Operator Theory
Advances and Applications
Vol. 75

Editor
I. Gohberg

Operator Theory in Function Spaces and Banach Lattices

**Essays dedicated to A.C. Zaanen
on the occasion of his 80th birthday**

Edited by

C.B. Huijsmans
M.A. Kaashoek
W.A.J. Luxemburg
B. de Pagter

Birkhäuser Verlag
Basel · Boston · Berlin

Volume Editorial Office:

Mathematisch Instituut
Rijksuniversiteit Leiden
Niels Bohrweg 1
Postbus 9512
2300 RA Leiden
The Netherlands

A CIP catalogue record for this book is available from the Library of Congress, Washington D.C., USA

Deutsche Bibliothek Cataloging-in-Publication Data
Operator theory in function spaces and banach lattices : essays
dedicated to A. C. Zaanen on the occasion of his 80th birthday
/ ed. by C. B. Huijsmans ... – Basel ; Boston ; Berlin :
Birkhäuser, 1995
  (Operator theory ; Vol. 75)
  ISBN-13:978-3-0348-9896-6   e-ISBN-13:978-3-0348-9076-2
  DOI: 10.1007/978-3-0348-9076-2

NE: Huijsmans, Charles B. [Hrsg.]; Zaanen, Adriaan C.: Festschrift;
  GT

Camera-ready copy prepared by the editors
Printed on acid-free paper produced from chlorine-free pulp
Cover design: Heinz Hiltbrunner, Basel
ISBN-13:978-3-0348-9896-6

9 8 7 6 5 4 3 2 1

Contents

Biographical Notes

During the first week of September 1993, a Symposium was held at the University of Leiden honoring Professor A.C. Zaanen on the occasion of his 80th birthday in June of the same year. In March 1993, Professor Zaanen also celebrated the 55th anniversary of receiving his Doctor's Degree in Philosophy at the University of Leiden, marking the beginning of his remarkable mathematical career.

His Ph.D. thesis was devoted to an important topic in the theory of the so-called Sturm-Liouville two point boundary value problems under the title "Over Reeksen van Eigenfuncties van zekere Randproblemen" ("Concerning eigenfunction expansions of certain boundary value problems"). His "promotor" (thesis supervisor) was the mathematical physicist J. Droste, Professor of Mathematics at the University of Leiden who had been a student at the University of Leiden of the famous Dutch theoretical physicist H.H. Lorentz. It may be of some interest to point out that Professor Droste's thesis dealt with the calculations of the field of gravitation of one or more bodies according to the theory of Einstein under the title "Het zwaartekrachtveld van één of meer lichamen volgens de theorie van Einstein".

In his thesis Zaanen investigated in detail the asymptotic behavior of the eigenvalues and eigenfunctions determined by various types of two-point boundary value problems and the nature of the convergence of the eigenfunction expansions. These investigations led in a natural way to a detailed study of the properties of the solutions of certain types of linear integral equations and their eigenfunction expansions. In particular, Zaanen showed that many of the classical L^2-type results have a natural counter part in the emerging more general theory of the L^p-spaces ($1 \leq p \leq \infty$) of measurable functions. A detailed account of these results can be found in the series of papers [8] through [14] as numbered in his list of publications. It is worth to observe that these papers were written during a period that included the years from 1938 through 1947 when Zaanen taught mathematics in the secondary school system in the Netherlands.

In the late thirties, primarily through the work of Professor A. Zygmund and his pupils, Orlicz spaces of measurable functions made their entrée in the theory of Fourier series. By modifying the definition of an Orlicz space so as to include the L^1 and L^∞- type space (see

[15] and [21]), Zaanen presented for the first time results in the theory of linear integral equations referred to above, in the more general setting of the theory of Orlicz spaces (see [17], [18], [19], [20] and [26]). Significant adjustments had to be made, in part due to the incomplete duality between a general Orlicz space and its complementary space, akin to the pair of L^1 and L^∞-spaces.

These fundamental publications and others that appeared during the late forties and early fifties formed the basis of the first real textbook on functional analysis entitled "Linear Analysis" which appeared in 1953. At the time of appearance there were only still a few books dealing with the subject of linear analysis available. Of course there was Banach's classic "Théorie des Operations Linéaires" that had just been reprinted, the book of Marshall Stone, "Linear Transformations in Hilbert Space" from 1932 and the book by F. Riesz and B. Sz.-Nagy "Leçons d'Analyse Fonctionnelle" that appeared about the same time as Linear Analysis and, in a sense, complemented Zaanen's book in a nice way. Zaanen divided his book into three parts. Part I contains a comprehensive and very elegant treatment of the theory of measures based on the so-called Carathéodory extension procedure for countably additive measures. Following von Neumann's idea, measures are defined on the simpler classes of sets called semi-rings rather than on rings or algebras of sets to facilitate the introduction of special examples of measures such as the Lebesgue measure in Euclidean spaces. In this part of the book the theory of Orlicz spaces appeared for the first time in book form. Special attention is given to the definition of the class of the so-called Young functions that are used to define the Orlicz spaces of measurable functions. The extended Young's inequality and Hölder's inequality are presented with elegant proofs and the first part concludes with the important duality theory for the classes of Orlicz spaces.

Part II is devoted to the main principles of functional analysis. It begins with a discussion of the definition and basic results of the theory of Hilbert and Banach spaces and of their bounded linear operators. The introduction of this material in analysis is motivated by a comprehensive treatment of the finite dimensional case. This approach also shows in a natural way the importance of the so-called Fredholm determinant theory in the theory of linear operators. The basic principles of functional analysis such as the Hahn-Banach extension theorem, the Steinhaus uniform boundedness principle, Banach's open mapping theorem and the closed graph theorem are treated, and their power is illustrated in many examples. In passing we may mention here that, as a sign of the time concerning one's view of the validity of the use of the axiom of choice or versions of it, Zaanen preferred to introduce the Hahn-Banach extension principle as a general property that Banach spaces could or could not possess.

Part III is entirely devoted to the general theory of linear integral equations in the setting of the theory of Orlicz spaces. It is the first of its kind that appeared in book form and even today it is still one of the most informative accounts of this theory available in textbook form. It contains all the basic results concerning eigenfunction expansions of linear integral equations. In particular, the spectral properties of symmetrizable kernel type operators are treated in detail. Zaanen's more general version of the famous Mercer expansion theorem [27] and its applications are discussed and many other important aspects of this theory can be found in this part of the book. A special feature is its large number of worked out examples and the numerous exercises, some with detailed hints for their solutions that supplement the text and facilitate greatly the study of the material presented in the book. Linear Analysis (counting more than 600 pages) was an immediate success. It saw two reprintings one in 1957 and one in 1960. There is no question that the book influenced the development of functional analysis. One may say that a generation of analysts was brought up on this work.

In the years that followed the appearance of Linear Analysis, Zaanen, strongly dedicated to the educational aspects of mathematics, found the time to write another important textbook now devoted entirely to the theory of integration and its applications; it appeared some five years after the appearance of Linear Analysis. In this textbook Zaanen introduced in the treatment of the theory of integration a new approach by defining directly the integral of a function as the outer product measure of the set under the graph of a function. This approach proved to possess a number of advantages. Its links with the previously developed methods of integration, such as the one given by Daniell in the twenties, are clearly explained. As was to be expected the book was again a great success. It saw reprintings in 1961 and 1965. A completely new revised and greatly enlarged edition counting more than 600 pages appeared in 1967. The new version not only contains the first comprehensive treatment of the theory of Banach function spaces, but also Plancherel's treatment of the L^2-theory of the Fourier transform and ergodic theory. This book also contains a wealth of worked out examples and exercises related to the theory of special functions, such as Euler's Gamma and Beta functions, as applications of the theory of measure and integration. Even today it is hard to find a book devoted to the theory of measure and integration that contains a treatment of so many of the important classical results from the theory of special functions.

A new research program started in the early 60ths jointly with his first Ph.D. student W.A.J. Luxemburg. This was devoted to a systematic study of those aspects of theory of lattices and vector lattices (or Riesz spaces, a terminology introduced in the fifties by N. Bourbaki in his treatise on the theory of integration) that play a fundamental role in functional analysis. To some extent this project was started by the results obtained in the paper [39] in which the authors presented necessary and sufficient conditions of a measure theoretic nature for a kernel operator to be compact. These results later played a role in the important theorems concerning compactly dominated operators. This research project resulted in a long series of papers called "Notes on Banach Function Spaces" (see [40] - [52]) and led to the publication of a two-volume monograph entitled "The Theory of Riesz Spaces". Volume I, jointly with W.A.J. Luxemburg, appeared in 1971. It deals to a large extent with the more algebraic aspects of the theory of vector lattices and their representations. Volume II, which appeared in 1983, is devoted to the more functional analytic aspects of the theory of vector lattices. The second volume, in particular, contains a wealth of material and includes many of the results of his students. We may single out here an up-to-date version of theory of order bounded linear operators on Banach lattices, a treatment of the theory of kernel operators on Banach function spaces and many other aspects of the theory of Riesz spaces and their order dual.

The research activities of Zaanen presented plenty of opportunities for his Ph.D. students to work on. In a time space of 26 years thirteen students received their Ph.D. degrees under the supervision of Zaanen. Their research projects range over many areas of functional analysis.

During his mathematical career Zaanen taught many different courses in analysis at the University of Indonesia at Bandung, the Technical University at Delft and the University of Leiden. A course clear to his heart was the one that introduced the beginners to the wonderful world of Fourier Analysis. This course led to the publication of a unique kind of textbook entitled "Continuity, Integration and Fourier Theory" that appeared recently in 1989. It is already a bestseller and will be so for many years to come.

Professor Zaanen: the collection of articles contained in this "Festschrift", written by your admirers, among them many of your former students, is a token of our gratitude and esteem. It should also be seen as a special thanks for your help and guidance in our research that we received from you over the years. We hope that it may give you many pleasant hours of reading. We wish you also many more years of research and study in the fields of analysis that you love so much.

C.B. Huijsmans

M.A. Kaashoek

W.A.J. Luxemburg

B. de Pagter

List of Publications of A.C. Zaanen

I. Books

1. Analytische Meetkunde 1 (Delft Manuel, a-1) (1952) 117 p.

2. Analytische Meetkunde 2 (Delft Manuel, a-5) (1953) 73 p.

3. Linear Analysis, North-Holland Publ. Comp., Amsterdam and P. Noordhoff, Groningen (Bibliotheca Mathematica II), (1953, 1957, 1960) 600 p.

4. An Introduction to the Theory of Integration, North-Holland Publ. Comp., Amsterdam (1958, 1961, 1965), 254 p.

5. Integration (revised and enlarged edition of: An Introduction to the Theory of Integration), North-Holland Publ. Comp., Amsterdam (1967) 604 p.

6. (with W.A.J. Luxemburg) Riesz Spaces I, North-Holland Publ. Comp., Amsterdam (1971) 525 p.

7. Riesz Spaces II, North-Holland Publ. Comp., Amsterdam (1983) 731 p.

8. Continuity, Integration and Fourier theory, Universitext, Springer Verlag, Berlin etc. (1989) 259 p.

II. Papers

1. Over reeksen van eigenfuncties van zekere randproblemen, proefschrift, R.U. Leiden, 10 maart 1938.

2. On some orthogonal systems of functions, Comp. Math. $\underline{7}$ (1939) p. 252-282.

3. A theorem on a certain orthogonal series and its conjugate series, Nieuw Arch. voor Wiskunde (2nd series) $\underline{20}$ (1940) p. 244-252.

4. Ueber die Existenz der Eigenfunktionen eines symmetrisierbaren Kernes, Proc. Netherl. Acad. Sc. $\underline{45}$ (1942) p. 973-977.

5. Ueber vollstetige symmetrische und symmetrisierbare Operatoren, Nieuw Arch. voor Wiskunde (2nd series) $\underline{22}$ (1943) p. 57-80.

6. Transformaties in de Hilbertsche ruimte, die van een parameter afhangen, Mathematica B $\underline{13}$ (1944) p. 13-22.

7. On the absolute convergence of Fourier series, Proc. Netherl. Acad. Sc. $\underline{48}$ (1945) p. 211-215.

8. On the theory of linear integral equations I, Proc. Netherl. Acad. Sc. $\underline{49}$ (1946) p. 194-204.

9. On the theory of linear integral equations II, Proc. Netherl. Acad. Sc. $\underline{49}$ (1946) p. 205-212.

10. On the theory of linear integral equations III, Proc. Netherl. Acad. Sc. $\underline{49}$ (1946) p. 292-301.

11. On the theory of linear integral equations IV, Proc. Netherl. Acad. Sc. $\underline{49}$ (1946) p. 409-416.

12. On the theory of linear integral equations IVa, Proc. Netherl. Acad. Sc. $\underline{49}$ (1946) p. 417-423.

13. On the theory of linear integral equations V, Proc. Netherl. Acad. Sc. $\underline{49}$ (1946) p. 571-585.

14. On the theory of linear integral equations VI, Proc. Netherl. Acad. Sc. $\underline{49}$ (1946) p. 608-621.

15. On a certain class of Banach spaces, Annals of Math. $\underline{47}$ (1946) p. 654-666.

16. Eenige karakteristieke kenmerken der moderne wiskunde, openbare les aanvaarding privaatdocentschap R.U. Leiden, 22 oktober 1946.

17. On the theory of linear integral equations VII, Proc. Netherl. Acad. Sc. $\underline{50}$ (1947) p. 357-368.

18. On the theory of linear integral equations VIII, Proc. Netherl. Acad. Sc. $\underline{50}$ (1947) p. 465-473.

19. On the theory of linear integral equations VIIIa, Proc. Netherl. Acad. Sc. $\underline{50}$ (1947) p. 612-617.

20. On linear functional equations, Nieuw Arch. voor Wiskunde (2nd series) $\underline{22}$ (1948) p. 269-282.

21. Note on a certain class of Banach spaces, Proc. Netherl. Acad. Sc. $\underline{52}$ (1949) p. 488-498.

22. Enige motieven die bij de beoefening der wiskunde ook een rol spelen, intreerede T.H. Delft, 24 januari 1951.

23. Normalisable transformations in Hilbert space and systems of linear integral equations, Acta Math. $\underline{83}$ (1950) p. 197-248.

24. Characterization of a certain class of linear transformations in an arbitrary Banach space, Proc. Netherl. Acad. Sc. $\underline{54}$ (1951) p. 87-93.

25. (with C. Visser) On the eigenvalues of compact linear transformations, Proc. Netherl. Acad. Sc. $\underline{55}$ (1952) p. 71-78.

26. Integral transformations and their resolvents in Orlicz and Lebesgue spaces, Comp. Math. 10 (1952) p. 56-94.

27. An extension of Mercer's theorem on continuous kernels of positive type, Simon Stevin 29 (1952) p. 113-124.

28. (with N.G. de Bruijn) Non σ-finite measures and product measures, Proc. Netherl. Acad. Sc. 57 (1954) p. 456-466.

29. (with W.A.J. Luxemburg) Some remarks on Banach function spaces, Proc. Netherl. Acad. Sc. 59 (1956) p. 110-119.

30. (with W.A.J. Luxemburg) Conjugate space of Orlicz spaces, Proc. Netherl. Acad. Sc. 59 (1956) p. 217-228.

31. Het kleed der wiskunde, intreerede R.U. Leiden, 15 november 1957.

32. A note on measure theory, Nieuw Arch. voor Wiskunde (3rd series) 6 (1958) p. 58-65.

33. A note on perturbation theory, Nieuw Arch. voor Wiskunde (3rd series) 7 (1959) p. 61-65.

34. A note on the Daniell-Stone integral, Colloque sur l'Analyse fonctionelle, Louvain 25-28 Mai, 1960, p. 63-69.

35. Banach function spaces, Proceedings International Symposium on Linear Spaces, July 5-12, 1960, p. 448-452.

36. The Radon-Nikodym theorem I, Proc. Netherl. Acad. Sc. 64 (1961) p. 157-170.

37. The Radon-Nikodym theorem II, Proc. Netherl. Acad. Sc. 64 (1961) p. 171-187.

38. Some examples in weak sequential convergence, Amer. Math. Monthly 69 (1962) p. 85-93.

39. (with W.A.J. Luxemburg) Compactness of integral operators in Banach function spaces, Math. Ann. 149 (1963) p. 150-180.

40. (with W.A.J. Luxemburg) Notes on Banach function spaces I, Proc. Netherl. Acad. Sc. 66 (1963) p. 135-147.

41. (with W.A.J. Luxemburg) Notes on Banach function spaces II, Proc. Netherl. Acad. Sc. 66 (1963) p. 148-153.

42. (with W.A.J. Luxemburg) Notes on Banach function spaces III, Proc. Netherl. Acad. Sc. 66 (1963) p. 239-250.

43. (with W.A.J. Luxemburg) Notes on Banach function spaces IV, Proc. Netherl. Acad. Sc. 66 (1963) p. 251-263.

44. (with W.A.J. Luxemburg) Notes on Banach function spaces V, Proc. Netherl. Acad. Sc. 66 (1963) p. 496-504.

45. (with W.A.J. Luxemburg) Notes on Banach function spaces VI, Proc. Netherl. Acad. Sc. 66 (1963) p. 655-668.

46. (with W.A.J. Luxemburg) Notes on Banach function spaces VII, Proc. Netherl. Acad. Sc. 66 (1963) p. 669-681.

47. (with W.A.J. Luxemburg) Notes on Banach function spaces VIII, Proc. Netherl. Acad. Sc. 67 (1964) p. 104-119.

48. (with W.A.J. Luxemburg) Notes on Banach function spaces IX, Proc. Netherl. Acad. Sc. 67 (1964) p. 360-376.

49. (with W.A.J. Luxemburg) Notes on Banach function spaces X, Proc. Netherl. Acad. Sc. 67 (1964) p. 493-506.

50. (with W.A.J. Luxemburg) Notes on Banach function spaces XI, Proc. Netherl. Acad. Sc. 67 (1964) p. 507-518.

51. (with W.A.J. Luxemburg) Notes on Banach function spaces XII, Proc. Netherl. Acad. Sc. 67 (1964) p. 519-529.

52. (with W.A.J. Luxemburg) Notes on Banach function spaces XIII, Proc. Netherl. Acad. Sc. 67 (1964) p. 530-543.

53. (with W.A.J. Luxemburg) Some examples of normed Köthe spaces, Math. Ann. 162 (1966) p. 337-350.

54. Stability of order convergence and regularity in Riesz spaces, Studia Math. 31 (1968) p. 159-172.

55. (with W.A.J. Luxemburg) The linear modulus of an order bounded linear transformation I, Proc. Netherl. Acad. Sc. 74 (1971) p. 422-434.

56. (with W.A.J. Luxemburg) The linear modulus of an order bounded linear transformation II, Proc. Netherl. Acad. Sc. 74 (1971) p. 435-447.

57. The linear modulus of an integral operator, Mémoire 31-32 du Bulletin de la Soc. Math. France (1971) p. 399-400.

58. Representation theorems for Riesz spaces, Proceedings Conference on Linear Operators and Approximation, Oberwolfach, August 14-22, 1971, p. 122-128.

59. Ideals in Riesz spaces, Troisième Colloque sur l'Analyse fonctionelle, (Liège, 1970) p. 137-146, Vander, Louvain (1971).

60. Examples of orthomorphisms, J. Approximation Theory 13 (1975) p. 192-204.

61. De ontwikkeling van het integraalbegrip, Verslag van de gewone vergadering van de Afd. Natuurkunde van de Koninkl. Nederl. Akademie van Wetenschappen 84 (1975) p. 49-54.

62. Riesz spaces and normed Köthe spaces, Proceedings Symposium Potchefstroom University and South African Math. Soc., July 23-24, 1974, p. 1-21.

63. (with E. de Jonge) The semi-M property for normed Riesz spaces, Measure Theory, Proceedings of the Conference held at Oberwolfach, 15-21 June 1975, p. 299-302 (Lecture Notes in Mathematics 541).

64. Kernel operators, Proceedings Conference on Linear Spaces and Approximation, Oberwolfach, 20-27 August 1977, p. 23-31.

65. (with W.J. Claas) Orlicz lattices, Commentationes Math., Tomus Specialis in Honorem Ladislai Orlicz I (1978) p. 77-93.

66. Some remarks about the definition of an Orlicz space, Measure theory, Proceedings of the Conference held at Oberwolfach, 21-27 June 1981, p. 263-268. (Lecture Notes in Mathematics 945).

67. Terugzien, Nieuw Arch. voor Wiskunde (4th series) 1 (1983) p. 224-240.

68. The universal completion of an Archimedean-Riesz space, Indag. Math. 45 (1983) p. 435-441.

69. Integral operators, in: Algebra and Order, Proceedings of the First International Symposium on Ordered Algebraic Structures held at Luminy, 11-15 June 1984, p. 197-202. (Research and Exposition in Mathematics 14, Heldermann Verlag).

70. Measurable functions and integral operators, Nieuw Arch. voor Wiskunde (4th series) 3 (1985), p. 167-205.

71. Some recent results in operator theory, Preprint, R.U. Leiden 2 (1985).

72. Continuity of measurable functions, Am. Math. Monthly 93 (1986), p. 128-130.

Curriculum Vitae of A.C. Zaanen

Adriaan Cornelis Zaanen was born June 14, 1913, Rotterdam, The Netherlands. He is married with Ada Jacoba van der Woude. They have four sons, born in 1946, 1948, 1957 and 1959.

Zaanen studied Mathematics at the University of Leiden, 1930 - 1938. He received his Doctor's degree in Mathematics at the University of Leiden, March 10, 1938 on a thesis entitled: "Over reeksen van eigenfuncties van zekere randproblemen". His thesis supervisor was J. Droste.

Positions held:

1938 - 1947	Highschool teaching.
1946 - 1947	Wiskunde Docent TH Delft.
1946 - 1947	Privaat-Docent RU Leiden.
1947 - 1950	Professor of Mathematics at the Faculty of Technology, University of Indonesia in Bandung.
1950 - 1956	Professor of Mathematics at the Delft University of Technology.
1956 - 1981	Professor of Mathematics at the University of Leiden.
1982	Teaching assignment at the University of Leiden.
1960 - 1961, 1968 - 1969	Visiting Professor at the California Institute of Technology (Pasadena, USA).

Memberships:

Dutch Mathematical Society (Wiskundig Genootschap), since 1937; honorary member since 1988.

American Mathematical Society, since 1948.

Royal Dutch Academy of Science (Koninklijke Nederlandse Akademie van Wetenschappen), since 1960.

Other Activities:

1. Examination Commitee "M.O.-wiskunde"; 1951 - 1982: member; 1974 - 1975: chairman; 1975 - 1983: vice-chairman.

2. Editor of "Nieuw Archief van Wiskunde", 1953 - 1982.

3. Member of the board of the Mathematical Centre (Amsterdam), 1950 - 1979.

4. President of the Dutch Mathematical Society, 1970 - 1972; member of the board, 1970 - 1973.

5. Member of the board of advisary editors of the series "North-Holland Mathematical Library".

6. Member of the committee of exact sciences, 1965 - 1967; occasionally secretary of Z.W.O.

7. Member of the Mathematics Section of the "Akademische Raad" for several years.

8. Chairman of the Department of Mathematics (Leiden University), 1965 - 1971; Member of the Faculty Board, 1966 - 1973; Chairman of the Subfaculty Board, 1976 - 1979; Member of the Subfaculty Board, 1976 - 1980.

Order of Knighthood:

Ridder in de Orde van de Nederlandse Leeuw, since 1982.

List of A.C. Zaanen's Ph.D. Students:

W.A.J. Luxemburg	: Banach function spaces, October 12, 1955.
B.C. Strydom	: Abstract Riemann integration, May 20, 1959.
M.A. Kaashoek	: Closed linear operators on Banach spaces, February 26, 1964.
A.C. van Eijnsbergen	: Beurling spaces, a class of normed Köthe spaces, July 3, 1967.
J.J. Grobler	: Non-singular linear integral equations in Banach function spaces, May 6, 1970.
N.A. van Arkel	: Algebras of holomorphic functions of n complex variables, June 24, 1970.
C.B. Huijsmans	: Prime ideals in commutative rings and Riesz spaces, May 17, 1973.
E. de Jonge	: Singular functionals on Köthe spaces, September 19, 1973.
P. Maritz	: Integration of set-valued functions, June 11, 1975.

W.J. Claas   : Orlicz lattices, June 1, 1977.

A.R. Schep   : Kernel operators, June 8, 1977.

W.K. Vietsch   : Abstract kernel operators and compact operators, June 13,
                  1979.

B. de Pagter   : f-Algebras and orthomorphisms, June 3, 1981.

Operator Theory:
Advances and Applications, Vol. 75
© 1995 Birkhäuser Verlag Basel/Switzerland

ANOTHER CHARACTERIZATION OF
THE INVARIANT SUBSPACE PROBLEM

Y. A. Abramovich, C. D. Aliprantis, and O. Burkinshaw

Dedicated to Professor A. C. Zaanen on the occasion of his 80th birthday

Recently, L. de Branges [6], building upon the work of V. I. Lomonosov [12], presented a characterization of the dual invariant subspace problem in terms of a denseness property of a certain subspace of vector valued functions. In this paper, we present two companion characterizations—one for the invariant subspace problem and one for its dual—also in terms of denseness properties of some appropriately chosen spaces of vector valued functions and linear topologies on them.

1. INTRODUCTION

The question

- *Does a continuous linear operator $T: X \to X$ on a Banach space have a non-trivial closed invariant subspace?*

is known as the **invariant subspace problem**. If the Banach space X is non-separable, then the closed vector subspace generated by the orbit $\{x, Tx, T^2x, \ldots\}$ of any non-zero vector x is a non-trivial closed invariant subspace for T. In other words, every operator on a non-separable Banach space has a non-trivial closed invariant subspace. However, if we consider not a single operator but rather an algebra of operators on $L(X)$, then the existence of a common non-trivial closed invariant subspace is of interest even in the non-separable case.

Very few affirmative results are known regarding the invariant subspace problem. The most prominent is the following invariant subspace theorem due to V. Lomonosov [11].

- **Lomonosov's Theorem:** *If a continuous operator T on a Banach space commutes with an operator S which is not a multiple of the identity and S in turn commutes with a non-zero compact operator, then T has a non-trivial closed invariant subspace.*

For more results on and the history of the invariant subspace problem see [3, 4, 9, 13, 14, 16] and the references given there. For some recent invariant subspace theorems for positive operators see [1, 2].

P. Enflo [8] was the first to exhibit a bounded linear operator on a separable Banach space without a non-trivial closed invariant subspace, and thus he gave a negative answer to the invariant subspace problem in its general formulation. Later on, C. J. Read [15] constructed a bounded linear operator on ℓ_1 without non-trivial closed invariant subspaces. In spite of these counterexamples, V. I. Lomonosov [12] conjectured that the invariant subspace problem may have an affirmative answer for adjoint operators. That is, he suggested the following "Dual Invariant Subspace Problem."

- **Lomonosov's Conjecture:** *The adjoint of a bounded linear operator on a Banach space has a non-trivial closed invariant subspace.*

In the same paper [12], V. I. Lomonosov also presented some characterizations of the dual invariant subspace problem.

Subsequently, in a recent paper, L. de Branges [6] has initiated a study aimed at proving Lomonosov's conjecture. Specifically, he reduced Lomonosov's conjecture to proving that a certain vector subspace of a natural space of vector functions is not dense with respect to a linear topology introduced by L. de Branges.

The purpose of this work is to present an alternate approach to the characterization of the invariant subspace problem for adjoint operators by introducing two other topologies, which are the analogues of the weak and strong operator topologies. As in the Lomonosov–de Branges approach, we study the invariant subspace problem for algebras of operators. We consider a Banach space X and denote the closed unit ball of its norm dual X' by S. The Banach space of all weak* continuous functions from S into X' is denoted by $C(S, X')$. For an algebra $\mathcal{A}$ of continuous operators on X, we establish the following two results.

1. *There is a non-trivial closed $\mathcal{A}$-invariant subspace of X if and only if there exist an operator B and a compact operator K on X such that $K'B'$ does not belong to the norm closed vector subspace of $C(S, X')$ generated by the collection*

$$\{\alpha K'T'\colon \ \alpha \in C(S) \ \text{ and } \ T \in \mathcal{A}\}.$$

2. *There is a non-trivial closed $\mathcal{A}'$-invariant subspace of X' if and only if there exist an operator B and a compact operator K on X such that $B'K'$ does not belong to the norm closure of the vector subspace generated by the collection*

$$\{\alpha T'K'\colon \ \alpha \in C(S) \ \text{ and } \ T \in \mathcal{A}\}.$$

The above characterizations of the invariant subspace problem show that the existence of invariant subspaces is closely related with the presence of compact operators.

2. PRELIMINARIES

In this work X will denote a complex Banach space of dimension greater than one and X' its norm dual. Also, $\mathcal{A}$ will designate a subalgebra of $L(X)$, the Banach algebra of all bounded linear operators on X. The **dual algebra** of $\mathcal{A}$ is the subalgebra of $L(X')$ defined by $\mathcal{A}' = \{T'\colon T \in \mathcal{A}\}$, where as usual T' denotes the norm adjoint of T.

A subspace V of X is said to be $\mathcal{A}$-**invariant** whenever V is invariant under every operator of $\mathcal{A}$, i.e., $T(V) \subseteq V$ holds for each $T \in \mathcal{A}$. A subspace V of X is, of course, invariant under an operator $T \in L(X)$ if and only if V is invariant under the algebra (unital or not) generated by T in $L(X)$. As usual, a subspace V of X is called **non-trivial** if $V \neq 0$ and $V \neq X$.

The next result is a folklore characterization of the existence of non-trivial closed $\mathcal{A}$-invariant subspaces.

Proposition 2.1. *A subalgebra $\mathcal{A}$ of $L(X)$ admits a non-trivial closed $\mathcal{A}$-invariant subspace if and only if there exist a nonzero vector $x \in X$ and a nonzero linear functional $x' \in X'$ satisfying*

$$\langle x', Tx \rangle = 0$$

for each $T \in \mathcal{A}$.

Proof. Let V be a non-trivial closed $\mathcal{A}$-invariant subspace. Fix a nonzero vector $x \in V$ and consider the closed subspace $\mathcal{A}x = \overline{\{Tx \colon T \in \mathcal{A}\}}$. Clearly, $\mathcal{A}x \subseteq V$, and so $\mathcal{A}x \neq X$. Therefore, there exists some nonzero $x' \in X'$ that annihilates $\mathcal{A}x$, i.e., we have

$$\langle x', Tx \rangle = 0$$

for all $T \in \mathcal{A}$.

For the converse, assume that $\langle x', Tx \rangle = 0$ holds for all $T \in \mathcal{A}$ and some nonzero vectors $x \in X$ and $x' \in X'$. It follows that the closed subspace $\mathcal{A}x = \overline{\{Tx \colon T \in \mathcal{A}\}}$ is not norm dense in X. If $\mathcal{A}x \neq 0$, then $\mathcal{A}x$ is a non-trivial closed $\mathcal{A}$-invariant subspace. In case $\mathcal{A}x = \{0\}$, the non-trivial closed subspace $V = \{\lambda x \colon \lambda \in \mathbf{C}\}$ is $\mathcal{A}$-invariant. ∎

From the identity $\langle x', Tx \rangle = \langle T'x', x \rangle$, it follows immediately that if there exists a non-trivial closed $\mathcal{A}$-invariant subspace, then there is also a non-trivial closed $\mathcal{A}'$-invariant subspace.

In accordance with de Branges' notation [6], we shall denote by S the closed unit ball of X'. That is,

$$S = \{x' \in X' \colon \|x'\| \leq 1\}.$$

As usual, S will be equipped with its weak* topology, and hence S is a weak* compact subset of X'.

Definition 2.2. *The vector space of all continuous functions from S into X', when both S and X' are equipped with the weak* topology, will be denoted by $C(S, X')$. Occasionally, $C(S, X')$ will also be denoted by Y, i.e., $Y = C(S, X')$.*

Observe that for each $T \in L(X)$ the restriction of the adjoint $T': S \to X'$ is an element of $C(S, X')$. Clearly, the vector space $C(S, X')$ equipped with the norm

$$\|f\| = \sup_{s \in S} \|f(s)\|, \quad f \in C(S, X'),$$

is a Banach space.

The Banach space $C(S, X')$ was the main object of study in [12] and [6]. V. I. Lomonosov [12], inspired by L. de Branges' proof of the Stone–Weierstrass theorem [5], characterized the extreme points of the closed unit ball of the norm dual of $C(S, X')$. Subsequently, L. de Branges [6] presented a deep analysis of the behavior of these extreme points and obtained an abstract version of the Stone–Weierstrass theorem. The Lomonosov–de Branges analysis will be employed later on in our characterization of the invariant subspace problem.

As always, the symbol $C(S)$ denotes the Banach space of all continuous complex valued functions defined on S. It is worth mentioning that each function $\alpha \in C(S)$ defines an "action" (or equivalently, an operator) on $C(S, X')$ via the formula

$$(\alpha f)(s) = \alpha(s) f(s), \quad f \in C(S, X'), \ s \in S.$$

Clearly,

$$\|\alpha f\| \le \|\alpha\|_\infty \|f\|.$$

In algebraic terminology this means that $C(S, X')$ is a $C(S)$-module and our discussion can be formulated in terms of modules. However, we shall not pursue this terminology any further.

The second norm dual of X will be denoted by X''. Every $x'' \in X''$ and every $s \in S$ give rise to a continuous linear functional $x'' \otimes s$ on $C(S, X')$ via the formula

$$\langle x'' \otimes s, f \rangle = (x'' \otimes s)(f) = \langle x'', f(s) \rangle = x''(f(s)), \quad f \in C(S, X').$$

Clearly, $x'' \otimes s$ is a norm continuous linear functional on $C(S, X')$. The vector space generated by the family $\{x'' \otimes s \colon x'' \in X'' \text{ and } s \in S\}$ in the norm dual of $Y = C(S, X')$ will be denoted by $Y^{\#}$, i.e.,

$$Y^{\#} = \Big\{\sum_{i=1}^{n} x_i'' \otimes s_i \colon x_i'' \in X'' \text{ and } s_i \in S \text{ for each } i = 1, \ldots, n\Big\}.$$

Obviously, $Y^{\#}$ separates the points of Y, and so $\langle Y, Y^{\#}\rangle$ with its natural duality is a dual system. Apart from the norm topology, we shall consider on the Banach space $C(S, X')$ two other easily accessible topologies and use them in our study of the invariant subspace problem. They are defined as follows.

1. The topology τ_w on $C(S, X')$ is the locally convex topology generated by the family of seminorms $\{\rho_{x'',s} \colon x'' \in X'' \text{ and } s \in S\}$, where

$$\rho_{x'',s}(f) = \big|(x'' \otimes s)(f)\big| = \big|\langle x'', f(s)\rangle\big|$$

for each $f \in C(S, X')$. Note that τ_w is simply the weak topology $\sigma(Y, Y^{\#})$.

2. The topology τ_s on $C(S, X')$ is the locally convex topology generated by the family of seminorms $\{\rho_s \colon s \in S\}$, where

$$\rho_s(f) = \big\|f(s)\big\|, \quad f \in C(S, X').$$

The alert reader will recognize immediately that these topologies are similar to the usual weak and strong operator topologies on $L(X)$ and this justifies our choice of the subscripts w and s. Moreover, in analogy with the classical weak and strong operator topologies [7, Theorem 4, p. 477], the topologies τ_w and τ_s have the same continuous linear functionals. That is, they are both consistent with the dual system $\langle Y, Y^{\#}\rangle$. The details follow.

Theorem 2.3. *The locally convex topology τ_s is consistent with the dual system $\langle Y, Y^{\#}\rangle$. That is, we have the inclusions*

$$\sigma(Y, Y^{\#}) \subseteq \tau_s \subseteq \tau(Y, Y^{\#}),$$

where as usual $\tau(Y, Y^{\#})$ denotes the Mackey topology.

Proof. Clearly, $\sigma(Y, Y^\#) \subseteq \tau_s$. So, it suffices to establish that $\tau_s \subseteq \tau(Y, Y^\#)$.

To this end, fix $s \in S$. We must show that the set $\{f \in Y: \rho_s(f) = \|f(s)\| \leq 1\}$ is a $\tau(Y, Y^\#)$-neighborhood of zero. Let U'' denote the closed unit ball of X''. Next, note that the operator $R: (X'', \sigma(X'', X')) \to (Y^\#, \sigma(Y^\#, Y))$, defined by $Rx'' = x'' \otimes s$, is continuous. Since U'' is a $\sigma(X'', X')$-compact set, it follows that the convex circled set $D = R(U'') = \{x'' \otimes s: x'' \in U''\}$ is $\sigma(Y^\#, Y)$-compact. So, its polar

$$D^\circ = \{f \in Y: |\langle x'' \otimes s, f \rangle| = |x''(f(s))| \leq 1 \text{ for all } x'' \in U''\}$$
$$= \{f \in Y: \|f(s)\| \leq 1\}$$

is a $\tau(Y, Y^\#)$-neighborhood of zero, and the proof is finished. ∎

The preceding theorem can be reformulated as follows.

Theorem 2.4. *For a linear functional ϕ defined on $C(S, X')$ the following statements are equivalent.*

1. *$\phi = \sum_{i=1}^n x_i'' \otimes s_i$, where $s_1, \ldots, s_n \in S$ and $x_1'', \ldots, x_n'' \in X''$.*

2. *ϕ is τ_w-continuous.*

3. *ϕ is τ_s-continuous.*

Now the standard duality theory yields the following result.

Corollary 2.5. *The topologies τ_w and τ_s on $C(S, X')$ have the same closed convex sets.*

We are now ready to establish that if a subspace $\mathcal{M}$ of $C(S, X')$ is $C(S)$-invariant, then $\mathcal{M}$ satisfies a nice separation property that elements outside of the closure of $\mathcal{M}$ can be separated by a linear functional of the form $x'' \otimes s$.

Lemma 2.6. *Let $\mathcal{M}$ be a vector subspace of $C(S, X')$ which is invariant under multiplication by elements of $C(S)$. Then an element $f_0 \in C(S, X')$ does not belong to the τ_s-closure of $\mathcal{M}$ if and only if there exist $x'' \in X''$ and $s \in S$ such that*

$$\langle x'' \otimes s, f_0 \rangle = 1 \quad and \quad \langle x'' \otimes s, f \rangle = 0$$

for all $f \in \mathcal{M}$.

Proof. The "only if" part needs verification. So, suppose that f_0 does not belong to the τ_s-closure of $\mathcal{M}$. Then, there exists some τ_s-continuous linear functional ϕ on $C(S, X')$ such that $\phi(f_0) \neq 0$ and $\phi(f) = 0$ for each $f \in \mathcal{M}$. By Theorem 2.4, ϕ is of the form $\phi = \sum_{i=1}^{n} x_i'' \otimes s_i$, where $s_i \neq s_j$ for $i \neq j$. From $\phi(f_0) = \sum_{i=1}^{n} \langle x_i'', f_0(s_i) \rangle \neq 0$, it follows that there exists some k satisfying $x_k'' \neq 0$ and $\langle x_k'', f_0(s_k) \rangle \neq 0$. We can suppose $\langle x_k'', f_0(s_k) \rangle = 1$.

Next, by Urysohns' Lemma, pick some $\alpha \in C(S)$ such that $\alpha(s_k) = 1$ and $\alpha(s_i) = 0$ for $i \neq k$. Since $\alpha f \in \mathcal{M}$ for each $f \in \mathcal{M}$, we have

$$\langle x_k'' \otimes s_k, f \rangle = \langle x_k'', f(s_k) \rangle = \sum_{i=1}^{n} \langle x_i'', \alpha(s_i) f(s_i) \rangle = \phi(\alpha f) = 0$$

for all $f \in \mathcal{M}$, and the proof is finished. ∎

3. THE INVARIANT SUBSPACE PROBLEM

As before, $\mathcal{A}$ denotes a subalgebra of $L(X)$. The next result gives a necessary and sufficient condition for the existence of a common non-trivial closed invariant subspace of X' for all the adjoint operators T' with $T \in \mathcal{A}$. As we shall see, this condition is closely related to the properties of the vector subspace generated in $C(S, X')$ by the collection of functions $\{\alpha T': \alpha \in C(S), T \in \mathcal{A}\}$.

Theorem 3.1. *For an arbitrary subalgebra $\mathcal{A}$ of $L(X)$, the following two statements are equivalent.*

1. *There exists a non-trivial closed $\mathcal{A}'$-invariant subspace.*

2. *There exists an operator $B \in L(X)$ such that the adjoint operator B' does not belong to the τ_s-closure in $C(S, X')$ of the vector space generated by the set*

$$\{\alpha T': \alpha \in C(S) \text{ and } T \in \mathcal{A}\}.$$

Proof. Let $\mathcal{M}$ denote the vector subspace of $C(S, X')$ generated by the collection of functions $\{\alpha T': \alpha \in C(S) \text{ and } T \in \mathcal{A}\}$, and let $\overline{\mathcal{M}}$ denote the closure of $\mathcal{M}$ in the topology τ_s.

$(1) \Longrightarrow (2)$ By Proposition 2.1 there exist non-zero $x'' \in X''$ and $s \in X'$ satisfying $\langle x'', T's \rangle = 0$ for each $T \in \mathcal{A}$. Without loss of generality we can suppose that $s \in S$.

Pick $b' \in X'$ and $b \in X$ such that $\langle x'', b' \rangle = 1$ and $\langle b, s \rangle = 1$. We claim that the rank-one operator $B = b' \otimes b \in L(X)$ satisfies $B' \notin \overline{\mathcal{M}}$. To see this, note that

$$\langle x'' \otimes s, \alpha T' \rangle = \alpha(s) \langle x'', T's \rangle = 0$$

for each $T \in \mathcal{A}$ and all $\alpha \in C(S)$. That is, the τ_s-continuous linear functional $x'' \otimes s$ vanishes on $\mathcal{M}$. On the other hand, the relation

$$\langle x'' \otimes s, B' \rangle = \langle x'' \otimes s, b \otimes b' \rangle = \langle x'', b' \rangle \langle b, s \rangle = 1,$$

implies that $B' \notin \overline{\mathcal{M}}$.

$(2) \Longrightarrow (1)$ Pick some $B \in L(X)$ such that $B' \notin \overline{\mathcal{M}}$. Since $\mathcal{M}$ is invariant under multiplication by elements of $C(S)$, it follows from Lemma 2.6 that there exist $x'' \in X''$ and $s \in S$ such that

$$\langle x'', B's \rangle = 1 \quad \text{and} \quad \langle x'', T's \rangle = 0$$

for all $T \in \mathcal{A}$. Since $\langle x'', B's \rangle = 1$, it follows that $x'' \neq 0$ and $s \neq 0$, and by Proposition 2.1 the proof is finished. ∎

Our next goal is to obtain a similar characterization for the invariant subspace problem in terms of the norm topology. To accomplish this, we need to introduce the class of completely continuous functions.

Definition 3.2. *A function $f \in C(S, X')$ is said to be **completely continuous** if it is continuous for the weak* topology on S and the norm topology on X'.*

The vector subspace of all completely continuous functions of $C(S, X')$ will be denoted by $\mathcal{K}(S, X')$.

In other words, a function $f \in C(S, X')$ is completely continuous if and only if $s_\alpha \xrightarrow{w^*} s$ in S implies $\|f(s_\alpha) - f(s)\| \to 0$. Observe that if $T: X \to X$ is a compact operator, then $T': S \to X'$ is completely continuous, i.e., $T' \in \mathcal{K}(S, X')$.

Clearly, $\mathcal{K}(S, X')$ is a normed closed subspace of $C(S, X')$. As a vector subspace of $C(S, X')$, the space $\mathcal{K}(S, X')$ inherits the three topologies considered on $C(S, X')$; the norm topology, the τ_w-topology, and the τ_s-topology. It is obvious that neither τ_w nor τ_s is consistent with the norm topology on $\mathcal{K}(S, X')$. Nevertheless, for $C(S)$-invariant subspaces of $\mathcal{K}(S, X')$ the situation is different. Using the Lomonosov–de Branges technique, we are now ready to characterize the closures of the $C(S)$-invariant subspaces of $\mathcal{K}(S, X')$ under these topologies.

Theorem 3.3. *For a vector subspace $\mathcal{M}$ of $\mathcal{K}(S, X')$ which is invariant under multiplication by elements of $C(S)$ the following statements are equivalent.*

1. *The vector space $\mathcal{M}$ is τ_w-closed in $\mathcal{K}(S, X')$.*

2. *The vector space $\mathcal{M}$ is τ_s-closed in $\mathcal{K}(S, X')$.*

3. *The vector space $\mathcal{M}$ is norm closed.*

Proof. Clearly, $(1) \Longrightarrow (2) \Longrightarrow (3)$. It remains to establish that $(3) \Longrightarrow (1)$.

To this end, let $f_0 \in \mathcal{K}(S, X')$ belong to the τ_w-closure of $\mathcal{M}$. To show that $f_0 \in \mathcal{M}$, it suffices to prove that f_0 belongs to the norm closure of $\mathcal{M}$. By the Hahn–Banach Theorem, this will be established if we verify that for a norm continuous linear functional ϕ on $C(S, X')$ the identity $\langle \phi, f \rangle = 0$ for all $f \in \mathcal{M}$ implies that $\langle \phi, f_0 \rangle = 0$. So, let $\langle \phi, f \rangle = 0$ for all $f \in \mathcal{M}$. We can assume that ϕ has norm one.

For each $\alpha \in C(S)$, we define on $Y = C(S, X')$ the continuous linear functional ϕ_α by $\phi_\alpha(f) = \phi(\alpha f)$ where $f \in C(S, X')$. Since $\mathcal{M}$ is invariant under multiplication by elements of $C(S)$, it follows that $\langle \phi_\alpha, f \rangle = 0$ for all $f \in \mathcal{M}$. This certainly implies that $\langle \theta, f \rangle = 0$ for all $\theta \in V$ and $f \in \mathcal{M}$, where V is the weak* closed subspace generated by the ϕ_α in the norm dual Y'. We denote by $\mathcal{U}$ the intersection of V with the closed unit ball of Y'. Obviously $\phi = \phi_1 \in \mathcal{U}$.

Let ψ be any extreme point of $\mathcal{U}$. By the characterization of extreme points obtained in [6, Theorem 1, p. 164], it follows that there exist an element $s \in S$ and an element $x'' \in X''$ such that

$$\psi(f) = \langle x'' \otimes s, f \rangle = \langle x'', f(s) \rangle$$

holds for every $f \in \mathcal{K}(S, X')$. In particular, $\psi(f_0) = \langle x'', f_0(s) \rangle$ and $\psi(f) = \langle x'', f(s) \rangle = 0$ for all $f \in \mathcal{M}$. Since f_0 is in the τ_w-closure of $\mathcal{M}$, it follows that $\psi(f_0) = \langle x'', f_0(s) \rangle = 0$.

That is, we have proved that $\psi(f_0) = 0$ for each extreme point ψ of $\mathcal{U}$. Since $\phi \in \mathcal{U}$ and since, by the Krein–Milman theorem, $\mathcal{U}$ is the weak* closed convex hull of its extreme points, we can conclude that $\phi(f_0) = 0$. This completes the proof. ■

We are now ready to state the main result of this work which considerably improves Theorem 3.1 by replacing the topology τ_s with the norm topology.

Theorem 3.4. *If $\mathcal{A}$ is a subalgebra of $L(X)$, then the following statements are equivalent.*

1. *There exists a non-trivial closed $\mathcal{A}'$-invariant subspace.*

2. *There exist operators $B, K \in L(X)$ with K compact such that the operator $B'K'$ does not belong to the norm closure in $C(S, X')$ of the vector space $\mathcal{M}$ generated by the set*

$$\{\alpha T' K' : \alpha \in C(S) \text{ and } T \in \mathcal{A}\}.$$

Proof. (1) $\Longrightarrow$ (2) By Proposition 2.1, there exist non-zero $x'' \in X''$ and $s \in X'$ satisfying $\langle x'', T's \rangle = 0$ for each $T \in \mathcal{A}$. We can suppose that $s \in S$. Now, pick elements $b \in X$ and $b' \in X'$ such that $\langle s, b \rangle = 1$ and $\langle x'', b' \rangle = 1$. Next, consider the rank-one operators $K = s \otimes b$ and $B = b' \otimes b$, and note that $K's = \langle s, b \rangle s = s$ and $B's = \langle s, b \rangle b' = b'$. We claim that $B'K'$ does not belong to $\overline{\mathcal{M}}$, the norm closure of $\mathcal{M}$.

To see this, note that the norm continuous linear functional $\phi = x'' \otimes s$ on $C(S, X')$ satisfies

$$\langle \phi, \alpha T' K' \rangle = \alpha(s) \langle x'', T'K's \rangle = \alpha(s) \langle x'', T's \rangle = 0$$

for each $T \in \mathcal{A}$ and all $\alpha \in C(S)$. That is, ϕ vanishes on $\mathcal{M}$. On the other hand, $\phi(B'K') = \langle x'', B'K's \rangle = \langle x'', b' \rangle = 1$ shows that $B'K' \notin \overline{\mathcal{M}}$.

(2) $\Longrightarrow$ (1) Assume that $B, K \in L(X)$ satisfy the stated properties. Clearly, $B'K'$ belongs to $\mathcal{K}(S, X')$ and $\mathcal{M} \subseteq \mathcal{K}(S, X')$. Also, $\mathcal{M}$ is $C(S)$-invariant.

Since $B'K'$ does not belong to the norm closure of $\mathcal{M}$, it follows from Theorem 3.3 that $B'K'$ is not in the τ_w-closure of $\mathcal{M}$ in $\mathcal{K}(S, X')$. Therefore, by Lemma 2.6, there exist $x'' \in X''$ and $s \in S$ satisfying

$$\langle x'', B'K's \rangle = 1 \quad \text{and} \quad \langle x'', T'K's \rangle = 0$$

for all $T \in \mathcal{A}$. The former condition implies that $K's \neq 0$, and hence the latter condition shows that Proposition 2.1 is applicable to $\mathcal{A}'$. ∎

The approach suggested by L. de Branges in [6] was aimed at the dual invariant subspace problem. As we shall see next, our arguments above allow us to obtain also a necessary and sufficient condition for the existence of a common non-trivial closed invariant subspace for the algebra $\mathcal{A}$ itself. The only difference is that we must apply the compact operator on the left, as opposed to the multiplication on the right in the preceding theorem.

Theorem 3.5. *For a subalgebra $\mathcal{A}$ of $L(X)$ the following statements are equivalent.*

1. *There exists a non-trivial closed $\mathcal{A}$-invariant subspace.*

2. *There exist operators $B, K \in L(X)$ with K compact such that the operator $K'B'$ does not belong to the norm closure in $C(S, X')$ of the subspace generated by the set*

$$\{\alpha K'T' \colon \alpha \in C(S) \text{ and } T \in \mathcal{A}\}.$$

Proof. (1) $\Longrightarrow$ (2) By Proposition 2.1 there exist non-zero $x \in X$ and $x' \in X'$ satisfying $\langle x', Tx \rangle = 0$ for each $T \in \mathcal{A}$. We can suppose that $x' \in S$.

Take any $b' \in X'$ such that $\langle b', x \rangle = 1$, and then consider the rank-one operator $K = b' \otimes x \in L(X)$. Observe that $Kx = \langle b', x \rangle x = x$. Next choose any element b in X such that $\langle x', b \rangle = 1$. Define now the operator $B = b' \otimes b \in L(X)$. Clearly, the adjoint operator B' satisfies $B'x' = \langle x', b \rangle b' = b'$.

The vector space generated by the collection $\{\alpha K'T' \colon \alpha \in C(S), T \in \mathcal{A}\}$ will be denoted by $\mathcal{N}$. We claim that the operator $K'B'$ is not in the norm closure of $\mathcal{N}$. To see this, note that the linear functional $\phi = x \otimes x'$ is norm continuous on $C(S, X')$ and satisfies

$$\langle \phi, \alpha K'T' \rangle = \alpha(x')\langle x, K'T'x' \rangle = \alpha(x')\langle TKx, x' \rangle = \alpha(x')\langle Tx, x' \rangle = 0$$

for each $T \in \mathcal{A}$ and all $\alpha \in C(S)$. That is, $\phi = x \otimes x'$ vanishes on $\mathcal{N}$. On the other hand, the equality $B'x' = b'$ yields

$$\phi(K'B') = \langle x, K'B'x' \rangle = \langle x, K'b' \rangle = \langle Kx, b' \rangle = \langle x, b' \rangle = 1,$$

which shows that $K'B'$ does not belong to the norm closure of $\mathcal{N}$.

$(2) \Longrightarrow (1)$ Assume that the operators B and K satisfy the stated properties. Again, let $\mathcal{N}$ denote the vector space generated in $C(S, X')$ by the collection of functions

$$\{\alpha K'T' \colon \alpha \in C(S),\ T \in \mathcal{A}\}.$$

Clearly, $K'B' \in \mathcal{K}(S, X')$ and $\mathcal{N} \subseteq \mathcal{K}(S, X')$. Also, $\mathcal{N}$ is $C(S)$-invariant.

Since the compact operator $K'B'$ does not belong to the norm closure of $\mathcal{N}$, it follows from Theorem 3.3 that $K'B'$ is not in the τ_s-closure of $\mathcal{N}$ in $\mathcal{K}(S, X')$. Consequently, by Lemma 2.6, there exist $x'' \in X''$ and $s \in S$ such that

$$\langle x'', K'B's \rangle = 1 \quad \text{and} \quad \langle x'', K'T's \rangle = 0$$

for all $T \in \mathcal{A}$. Since $\langle x'', K'B's \rangle = 1$, it follows that $x_0 = K''x'' \neq 0$. Moreover, the compactness of K implies that $x_0 = K''x'' \in X$. Therefore, for each $T \in \mathcal{A}$, we have

$$\langle Tx_0, s \rangle = \langle TK''x'', s \rangle = \langle K''x'', T's \rangle = \langle x'', K'T's \rangle = 0,$$

and by Proposition 2.1 the proof is finished. ∎

We close the section by emphasizing once more the following remarkable fact that has been present throughout our discussion. As soon as we start asking about the existence of a non-trivial closed $\mathcal{A}$-invariant subspace of X (or about the existence of a non-trivial closed $\mathcal{A}'$-invariant subspace of X'), a compact operator emerges though the given algebra $\mathcal{A}$ need not be connected with compactness in any way!

4. AN APPLICATION

The objective of this section is to illustrate our results by presenting an invariant subspace existence theorem for dual algebras acting on ℓ_∞. Before we state the result, we need a lemma.

Lemma 4.1. *Assume that for a subalgebra $\mathcal{A}$ of $L(X)$ there exists a non-zero operator $T_0 \in L(X)$ such that:*

1. *T_0 commutes with each member of $\mathcal{A}$; and*

2. *for some $B \in L(X)$ the adjoint operator B' does not belong to the τ_s-closure of the vector subspace generated by $\{\alpha T'T_0' \colon \alpha \in C(S) \text{ and } T \in \mathcal{A}\}$ in $C(S, X')$.*

Then there exists a non-trivial closed $\mathcal{A}'$-invariant subspace.

Proof. By Lemma 2.6, there exist two vectors $x'' \in X''$ and $s \in S$ satisfying

$$\langle x'' \otimes s, B' \rangle = x''(B's) = 1 \quad \text{and} \quad \langle x'' \otimes s, T'T_0' \rangle = \langle x'', T'T_0's \rangle = 0$$

for all $T \in \mathcal{A}$. If $x' = T_0's \neq 0$, then the relation $\langle x'', T'x' \rangle = 0$ for each $T \in \mathcal{A}$ coupled with Proposition 2.1 shows that $\mathcal{A}'$ admits a non-trivial closed $\mathcal{A}'$-invariant subspace.

Now assume that $T_0's = 0$, and let N denote the null space of T_0'. Since T_0 commutes with every member of $\mathcal{A}$, it follows that N is a closed $\mathcal{A}'$-invariant subspace. From $x''(B's) = 1$, we infer that $s \neq 0$. Since $s \in N$, we see that N is non-zero. On the other hand, if $N = X'$, then $T_0' = 0$, and so $T_0 = 0$, a contradiction. Therefore N is a non-trivial closed $\mathcal{A}'$-invariant subspace, and the proof is finished. ∎

As mentioned earlier, every continuous linear operator on the (non-separable) Banach space ℓ_∞ admits a non-trivial closed invariant subspace. It is far less obvious that the dual algebra of a commutative algebra of bounded operators on ℓ_1 also admits a non-trivial closed invariant subspace. This was proven in [10]. We are now ready to present an alternate proof of this interesting result using our techniques.

Theorem 4.2. *If $\mathcal{A}$ is a commutative algebra of bounded linear operators on ℓ_1, then the dual algebra $\mathcal{A}'$ of bounded linear operators on ℓ_∞ admits a non-trivial closed $\mathcal{A}'$-invariant subspace.*

Proof. Let $\mathcal{A}$ be a commutative algebra of bounded linear operators on ℓ_1. If $\mathcal{A}$ coincides with the algebra generated by the identity operator I on ℓ_1, then the conclusion is trivial. So, we can assume that there exists some $T \in \mathcal{A}$ such that $T \neq \lambda I$ for each complex number λ. Fix some λ_0 in the approximate point spectrum of T and consider the operator $T_0 = T - \lambda_0 I$. Also, select a sequence $\{x_n\}$ of unit vectors of ℓ_1 (that is, $\|x_n\|_1 = 1$ for each n) satisfying

$$\|T_0 x_n\|_1 = \|T x_n - \lambda_0 x_n\|_1 \to 0. \qquad (\star)$$

Clearly, T_0 commutes with every member of $\mathcal{A}$.

Next, consider the subspace $\mathcal{M}$ of $C(S, X')$ generated by the collection of functions $\{\alpha T' T_0' : \alpha \in C(S) \text{ and } T \in \mathcal{A}\}$. We claim, that the identity operator $I' : \ell_\infty \to \ell_\infty$ (restricted to S) does not belong to the τ_s-closure of $\mathcal{M}$.

To see this, assume by way of contradiction that I' belongs to the τ_s-closure of $\mathcal{M}$. So, if $\epsilon > 0$ and $s \in S$ are fixed, then there exist $\alpha_i \in C(S)$ and $T_i \in \mathcal{A}$ $(i = 1, \ldots, m)$ satisfying $\left\|(I' - \sum_{i=1}^m \alpha_i T_i' T_0')(s)\right\| < \epsilon$. This implies

$$\left| \langle x, s \rangle - \sum_{i=1}^m \alpha_i(s) \langle x, T_i' T_0' s \rangle \right| < \epsilon \qquad (\star\star)$$

for all $x \in X$ with $\|x\| \leq 1$. Taking into account that $\mathcal{A}$ is a commutative algebra, it follows from $(\star\star)$ that

$$\left| \langle x, s \rangle - \sum_{i=1}^m \alpha_i(s) \langle T_0 x, T_i' s \rangle \right| < \epsilon$$

for each $x \in X$ with $\|x\| \leq 1$. In particular, we have

$$\left| \langle x_n, s \rangle - \sum_{i=1}^m \alpha_i(s) \langle T_0 x_n, T_i' s \rangle \right| < \epsilon$$

for each n, which, in view of $(\star)$, yields $|\langle x_n, s \rangle| < \epsilon$ for all sufficiently large n. In other words, the sequence $\{x_n\}$ converges weakly to zero in ℓ_1. Since ℓ_1 has the Schur property, it follows that $\|x_n\|_1 \to 0$, contrary to $\|x_n\|_1 = 1$ for each n. Therefore, the identity operator I' does not belong to the τ_s-closure of $\mathcal{M}$ and the conclusion follows from Lemma 4.1. ∎

The careful reader should notice that the preceding proof in actuality yields the following general result.

Theorem 4.3. *If X is a Banach space with the Schur property and $\mathcal{A}$ is a commutative subalgebra of $L(X)$, then the dual algebra $\mathcal{A}'$ of $L(X')$ admits a non-trivial closed $\mathcal{A}'$-invariant subspace.*

Acknowledgment. The authors express their sincerest thanks to Professor Louis de Branges for several fruitful and inspiring discussions on the invariant subspace problem. They also thank the referee for useful suggestions.

REFERENCES

1. Y. A. Abramovich, C. D. Aliprantis, and O. Burkinshaw, Invariant subspaces of operators on ℓ_p-spaces, *J. Funct. Anal.* **115** (1993), 418–424.

2. Y. A. Abramovich, C. D. Aliprantis, and O. Burkinshaw, Invariant subspace theorems for positive operators, *J. Funct. Anal.*, forthcoming.

3. H. Bercovici, Notes on invariant subspaces, *Bull. Amer. Math. Soc. (N.S.)* **23** (1990), 1–36.

4. S. Brown, Invariant subspaces for subnormal operators, *Integral Equations Operator Theory* **1** (1978), 310–333.

5. L. de Branges, The Stone–Weierstrass theorem, *Proc. Amer. Math. Soc.* **10** (1959), 822–824.

6. L. de Branges, A construction of invariant subspaces, *Math. Nachr.* **163** (1993), 163–175.

7. N. Dunford and J. T. Schwartz, *Linear Operators, I*, Wiley (Interscience), New York, 1958.

8. P. Enflo, On the invariant subspace problem for Banach spaces, *Seminaire Maurey–Schwarz* (1975–1976); *Acta Math.* **158** (1987), 213–313.

9. P. R. Halmos, *A Hilbert Space Problem Book*, 2nd Edition, Graduate Texts in Mathematics #19, Springer–Verlag, New York & Heidelberg, 1982.

10. R. B. Honor, Density and transitivity results on ℓ^∞ and ℓ^1, *J. London Math. Soc.* **32** (1985), 521–527.

11. V. I. Lomonosov, Invariant subspaces for operators commuting with compact operators, *Funktsional. Anal. i Prilozhen* **7** (1973), 55–56 (Russian); *Functional Anal. Appl.* **7** (1973), 213–214 (English).

12. V. I. Lomonosov, An extension of Burnside's theorem to infinite-dimensional spaces, *Israel J. Math.* **75** (1991), 329–339.

13. C. Pearcy and A. L. Shields, A survey of the Lomonosov technique in the theory of invariant subspaces, in *Topics in Operator Theory*, 219–229, *Math. Surveys No.* 13, Amer. Math. Soc., Providence, R.I., 1974.

14. H. Radjavi and P. Rosenthal, *Invariant Subspaces*, Springer–Verlag, Berlin & New York, 1973.

15. C. J. Read, A solution to the invariant subspace problem on the space ℓ_1, *Bull. London Math. Soc.* **17** (1985), 305–317.

16. P. Rosenthal, Equivalents of the invariant subspace problem, in *Paul Halmos: Celebrating 50 Years of Mathematics*, Springer–Verlag, Berlin & New York, 1991, 179–188.

1991 AMS Mathematics Subject Classification: 47A15, 47D30, 47D35

Department of Mathematical Sciences
IUPUI
402 N. Blackford Street
Indianapolis, IN 46202-3216
USA

Operator Theory:
Advances and Applications, Vol. 75
© 1995 Birkhäuser Verlag Basel/Switzerland

MATRIX YOUNG INEQUALITIES

T. Ando

Dedicated to Professor A. C. Zaanen on the occasion of his 80-th birthday

Let p, $q > 0$ satisfy $1/p + 1/q = 1$. We prove that for any pair A, B of $n \times n$ complex matrices there is a unitary matrix U, depending on A, B, such that

$$U^*|AB^*|U \leq |A|^p/p + |B|^q/q.$$

1. The most important case of the Young inequalities (see [1] Chap. 2) says that if $1/p + 1/q = 1$ with p, $q > 0$ then

$$|ab| \leq |a|^p/p + |b|^q/q \quad \text{for} \quad a, b \in \mathbb{C}. \tag{1}$$

In considering a matrix generalization of (1), the order relation $\leq$ should be defined in accordance with positive semi-definiteness; $A \geq B$ for Hermitian A, B means that $A - B$ is positive semi-definite. In particular, $A \geq 0$ means positive semi-definiteness of A. Let us write $A > 0$ when A is positive definite. $|A|$ should be understood as the modulus $|A| = (A^*A)^{1/2}$.

A direct matrix generalization of (1)

$$|AB| \leq |A|^p/p + |B|^q/q$$

does not hold in general. If A, $B \geq 0$ is a commuting pair, however, $AB \geq 0$ and via simultaneous diagonalization, it is proved that

$$AB \leq |A|^p/p + |B|^q/q.$$

Recently Bhatia-Kittaneh [2] established a matrix version of (1) for the special case $p = q = 1/2$ in the following form.

THEOREM (Bhatia and Kittaneh). *For any pair A, B of $n \times n$ complex matrices there is a unitary matrix U, depending on A, B, such that*

$$U^*|AB^*|U \leq |A|^2/2 + |B|^2/2. \tag{2}$$

In the present paper we shall extend this result to the case of general p. The proof of Bhatia-Kittaneh is based on the special situation that $|A|^2 = A^*A$ and $|B|^2 = B^*B$ while our proof is based on pinching inequalities for the map $X \geq 0 \mapsto X^r$ with $r > 0$ (see Lemma 2).

2. THEOREM. *Let p, $q > 0$ be mutually conjugate exponents, that is, $1/p + 1/q = 1$. Then for any pair A, B of $n \times n$ complex matrices there is a unitary matrix U, depending on A, B, such that*

$$U^*|AB^*|U \leq |A|^p/p + |B|^q/q. \tag{3}$$

Before going into proof, let us present some alternates of the assertion. Since each Hermitian matrix is unitarily similar to a diagonal matrix with its eigenvalues on the diagonal, the assertion is equivalent to the following system of inequalities for eigenvalues;

$$\lambda_i(|AB^*|) \leq \lambda_i(|A|^p/p + |B|^q/q) \quad (i = 1, 2, \cdots, n). \tag{4}$$

where, for a Hermitian matrix X, $\lambda_1(X) \geq \lambda_2(X) \geq \cdots \geq \lambda_n(X)$ are its eigenvalues, arranged in decreasing order.

If $A = V|A|$ and $B = W|B|$ are the polar representations of A and B respectively, with unitary V, W then

$$|AB^*| = W\big||A| \cdot |B|\big|W^*$$

and

$$\lambda_i(|AB^*|) = \lambda_i\big(\big||A| \cdot |B|\big|\big) \quad (i = 1, 2, \cdots, n).$$

Therefore, by replacing A and B by $|A|$ and $|B|$, respectively, it suffices to prove (4) only for A, $B \geq 0$.

3. Now let us enter a proof of (4) for A, $B \geq 0$. Fix k and let us prove

$$\lambda_k\big((BA^2B)^{1/2}\big) \leq \lambda_k(|A|^p/p + |B|^q/q). \tag{5}$$

Since

$$\lambda_k\big((BA^2B)^{1/2}\big) = \lambda_k\big((AB^2A)^{1/2}\big),$$

by exchanging the roles of A and B if necessary, we may assume that $1 < p \leq 2$, hence $2 \leq q < \infty$. Further by considering $B + \varepsilon I$ with $\varepsilon > 0$ and taking limit as $\varepsilon \to 0$, we may assume $B > 0$.

Write $\lambda = \lambda_k((BA^2B)^{1/2})$ and denote by P the orthoprojection (of rank k) to the spectral subspace, spanned by the eigenvectors corresponding to $\lambda_i((BA^2B)^{1/2})$ for $i = 1, 2, \cdots, k$. Denote by Q the orthoprojection (of rank k) to the subspace $\mathcal{M} = \mathrm{ran}(B^{-1}P)$. In view of the min-max characterization of eigenvalues of a Hermitian matrix (see [1] Chap. 2, §26) for inequality (5) it suffices to prove

$$\lambda Q \leq QA^pQ/p + QB^qQ/q. \tag{6}$$

By definition of Q we have

$$B^{-1}P = QB^{-1}P \quad \text{and} \quad PBQ = BQ,$$

which implies

$$PB^{-1}Q = PB^{-1} \quad \text{and} \quad QBP = QB.$$

Now it follows from these identities that

$$(QB^2Q) \cdot (B^{-1}PB^{-1}) = QB^2 \cdot QB^{-1}PB^{-1}$$
$$= QB^2 \cdot B^{-1}PB^{-1}$$
$$= QBPB^{-1} = Q,$$

and similarly

$$(B^{-1}PB^{-1}) \cdot (QB^2Q) = Q.$$

These together mean that $B^{-1}PB^{-1}$ and QB^2Q map $\mathcal{M}$ onto itself, vanish on its ortho-complement and are inverse to each other on $\mathcal{M}$.

4. For a proof of (6) we need some lemmas on the map $X \geq 0 \mapsto X^r$ with $r > 0$.

LEMMA 1. Let $0 < r \leq 1$. Then $0 \leq X \leq Y$ implies $X^r \leq Y^r$.

See Heinz [5] and Kato [6] for a proof of even more general cases of unbounded positive selfadjoint operators.

LEMMA 2. *Let Q be an orthoprojection. Then for $X \geq 0$*

$$QX^rQ \leq (QXQ)^r \quad \text{if} \quad 0 < r \leq 1,$$

and

$$QX^rQ \geq (QXQ)^r \quad \text{if} \quad 1 \leq r \leq 2.$$

See [3] and [4] for the proofs of these lemmas.

Let us return to the proof of (6). First by definition of P we have

$$(BA^2B)^{1/2} \geq \lambda P,$$

which implies, via commutativity of $(BA^2B)^{1/2}$ and P,

$$A^2 \geq \lambda^2 B^{-1}PB^{-1}.$$

Then by LEMMA 1 with $r = p/2$ we have

$$A^p \geq \lambda^p (B^{-1}PB^{-1})^{p/2},$$

hence

$$QA^pQ \geq \lambda^p (B^{-1}PB^{-1})^{p/2}.$$

Since $B^{-1}PB^{-1}$ is the inverse of QB^2Q on $\mathcal{M}$, this means that on $\mathcal{M}$

$$QA^pQ \geq \lambda^p (QB^2Q)^{-p/2}. \tag{7}$$

To prove (6), let us first consider the case $2 \leq q \leq 4$. Then by LEMMA 2 with $r = q/2$ we have

$$QB^qQ \geq (QB^2Q)^{q/2}. \tag{8}$$

Now it follows from (7) and (8) that on $\mathcal{M}$

$$QA^pQ/p + QB^qQ/q$$
$$\geq \lambda^p (QB^2Q)^{-p/2}/p + (QB^2Q)^{q/2}/q.$$

In view of the Young inequality for the commuting pair, $\lambda \cdot (QB^2Q)^{-1/2}$ and $(QB^2Q)^{1/2}$, this implies

$$(QA^pQ)/p + (QB^qQ)/q$$
$$\geq \lambda \cdot (QB^2Q)^{-1/2} \cdot (QB^2Q)^{1/2} = \lambda Q,$$

proving (6).

Let us next consider the case $4 < q < \infty$. Let $s = q/2$. Then $0 < 2/s \leq 1$ and $q/s = 2$. By LEMMA 2 with $r = q/s$ we have

$$QB^qQ \geq (QB^sQ)^{q/s}. \tag{9}$$

On the other hand, by LEMMA 2 with $r = 2/s$ we have

$$(QB^sQ)^{2/s} \geq QB^2Q,$$

and then by LEMMA 1 with $r = p/2$

$$(QB^sQ)^{p/s} \geq (QB^2Q)^{p/2},$$

hence on $\mathcal{M}$

$$(QB^sQ)^{-p/s} \leq (QB^2Q)^{-p/2}. \tag{10}$$

Now it follows from (7), (10) and (9) that

$$QA^pQ/p + QB^qQ/q$$
$$\geq \lambda^p \cdot (QB^sQ)^{-p/s}/p + (QB^sQ)^{q/s}/q.$$

In view of the Young inequality for the commuting pair $\lambda \cdot (QB^sQ)^{-1/s}$ and $(QB^sQ)^{1/s}$, this implies

$$QA^pQ/p + QB^qQ/q$$
$$\geq \lambda \cdot (QB^sQ)^{-1/s} \cdot (QB^sQ)^{1/s} = \lambda Q,$$

proving (6). This completes the proof.

REFERENCES

1.   E. F. Beckenbach and R. Bellman, *Inequalities*, Springer-Verlag, New York, 1965.

2.   R. Bhatia and F. Kittaneh, *On the singular values of a product of operators*, SIAM J. Matrix Anal. Appl. **11** (1990), 272–277.

3.   Ch. Davis, *Notions generalizing convexity for functions defined on spaces of matrices*, in Convexity: Proc. Symp. Pure Math. , 1963, 187–201.

4.   F. Hansen, *An operator inequality*, Math. Ann. **246** (1980), 249–250.

5.   E. Heinz, *Beiträge zur Störungstheorie der Spektralzerlegung*, Math. Ann. **123** (1951), 415–438.

6.   T. Kato, *Notes on some inequalities for linear operators*, Math. Ann. **125** (1952), 208–212.

Laboratory of Information Mathematics
Research Institute for Electronic Science
Hokkaido University
Sapporo 060, Japan

Operator Theory:
Advances and Applications, Vol. 75

PRINCIPAL EIGENVALUES AND PERTURBATION

Wolfgang Arendt and Charles J.K. Batty

Dedicated to Professor A.C. Zaanen on the occasion of his 80th birthday.

INTRODUCTION. In a classical article, Hess and Kato [HK] study the problem

$$(0.1) \qquad \begin{cases} Au + \lambda mu = 0 \\ 0 \leq u \in D(A), \quad u \neq 0, \end{cases}$$

where A is a strongly elliptic operator on a bounded open set Ω of $\mathbf{R}^n$ with Dirichlet boundary conditions and m is a continuous bounded function on Ω. They show that there exists a unique $\lambda > 0$ such that the problem (0.1) has a solution.

This result can be reformulated by saying that there is a unique $\lambda > 0$ such that the spectral bound $s(A + \lambda m)$ of the operator $A + \lambda m$ on $C_0(\Omega)$ is 0.

The motivation of Hess and Kato was to investigate bifurcation of a nonlinear problem and (0.1) is obtained by linearization.

There is also another reason to study the spectral bound : in many cases it determines the asymptotic behavior of the semigroup. In particular, if $s(A) = 0$, then, under suitable hypotheses, the semigroup generated by A converges to a rank-1-projection. We show a typical result of this sort in the first section.

In the second section we assume that A generates a positive semigroup such that $s(A) < 0$ and consider a positive compact perturbation $B : D(A) \to E$. We show that there exists a unique $\lambda > 0$ such that $s(A + \lambda B) = 0$. Continuity properties of the spectral bound play a role in this context. They are presented in the appendix.

In Section 3 we consider the case where m is no longer positive. A beautiful theorem due to Kato [K2] says that spectral bound $s(A + \lambda m)$ and type $\omega(A + \lambda m)$ are convex functions of $\lambda \in \mathbf{R}$ if $E = C_0(\Omega)$ and $m \in C^b(\Omega)$ or $E = L^p(\Omega)$ and $m \in L^\infty(\Omega)$.

We extend this result to the case where A is the generator of a positive C_0−semigroup on an arbitrary Banach lattice and m is in the centre of E.

In this context the lattice structure, and Kakutani's theorem in particular, play an important role. First of all, it allows one to define multiplication operators abstractly (in the form of the centre, see Section 3 and Zaanen [Z1], [Z2]). Secondly, in the proof of the convexity theorem we use an interesting approximation property in Banach lattices due to B. Walsh [W]. We give the proof of this apparently not very well-known result in Appendix B. Thus our presentation of the convexity theorem is selfcontained.

1. PERRON-FROBENIUS THEORY AND PRINCIPAL EIGENVALUES.

The aim of Perron-Frobenius theory is to deduce asymptotic behavior from the location of the spectrum. We describe one particular case.

Let E be a complex Banach lattice, e.g., $E = L^p$, $1 \le p < \infty$, or $E = C_0(\Omega)$, the space of all continuous functions on a locally compact space Ω which vanish at infinity.

Let A be the generator of a positive semigroup $T = (T(t))_{t \ge 0}$ on E (by this we mean a C_0−semigroup throughout). We denote by $\sigma(A)$ the spectrum of A and by

$$(1.1) \qquad\qquad s(A) = \sup \{\operatorname{Re} \lambda : \lambda \in \sigma(A)\}$$

the **spectral bound** of A. Then $s(A) < \infty$, and if $s(A) > -\infty$, then $s(A) \in \sigma(A)$. Moreover, $R(\mu, A) := (\mu - A)^{-1} \ge 0$ whenever $\mu > s(A)$ and conversely, if $\mu \in \rho(A) \cap \mathbf{R}$ such that $R(\mu, A) \ge 0$, then $\mu > s(A)$.

The spectral bound is the analogue of the spectral radius for unbounded operators. This is made precise by the following formula

$$(1.2) \qquad r(R(\mu, A)) = \frac{1}{\mu - s(A)} \quad (\mu > s(A)),$$

where $r(B)$ denotes the spectral radius of B. In particular, $s(A) = -\infty$ if and only if $R(\mu, A)$ is quasi-nilpotent. See [N] for these results. We use the following notation. An element $u \in E_+$ is called a **quasi-interior point** if $\varphi \in E'_+$, $\langle u, \varphi \rangle = 0$ implies $\varphi = 0$. Thus, u is quasi-interior if and only if the principal ideal $E_u = \{f \in E : \exists n \in \mathbf{N} \quad |f| \leq nu\}$ is dense in E. If $E = L^p$ $(1 \leq p < \infty)$ this is equivalent to $u > 0$ a.e. A positive linear form $\varphi \in E'_+$ is **strictly positive** if $\langle f, \varphi \rangle > 0$ for all $f \in E_+, f \neq 0$. We write $\varphi \gg 0$.

An operator $B \in \mathcal{L}(E)_+$ is **strictly positive** (we write $B \gg 0$) if Bf is a quasi-interior point for all $f \in E_+$, $f \neq 0$.

We say that the semigroup T is **irreducible** if $R(\mu, A) \gg 0$ for all $\mu > s(A)$.

The following definition is central for our purposes.

DEFINITION 1.1. We say that 0 is a **principal eigenvalue** if
(a) there exists $\epsilon > 0$ such that $\{\lambda \in \sigma(A) : \mathrm{Re}\,\lambda > -\epsilon\} = \{0\}$ and
(b) 0 is a pole of the resolvent.

Note that $s(A) = 0$ whenever 0 is a principal eigenvalue. If 0 is a principal eigenvalue and T satisfies a regularity condition, then T has a very specific asymptotic behavior : it converges to a rank-1-projection.

We say that T is **eventually norm continuous** if there exists $t_0 > 0$ such that $\lim_{t \downarrow 0} \|T(t_0 + t) - T(t_0)\| = 0$; e.g., a holomorphic semigroup satisfies this condition.

THEOREM 1.2. *Let T be irreducible and eventually norm continuous. Assume that 0 is a principal eigenvalue. Then there exists a unique quasi-interior point $u_0 \in D(A)$ such that $Au_0 = 0$, $\|u_0\| = 1$ and a unique $0 \ll \varphi_0 \in D(A')$ satisfying $A'\varphi_0 = 0$ and $\langle u_0, \varphi_0 \rangle = 1$. Moreover there exist $\delta > 0, M \geq 0$ such that*

$$(1.2) \qquad \|T(t) - \varphi_0 \otimes u_0\| \leq Me^{-\delta t} \quad (t \geq 0).$$

*We call u_0 the **principal eigenvector** of A and φ_0 the **principal eigenfunctional** of A'. By $P = \varphi_0 \otimes u_0$ we mean the rank-1-operator $Pf = \langle f, \varphi_0 \rangle u_0$.*

REMARK 1.3. Note that 0 is the only eigenvalue with a positive eigenvector. In fact, assume that $0 \leq v \in D(A)$, $Av = \mu v$. Then $\mu \langle v, \varphi_0 \rangle = \langle Av, \varphi_0 \rangle = \langle v, A'\varphi_0 \rangle = 0$. Hence $\mu = 0$ or $\langle v, \varphi_0 \rangle = 0$ (which implies $v = 0$ since $\varphi_0 \gg 0$).

Let $f \in D(A)$. Then $u(t) = T(t)f$ is the unique solution of the problem

$$(P) \quad \begin{cases} u \in C^1([0, \infty); E); \\ u(t) \in D(A) \quad (t \geq 0); \\ u'(t) = Au(t) \quad (t \geq 0); \\ u(0) = f. \end{cases}$$

Note that in the situation of Theorem 1.2 $T(t)u_0 \equiv u_0$. The estimate (1.2) implies that every solution converges to the stationary point u_0, i.e.

$$\lim_{t \to \infty} T(t)f = \langle f, \varphi_0 \rangle u_0.$$

Next we give a criterion for 0 to be a principal eigenvalue.

THEOREM 1.4. *Assume that T is irreducible and holomorphic. Assume that* $s(A) = 0$. *If*
(a) $R(\mu, A)$ is compact for (some) $\mu \in \rho(A)$ or
(b) $s(A - B) < 0$ for some compact operator $B : D(A) \to E$,
then 0 is a principal eigenvalue.

Here we consider $D(A)$ as a Banach space with the graph norm $\|f\|_A := \|f\| + \|Af\|$. Then $R(\mu, A)$ is an isomorphism from E onto $D(A)$ $(\mu > s(A))$. To say that $B : D(A) \to E$ is compact is equivalent to $BR(\mu, A) : E \to E$ being compact for one (equivalently all) $\mu \in \rho(A)$.

Note that, if $B : D(A) \to E$ is compact, then by a result of Desch-Schappacher [DS1] $A - B$ generates a holomorphic semigroup.

Theorems 1.2 and 1.4 are variants of the Perron-Frobenius theory developed in [N]. An essential argument is the cyclicity of the boundary spectrum, a result which is due to G. Greiner and is analogous to results of Lotz for bounded positive operators and of Perron-Frobenius for positive matrices (see [N] and [S] for details and bibliographical notes).
We give the proofs (based on [N]) in order to be complete.

PROOF OF THEOREM 1.2. It follows from [N, C-III, Prop. 3.5, p. 310] that 0 is a pole of order 1 and that the residue is of the form $P = \varphi_0 \otimes u_0$ with φ_0, u_0 as in the statement of the theorem. By [N, A-III, Theorem 3.3], P is the spectral projection corresponding to $\{0\}$. Let A_1 be the generator of $T_1(t) = T(t)_{|F}$ with $F = (I - P)E$. Then $s(A_1) < 0$. Since T_1 is eventually norm continuous one has $s(A_1) = \omega(A_1)$ (the type of T_1). Thus, if δ is such that $s(A_1) < -\delta < 0$, then there exists $M \geq 0$ such that $\|T_1(t)\| \leq Me^{-\delta t}$ $(t \geq 0)$. This implies (1.2). $\Diamond$

PROOF OF THEOREM 1.4. Let $\epsilon > 0$ be arbitrary in the first case and $s(A - B) < -\epsilon < 0$ in the second. It follows from [K1, IV §5.6, p. 242-244] that the set $H = \{\mu \in \sigma(A) : \text{Re } \mu \geq -\epsilon\}$ consists of isolated points which are poles of the resolvent of A. Since T is eventually norm-continuous the set H is compact [N, A-II, Theorem 1.20, p. 38]. Thus H is finite. By cyclicity [N, C-III, Theorem 3.12, p. 315] $\sigma(A) \cap i\mathbf{R}$ is unbounded or reduced to $\{0\}$. Since H is compact, it follows that $\sigma(A) \cap i\mathbf{R} = \{0\}$. Hence 0 is a principal eigenvalue. $\Diamond$

REMARK. Theorem 1.4 remains true if T is merely eventually norm-continuous. In that case $A - B$ is not necessarily the generator of a semigroup in the case (b), see Desch-Schappacher [DS2].

2. POSITIVE COMPACT PERTURBATION.

Let A be the generator of a positive irreducible holomorphic semigroup on a Banach lattice E. We consider perturbations of the form $A + \lambda B$ $(\lambda \geq 0)$, where $B : D(A) \to E$ is linear, compact $(D(A)$ being considered with the graph norm) and positive (i.e. $0 \leq f \in D(A)$ implies $Bf \geq 0$).

Then $A + \lambda B$ generates a holomorphic semigroup for all $\lambda > 0$ by a result of Desch-Schappacher [DS1] (see also [KS]). This semigroup is positive and irreducible (see proof of Theorem 2.4).

We assume throughout that $B \neq 0$. Under these conditions we show the following.

THEOREM 2.1. *Assume in addition to the hypotheses made above that $s(A) < 0$. Then there exists a unique $\lambda_0 > 0$ such that $s(A + \lambda_0 B)$ is a principal eigenvalue.*

COROLLARY 2.2. *Under the hypotheses of Theorem 2.1 there exists a* *unique $\lambda_0 > 0$ such that the problem*

$$\begin{cases} 0 \leq u \in D(A), \quad \|u\| = 1 \\ Au + \lambda_0\, Bu = 0 \end{cases}$$

has a solution. Moreover, this solution u is unique and u is a quasi interior point.

PROOF. This follows from Theorems 2.1, 1.2 and Remark 1.3. ◇

The proof of Theorem 2.1 consists of several steps.
We investigate the function

$$s : [0, \infty) \to [s(A), \infty)$$

$$\lambda \mapsto s(A + \lambda B).$$

Since B is positive, the function s is increasing :

(2.1) $$0 \leq \lambda_1 \leq \lambda_2 \text{ implies } s(\lambda_1) \leq s(\lambda_2).$$

PROOF. Let $\mu > \max\{s(\lambda_1), s(\lambda_2)\}$. Then

$$R(\mu, A + \lambda_2 B) - R(\mu, A + \lambda_1 B) = R(\mu, A + \lambda_2 B)(\lambda_2 - \lambda_1)BR(\mu, A + \lambda_1 B) \geq 0.$$

Hence $0 \leq R(\mu, A + \lambda_1 B) \leq R(\mu, A + \lambda_2 B)$ and by (1.2),

$$\frac{1}{\mu - s(A + \lambda_1 B)} = r(R(\mu, A + \lambda_1 B)) \leq r(R(\mu, A + \lambda_2 B)) = \frac{1}{\mu - s(A + \lambda_2 B)}. \quad ◇$$

Next we show that

(2.2) $$\lim_{\lambda \to \infty} s(A + \lambda B) = \infty.$$

This depends heavily on irreducibility (see Remark 2.7 below). In fact, we use the following deep theorem due to de Pagter [P] :

THEOREM 2.3. *Let K be a positive compact operator on E. If K is irreducible, then $r(K) > 0$.*

REMARK. A positive operator $K \in \mathcal{L}(E)$ is called **irreducible** if $(e^{tK})_{t \geq 0}$ is irreducible, i.e. if $R(\mu, K) \gg 0$ for $\mu > r(K)$. From the Neumann series one sees that K is irreducible whenever $K \gg 0$.

PROOF OF (2.2). Assume that there exists μ_0 such that $s(A + \lambda B) \leq \mu_0$ for all $\lambda > 0$. Let $\mu > \mu_0$. Then for $\lambda > 0$,

$$R(\mu, A + \lambda B) = R(\mu, A) + \lambda R(\mu, A + \lambda B)BR(\mu, A) \geq \lambda R(\mu, A)BR(\mu, A)$$

(see the proof of (2.1)). Hence

$$\frac{1}{\mu - \mu_0} \geq \frac{1}{\mu - s(A + \lambda B)} = r(R(\mu, A + \lambda B)) \geq \lambda r(K)$$

for all $\lambda > 0$, where $K = R(\mu, A)BR(\mu, A)$. Consequently, $r(K) = 0$. However, K is compact and strictly positive, and thus irreducible. This is impossible by de **Pagter**'s theorem.   ◊

Next we show that

(2.3) s is continuous on $[0, \infty)$.

PROOF. We know that s is increasing and $s(0) = s(A) < 0$. Let $\lambda_0 := \sup \{\lambda \geq 0 : s(\lambda) = s(A)\}$. It follows from Proposition A1 a) in Appendix A that $\varlimsup_{\lambda \downarrow \lambda_0} s(\lambda) \leq s(\lambda_0) = s(A)$. Thus s is continuous at λ_0 and on $[0, \lambda_0]$. If $\lambda > \lambda_0$, $s(A + \lambda B) > s(A)$. Thus $s(A + \lambda B)$ is an isolated point in the spectrum of $A + \lambda B$. So continuity in λ follows from Proposition A1 b).   ◊

It follows from (2.2), (2.3) and the assumption $s(A) < 0$ that there exists $\lambda_0 > 0$ such that $s(A + \lambda_0 B) = 0$. Then 0 is a principal eigenvalue of $A + \lambda_0 B$ by Theorem 1.4.

Finally, it follows from Proposition A2 that

(2.4) $s(\lambda_1) < s(\lambda_2)$ whenever $\lambda_0 < \lambda_1 < \lambda_2$.

This shows uniqueness of λ_0 and the proof of Theorem 2.1 is finished.

More generally, our arguments show the following.

THEOREM 2.4. *Let A be a resolvent positive densely defined operator (see Appendix A) such that*
(a) $s(A) < 0$;
(b) $\sup_{\mu \geq 0} \|\mu R(\mu, A)\| =: M < \infty$;
(c) $R(\mu, A) \gg 0$ $(\mu > 0)$.

Let $B : D(A) \to E$ be positive, compact and $\neq 0$.

Then $A + \lambda B$ is resolvent positive for all $\lambda \geq 0$. The function $s(\lambda) = s(A + \lambda B)$ is continuous on $[0, \infty)$, strictly increasing and $\lim_{\lambda \to \infty} s(\lambda) = \infty$. In particular, there exists a unique $\lambda_0 > 0$ such that $s(A + \lambda_0 B) = 0$.

PROOF. a) We show that $A + \lambda B$ is resolvent positive for all $\lambda \geq 0$. In fact, an argument given by Desch-Schappacher [DS1] can be adapted to the situation considered here. We can assume $\lambda = 1$. Since $s(A) < 0$, one may consider the equivalent norm $\|x\|_{D(A)} := \|Ax\|$ on $D(A)$. It follows from (b) that

$$\lim_{\mu \to \infty} \|R(\mu, A)x\|_{D(A)} = \lim_{\mu \to \infty} \|AR(\mu, A)x\| = \lim_{\mu \to \infty} \|\mu R(\mu, A)x - x\| = 0 \quad (x \in E).$$

Since

$$\sup_{\mu \geq 0} \|R(\mu, A)\|_{\mathcal{L}(E, D(A))} = \sup_{\mu \geq 0} \|AR(\mu, A)\| = \sup_{\mu \geq 0} \|\mu R(\mu, A) - I\| \leq M + 1 < \infty,$$

it follows that $\lim_{\mu \to \infty} \|R(\mu, A)x\|_{D(A)} = 0$ uniformly for x in compact subsets of E. Since $B : D(A) \to E$ is compact, we conclude that

$$\|R(\mu, A)B\|_{\mathcal{L}(D(A))} \to 0 \quad (\mu \to \infty).$$

Hence there exists $\mu_0 > 0$ such that $\|R(\mu, A)B\|_{\mathcal{L}(D(A))} \leq \frac{1}{2}$ whenever $\mu \geq \mu_0$. This implies that $(I - R(\mu, A)B)$ is invertible in $\mathcal{L}(D(A))$ and

$$(I - R(\mu, A)B)^{-1} = \sum_{n=0}^{\infty} (R(\mu, A)B)^n \geq 0.$$

Hence

$$(\mu - (A + B))^{-1} = [(\mu - A)(I - R(\mu, A)B)]^{-1} = (I - R(\mu, A)B)^{-1} R(\mu, A)$$

exists and is positive for all $\mu \geq \mu_0$.

b) Note that

$$R(\mu, A + B) = R(\mu, A) + \sum_{n=1}^{\infty} (R(\mu, A)B)^n R(\mu, A) \geq R(\mu, A) \gg 0 \quad (\mu > \mu_0).$$

The remaining arguments are the same as above.   ◊

REMARK 2.5. The proof also shows that

$$\overline{\lim_{\mu \to \infty}} \|\mu R(\mu; A + \lambda B)\| < \infty \quad (\lambda > 0).$$

REMARK 2.6. Also in the situation of Theorem 2.4 we can conclude that there exists a unique $\lambda_0 > 0$ such that the problem

(2.5)
$$\begin{cases} (A + \lambda_0 B)u = 0 \\ u \in D(A), u \geq 0, \|u\| = 1 \end{cases}$$

has a solution. And, as before, this solution u is unique and u is a quasi interior point. However, we do not know whether, in the general situation considered in Theorem 2.4, 0 is a principal eigenvalue. As before, the boundary spectrum $\sigma(A + \lambda_0 B) \cap i\mathbf{R}$ is cyclic (by the proofs given in [N]). But it is not clear whether $\sigma(A + \lambda_0 B) \cap i\mathbf{R}$ is bounded (which is needed to conclude that $\sigma(A + \lambda_0 B) \cap i\mathbf{R} = \{0\}$).

REMARK 2.7. In Theorem 2.1 and 2.4 irreducibility is essential. In fact, if $E = \mathbf{R}^2$, $A = \begin{pmatrix} -1 & 0 \\ 0 & -1 \end{pmatrix}$, $B = \begin{pmatrix} 0 & 1 \\ 0 & 0 \end{pmatrix}$, then $s(A + \lambda B) = -1$ for all $\lambda > 0$.

3. KATO'S CONVEXITY THEOREM.

If B is a multiplication operator, then the function $\lambda \mapsto s(A + \lambda B)$ is continuous even if no assumption of compactness and on the sign of B is made. This follows from the following theorem due to T. Kato [K2].

THEOREM 3.1. *Let A be the generator of a positive semigroup on $E = L^p$ $(1 \leq p < \infty)$ or $E = C_0(\Omega)$ (Ω locally compact).*

Then the functions

$$m \mapsto s(A + m) \quad and$$
$$m \mapsto \omega(A + m)$$

from L^∞ (resp. $C^b(\Omega)$) into $[-\infty, \infty)$ are convex.

Here we identify $m \in L^\infty$ (resp. $m \in C^b(\Omega)$) with the multiplication operator $f \mapsto mf$ on L^p (resp. $C_0(\Omega)$). Moreover, we use the following definition of convexity :

A function s defined on a vector space Z with values in $[-\infty, \infty)$ is called **convex** if either $s(m) = -\infty$ for all $m \in Z$ or $s(m) > -\infty$ for all $m \in Z$ and $s(\theta m_1 + (1 - \theta)m_2) \leq \theta s(m_1) + (1 - \theta) s(m_2)$ whenever $0 \leq \theta \leq 1$, $m_1, m_2 \in Z$.

The case $C_0(\Omega)$ is of particular interest. It seems to play a special role concerning de Pagter's theorem. In fact, in this case, it is very easy to see that every positive irreducible operator on $C_0(\Omega)$ has strictly positive spectral radius even without compactness assumptions (cf. [S, V. 6.1], [N, B-III. Prop. 3.5a] and the proof below). Thus, as a consequence of Kato's theorem, one obtains the same conclusions as in Theorem 2.4 in the case where $E = C_0(\Omega)$ and B is a positive multiplication operator :

COROLLARY 3.2. *Let A be the generator of a positive irreducible semigroup on $C_0(\Omega)$, Ω locally compact. Assume that $s(A) < 0$. Let $0 \leq m \in C^b(\Omega)$, $m \neq 0$. Then there exists a unique $\lambda > 0$ such that $s(A + \lambda m) = 0$.*

PROOF. a) If K is a strictly positive operator on $C_0(\Omega)$, then $r(K) > 0$. In fact, let $f \in C_0(\Omega)$ be of compact support such that $f \geq 0$, $\|f\| = 1$. Since $Kf \gg 0$, there exists $c > 0$ such that $Kf \geq cf$. Consequently, $K^n f \geq c^n f$ $(n \in \mathbf{N})$, and so $\|K^n\| \geq c^n$. It follows that $r(K) \geq c$.

b) Now the proof of (2.2) shows that $\lim\limits_{\lambda \to \infty} s(A + \lambda m) = \infty$. Since the function $\lambda \mapsto s(A + \lambda m)$ is convex and $s(A) < 0$, the claim follows. $\Diamond$

It has been pointed out by Kato, that Theorem 3.1 remains true on similar spaces where multiplication operators can be defined. The purpose of this section is to show that, indeed, the theorem remains true on arbitrary Banach lattices if multiplication operators are replaced by operators in the centre.

Let E be an arbitrary real Banach lattice. By

$$Z(E) := \{T \in \mathcal{L}(E) : \exists c \geq 0 \quad |Tx| \leq c|x| \quad (x \in E)\}$$

we denote **the centre** of E. It is remarkable that $Z(E)$ can be described purely in terms of orthogonality in the lattice sense :

$$Z(E) = \{T : E \to E \text{ linear } : |x| \wedge |y| = 0 \Rightarrow |Tx| \wedge |y| = 0\},$$

see Zaanen [Z2] for proofs and further results.

For our purposes the following theorem is important. It is a consequence of Kakutani's representation theorem (see Zaanen [Z2] and Meyer-Nieberg [MN]).

THEOREM 3.3. *The space $Z(E)$ is a closed subalgebra of $\mathcal{L}(E)$. Moreover, there exists a compact space K and a lattice and algebra isomorphism $\phi : C(K) \to Z(E)$.*

In the concrete cases considered in Theorem 3.1 the centre consists precisely of all multiplication operators :

EXAMPLE 3.4. a) Let Ω be locally compact, $E = C_0(\Omega)$. Then $Z(E)$ is isomorphic to $C^b(\Omega)$ identifying elements of $C^b(\Omega)$ with multiplication operators on E.

The space $C^b(\Omega)$ being isomorphic (as Banach lattice and algebra) to the space $C(\beta\Omega)$, this yields a proof of Theorem 3.3 in this concrete case.

b) Let (Ω, Σ, μ) be a $\sigma-$finite measure space and $E = L^p(\Omega)$, $1 \leq p \leq \infty$. Then $Z(E)$ is isomorphic to L^∞ (by action as multiplication operators). Again L^∞ is isomorphic to a space $C(K)$, K compact.

The results reported in Example 3.4 are due to Zaanen [Z1], see also Zaanen [Z2].

Now we can formulate the following more general version of Kato's theorem :

THEOREM 3.5. *Let A be the generator of a positive semigroup on a Banach lattice E. Then the functions*

$$M \mapsto s(A + M) \quad and$$
$$M \mapsto \omega(A + M)$$

from $Z(E)$ into $[-\infty, \infty)$ are convex.

For the proof we use the following defintion. Let $-\infty \leq a < b \leq \infty$; $\mathcal{L}(E)_+ = \{Q \in \mathcal{L}(E) : Q \geq 0\}$.

DEFINITION 3.6. (a) A function $q : (a, b) \to [0, \infty)$ is **superconvex** if $\log q$ is convex.

(b) A function $Q : (a, b) \to \mathcal{L}(E)_+$ is **superconvex** if $\langle Q(\cdot)x, x' \rangle$ is superconvex for all $x \in E_+$, $x' \in E'_+$.

The following properties are immediately clear.

LEMMA 3.7. *(a) Let $Q_1, Q_2 : (a, b) \to \mathcal{L}(E)_+$ be superconvex and let $\alpha, \beta \geq 0$. Then $\alpha Q_1 + \beta Q_2$ is superconvex.*

(b) Let $Q_n : (a, b) \to \mathcal{L}(E)_+$ be superconvex. Assume that $Q_n(\lambda)$ converges to $Q(\lambda)$ in the weak operator topology for all $\lambda \in (a, b)$. Then $Q(\lambda)$ is superconvex.

It is clear that the product of two numerical-valued superconvex functions is super-convex. Walsh's approximation theorem (Theorem B.1) allows one to extend this to operator-valued functions :

PROPOSITION 3.8. Let $Q_1, Q_2 : (a, b) \to \mathcal{L}(E)_+$ be superconvex. Then $\lambda \mapsto Q_1(\lambda) Q_2(\lambda)$ is superconvex.

PROOF. (a) Let $y \in E_+, y' \in E'_+$. Then for all $x \in E_+$, $x' \in E'_+$, $\lambda \mapsto \langle Q_1(\lambda)(y' \otimes y) Q_2(\lambda)x, x' \rangle = \langle Q_2(\lambda)x, y' \rangle \langle Q_1(\lambda)y, x' \rangle$ is superconvex.

(b) By Theorem B1 there exists a net R_α in P (see Appendix B for the definition) which converges strongly to the identity. It follows from (a) and Lemma 3.7 that $Q_1(\cdot) R_\alpha Q_2(\cdot)$ is superconvex for all α. Hence the limit $Q_1(\cdot) Q_2(\cdot)$ is superconvex by Lemma 3.7. $\Diamond$

COROLLARY 3.9. If $Q : (a, b) \to \mathcal{L}(E)_+$ is super-convex, then the spectral radius $r(Q(\cdot))$ is superconvex.

PROOF. Let $B_+ = \{x \in E_+ : \|x\| \leq 1\}$, $B'_+ = \{x' \in E'_+ : \|x'\| \leq 1\}$. By Proposition 3.8, the function $\lambda \mapsto \langle Q(\lambda)^n x, x' \rangle$ is superconvex for all $x \in B_+$, $x' \in B'_+$. Hence

$$\lambda \mapsto r(Q(\lambda)) = \lim_{n \to \infty} \|Q(\lambda)^n\|^{1/n}$$

$$= \lim_{n \to \infty} \left(\sup_{\substack{x \in B_+ \\ x' \in B'_+}} \langle Q(\lambda)^n x, x' \rangle \right)^{1/n}$$

is superconvex. $\Diamond$

PROOF OF THEOREM 3.5. Let $M \in Z(E)$.

(a) We show that the function $\lambda \mapsto e^{\lambda M} : \mathbb{R} \to \mathcal{L}(E)$ is superconvex. Let $x \in E_+$, $x' \in E'_+$. There exists an algebra isomorphism ϕ from $Z(E)$ onto $C(K)$. In particular, $e^{\phi(N)} = \phi(e^N)$ $(N \in Z(E))$.

By the Riesz representation theorem there exists a positive Borel measure μ on K such that

$$\langle N x, x' \rangle = \int_K \phi(N) d\mu   (N \in Z(E)).$$

Let $\lambda_1, \lambda_2 \in \mathbf{R}, 0 < \theta < 1, M_1 = \lambda_1 M, M_2 = \lambda_2 M$. Then by Hölder's inequality,

$$\langle e^{\theta M_1 + (1-\theta)M_2} x, x' \rangle = \int \phi(e^{\theta M_1 + (1-\theta)M_2}) \, d\mu$$

$$= \int (e^{\phi(M_1)})^\theta \cdot (e^{\phi(M_2)})^{1-\theta} \, d\mu$$

$$\leq \left(\int e^{\phi(M_1)} \, d\mu \right)^\theta \cdot \left(\int e^{\phi(M_2)} \, d\mu \right)^{1-\theta}$$

$$= (\langle e^{M_1} x, x' \rangle)^\theta (\langle e^{M_2} x, x' \rangle)^{1-\theta}.$$

(b) Let $t > 0$, $M \in Z(E)$. By (a) the function $\lambda \mapsto e^{\lambda t M}$ is superconvex. It follows from Lemma 3.7 and Proposition 3.8 that also the function $\lambda \mapsto e^{t(A+\lambda M)} = s - \lim_{n \to \infty} (e^{t/nA} e^{\lambda t/nM})^n$ is superconvex.

(c) It follows from (b) and Corollary 3.9 that the function $\lambda \mapsto r(e^{(A+\lambda M)}) = e^{\omega(A+\lambda M)}$ is superconvex. Hence $\omega(A + \lambda M)$ is convex in λ.

(d) Let $\lambda_0 > 0$. We show that $s(A+\lambda M)$ is convex in $\lambda \in (-\lambda_0, \lambda_0)$. Let $w > \omega(A)+\lambda_0 \|M\|$. There exists $c > 0$ such that $\|e^{t(A+\lambda M)}\| \leq c e^{wt}$ $(t \geq 0)$ for all $\lambda \in (-\lambda_0, \lambda_0)$.

It follows from (b) and Lemma 3.7 that

$$\lambda \mapsto R(\mu, A + \lambda M) = \int_0^\infty e^{-\mu t} e^{t(A+\lambda M)} \, dt$$

is super convex on $(-\lambda_0, \lambda_0)$ whenever $\mu > w$.

By Corollary 3.9, $r(R(\mu, A + \lambda M))$ is super convex and a fortiori convex in $\lambda \in (-\lambda_0, \lambda_0)$. Consequently, $\mu^2 r(R(\mu, A + \lambda M)) - \mu = \frac{\mu^2}{\mu - s(A+\lambda M)} - \mu = s(A + \lambda M)(1 - \frac{s(A+\lambda M)}{\mu})^{-1}$ is convex in $\lambda \in (-\lambda_0, \lambda_0)$ for all $\mu > w$. Letting $\mu \to \infty$ one concludes that $s(A + \lambda M)$ is convex.

e) Let $M_1, M_2 \in Z(E)$, $0 \leq \theta \leq 1$. It follows from (d) that $\lambda \mapsto s(A + B(\lambda))$ is convex where $B(\lambda) = M_2 + \lambda(M_1 - M_2)$. In particular,

$$s(A + \theta M_1 + (1 - \theta)M_2) = s(A + B(\theta 1 + (1 - \theta)0))$$
$$\leq \theta s(A + B(1)) + (1 - \theta) s(A + B(0))$$
$$= \theta s(A + M_1) + (1 - \theta) s(A + M_2).$$

This proves convexity of $M \mapsto s(A + M)$.

In the same way (c) implies convexity of $M \mapsto \omega(A + M)$.   ◇

REMARK : As pointed out by Kato [K2, p. 268], it follows from Theorem B1 that (weak) superconvexity as defined here (Definition 3.6 (b)) is equivalent to (strong) superconvexity in the sense of [K2, p. 26]. This simplifies the proof (but is restricted to Banach lattices). Otherwise, the arguments given here are those of Kato besides (a) in the proof of Theorem 3.5 which establishes the link with the abstract setting.

For positive perturbations which are not multiplication operators, the function $\lambda \mapsto s(A + \lambda B)$ is not convex, in general :

EXAMPLE 3.9. Let $E = \mathbf{R}^2$, $A = \begin{pmatrix} 1 & 1 \\ 1 & 1 \end{pmatrix}$, $B = \begin{pmatrix} 0 & 1 \\ 0 & 0 \end{pmatrix}$. Then $s(A + \lambda B) = \sqrt{2 + \lambda}$ is a concave function. $\Diamond$

REMARK 3.10. We have considered a real Banach lattice E. Of course, the spectrum is understood with respect to the corresponding complexifications. However, for a resolvent positive operator A on E one always has $s(A) = \inf \{\mu \in \mathbf{R} : (\lambda - A)^{-1}$ exists and $(\lambda - A)^{-1} \geq 0$ for all $\lambda > \mu\}$; this expression involves only the real space and real operators.

APPENDIX A. THE SPECTRAL BOUND OF RESOLVENT POSITIVE OPERATORS

Let A be a Banach lattice. An operator A on E is called **resolvent positive** if there exists $\lambda_0 \in \mathbf{R}$ such that $(\lambda_0, \infty) \subset \rho(A)$ and $R(\lambda, A) \geq 0$ for all $\lambda > \lambda_0$. The spectral bound of such an operator can be defined as in Section 1 and has the same properties (see [A]). We need the following continuity property of the spectral bound.

PROPOSITION A1. *Let A_n, A be resolvent positive operators. Assume that there exists $\mu > \sup (\{s(A_n) : n \in \mathbf{N}\} \cup \{s(A)\})$ such that $\lim_{n \to \infty} \|R(\mu, A_n) - R(\mu, A)\| = 0$. Then*

a) $\overline{\lim_{n \to \infty}} \, s(A_n) \leq s(A)$ *and,*
b) *if $s(A)$ is an isolated point in $\sigma(A)$, then $\lim_{n \to \infty} s(A_n) = s(A)$.*

PROOF. a) This follows from the upper continuity of the spectral radius [Au, Théorème 3, p. 6] and (1.2).

b) The assertion follows from a) and (1.2) together with the corresponding result for bounded operators [Au, Théorème 4, p. 8]. $\Diamond$

Next we formulate a result on strict monotonicity of the spectral bound. The argument is the same as in [AB, Theorem 1.3].

PROPOSITION A2. *Let A_1, A_2 be resolvent positive operators with dense domain such that*

$$0 \ll R(\lambda, A_1) \le R(\lambda, A_2) \text{ for } \lambda > \ max \ \{s(A_1), s(A_2)\}.$$

Assume that
(a) $A_1 \ne A_2$ and
(b) $s(A_i)$ is a pole of the resolvent of A_i, $i = 1, 2$.

Then $s(A_1) < s(A_2)$.

APPENDIX B. WALSH'S APPROXIMATION PROPERTY AND SUPERCONVEXITY.

Let E be a real Banach lattice. By P we denote the cone of all positive operators on E which are of the form $\sum_{i=1}^{n} x_i' \otimes x_i$ (i.e. $x \mapsto \sum_{i=1}^{n} \langle x, x_i' \rangle x_i$) with $x_i \in E_+$, $x_i' \in E_+'$ $i = 1, \ldots, n$.

The following approximation property is due to B. Walsh [W].

THEOREM B1. *The identity I is in the closure of P with respect to the strong operator topology.*

For completeness we include Walsh's proof. It depends on

LEMMA B2. *The assertion of Theorem B1 holds iff for every operator $R = \sum_{i=1}^{n} x_i' \otimes x_i$, $R \ge 0$ implies $\sum_{i=1}^{n} \langle x_i, x_i' \rangle \ge 0$.*

PROOF. The dual space of $\mathcal{L}_s(E)$ is $E' \otimes E$, the duality being defined by $\langle T, x' \otimes x \rangle = \langle Tx, x' \rangle$ (see [S1, p. 139]).

a) Assume that $I \notin \overline{P}$. Then there exists $R \in E' \otimes E$ such that $\langle p, R \rangle \ge 0$ for all $p \in P$ but $\langle I, R \rangle < 0$. Write $R = \sum_{j=1}^{n} x_j' \otimes x_j$. Then $0 > \langle I, R \rangle = \sum_{j=1}^{n} \langle x_j, x_j' \rangle$.
On the other hand let $p = y' \otimes y$ with $y' \in E_+'$, $y \in E_+$. Then $p \in P$. Thus $0 \le \langle p, R \rangle =$

$$\sum_{j=1}^{n} \langle px_j, x_j' \rangle = \sum_{j=1}^{n} \langle y, x_j' \rangle \langle x_j, y' \rangle = \langle Ry, y' \rangle.$$ Since $y' \in E_+'$, $y \in E_+$ are arbitrary this implies that R is a positive operator.

b) Conversely, suppose that $R = \sum_{j=1}^{n} x_j' \otimes x_j \geq 0$. Then $\langle Ry, y' \rangle \geq 0$ for all $y \in E_+$, $y' \in E_+'$. Thus $0 \leq \langle p, R \rangle$ for all $p \in P$. Hence, if $I \in \overline{P}$, then it follows that $0 \leq \langle I, R \rangle = \sum_{j=1}^{n} \langle x_j', x_j \rangle.$ $\Diamond$

PROOF OF THEOREM B1. a) One easily verifies that the theorem holds if $E = C(K)$, K compact ; see [S2, IV. Theorem 2.4, p. 239].

b) Let E be arbitrary. Let $R = \sum_{j=1}^{n} x_j' \otimes x_j$ be positive. By Lemma B1 it suffices to show that $\sum_{j=1}^{n} \langle x_j', x_j \rangle \geq 0$. Let $u = \sum_{j=1}^{n} |x_j|$. Then R leaves E_u invariant. By Kakutani's theorem E_u is isomorphic to a space $C(K)$. Thus we can assume that $E = C(K)$. Now the claim follows from a) and the other implication of Lemma B2. $\Diamond$

REMARK. Walsh [W] actually shows that Theorem B1 holds in an ordered Banach space E with normal and generating cone if and only if E' is a lattice.

REFERENCES

[Au]   Aupetit, B. : Propriétés Spectrales des Algèbres de Banach, Springer LN 375, Berlin 1979.

[A]   Arendt, W. : Resolvent positive operators, Proc. London Math. Soc. 54 (1987) 321-349.

[AB]   Arendt, W. and Batty, C. : Domination and ergodicity for positive semigroups, Proc. AMS 114 (1992) 743-747.

[DS1]   Desch, W. and Schappacher, W. : Some perturbation results for analytic semigroups, Math. Ann. 281 (1988)157-162.

[DS2]   Desch, W. and Schappacher, W. : On relatively bounded perturbations of linear C_0-semigroups, Ann. Scuola Normale Sup. Pisa XX (1984) 327-341.

[H]   Hess, P. and Kato, T. : On some linear and nonlinear eigenvalue problems with an indefinite weight function, Comm. PDE 5 (1980) 999-1030.

[KS]   Kappel, F. and Schappacher, W. : Strongly Continuous Semigroups,
            An Introduction, Book to appear.

[K1]   Kato, T. : Perturbation Theory, Springer, Berlin 1976.

[K2]   Kato, T. : Superconvexity of the spectral radius and convexity of the spectral
            bound and the type, Math. Z. 180 (1982) 265-273.

[LZ]   Luxemburg, W.A.J. and Zaanen, A.C. : Riesz Spaces I, North Holland,
            Amsterdam 1971.

[MN]   Meyer-Nieberg, P. : Banach lattices, Springer, Berlin 1991.

[N]   Nagel, R. : One-parameter semigroups of positive operators,
            Springer LN 1184, Berlin 1986.

[P]   de Pagter, B. : Irreducible compact operators, Math. Z. 192 (1986) 149-153.

[S1]   Schaefer, H.H. : Topological Vector Spaces, Springer, Berlin 1971.

[S2]   Schaefer, H.H. : Banach Lattices and Positive Operators, Springer, Berlin 1974.

[W]   Walsh, B. : Positive approximate identities and lattice-ordered dual space,
            Manuscripta Mathematica 14 (1974) 57-63.

[Z1]   Zaanen, A.C. : Examples of orthomorphisms, J. Approx. Theory 13 (1975)
            192-204.

[Z2]   Zaanen, A.C. : Riesz Spaces II, North-Holland, Amsterdam 1983.

AMS classification: 47D07, 47B65

Wolfgang ARENDT   Charles J.K. BATTY
URA CNRS 741   7St. John's College
Equipe de Mathématiques   OXFORD OX1 3JP
16, Route de Gray   Great Britain
F-25030 BESANCON Cedex

Operator Theory:
Advances and Applications, Vol. 75
© 1995 Birkhäuser Verlag Basel/Switzerland

INPUT-OUTPUT OPERATORS OF J-UNITARY

TIME-VARYING CONTINUOUS TIME SYSTEMS

J.A. Ball, I. Gohberg and M.A. Kaashoek

Dedicated to A.C. Zaanen on the occasion of his 80th birthday, with respect and admiration

This paper presents the construction of all integral operators T on the Hilbert space $L_2^\ell(\mathbb{R})$ which appear as input-output operators of J-unitary time-varying systems and for which the set of all operators TK, where K runs over the lower triangular Hilbert-Schmidt operators on $L_2^\ell(\mathbb{R})$, is prescribed. The problem solved here may be viewed as a nonstationary analogue of the problem of constructing a rational matrix function which is J-unitary on the imaginary axis and has a prescribed null-pole structure in the open right half plane.

0. INTRODUCTION

In interpolation problems (relative to the right half plane) of Nevanlinna Pick

type for rational matrix functions an important role is played by rational matrix functions,

$$\Theta(\lambda) = \begin{pmatrix} \Theta_{11}(\lambda) & \Theta_{12}(\lambda) \\ \Theta_{21}(\lambda) & \Theta_{22}(\lambda) \end{pmatrix},$$

which are J-unitary on the imaginary axis, where

$$J = \begin{pmatrix} I_{\mathbb{C}^m} & 0 \\ 0 & -I_{\mathbb{C}^r} \end{pmatrix},$$

and which have a prescribed null-pole structure in the open right half plane. In fact, the

entries Θ_{ij}, $1 \le i, j \le 2$, of such a matrix function appear as the coefficients in the linear

fractional representation of the solutions of interpolation problems of Nevanlinna-Pick,

Nudelman, or Nehari type (see, e.g., [BGR], Part V). The null-pole structure of a rational

matrix function relative to the open right half plane involves (see [BGR], Chapter 4) a triple $(C, A; Z, B; \Gamma)$, consisting of two pairs of matrices and a matrix Γ, of appropriate sizes, satisfying the following conditions:

(a) A and Z are square matrices with eigenvalues in the open right half plane only;

(b) $\cap_{j=1}^{n} \text{Ker } CA^{j-1} = \{0\}$ and $\mathbb{C}^k = \text{span } \{\text{Im } Z^{j-1}B \mid j = 1, \dots, k\}$, where n and k are the orders of A and Z, respectively;

(c) $\Gamma A - Z\Gamma = BC$.

Here (C, A) describes the pole structure of the rational matrix function involved, (Z, B) its null structure, and Γ is the so-called coupling matrix.

A basic problem is, given a signature matrix J (i.e., $J = J^* = J^{-1}$) and a triple $\omega = (C, A; Z, B; \Gamma)$ satisfying the conditions (a)–(c) above, construct a rational matrix function Θ such that Θ is J-unitary on the imaginary axis and has ω as its right half plane (RHP) null-pole triple. The latter can be expressed in different forms. One way is to say that

$$\Theta \mathcal{R}_+^{p \times 1} = S_\omega,$$

where $\mathcal{R}_+^{p \times 1}$ denotes the set of all rational $\mathbb{C}^p$-valued functions with poles only in the open left half plane and S_ω is the singular subspace associated with ω, i.e., the space consisting of all $\mathbb{C}^p$-valued functions of the form

$$C(\lambda - A)^{-1}x + h(\lambda),$$

where $x \in \mathbb{C}^n$ and $h \in \mathcal{R}_+^{p \times 1}$ are such that the total RHP residue of the matrix functions $(\lambda - Z)^{-1}Bh(\lambda)$ is equal to Γx. The solution of the above problem is known (see, e.g., [BGR], Section 6.3) and proceeds as follows. First one has to construct the matrix

$$\Lambda = \begin{pmatrix} K_1 & \Gamma^* \\ \Gamma & K_2 \end{pmatrix},$$

where K_1 and K_2 are the unique solutions of the Lyapunov equations

$$K_1 A + A^* K_1 = -C^* JC, \quad K_2 Z^* + ZK_2 = BJB^*.$$

Then there exists a J-unitary Θ which has $\omega = (C, A; Z, B; \Gamma)$ as its RHP null-pole triple if and only if the matrix Λ is nonsingular, and in this case any such Θ has the form

$$\Theta(\lambda) = D + (C \quad -JB^*) \begin{pmatrix} (\lambda - A)^{-1} & 0 \\ 0 & (\lambda + Z^*)^{-1} \end{pmatrix} \Lambda^{-1} \begin{pmatrix} -C^*J \\ B \end{pmatrix} D,$$

where D is any J-unitary matrix.

In interpolation problems of nonstationary type (see the papers in [G], [DKV]) one meets an analogous problem. To state this analogous version, first note that the rational matrix function Θ described above is the transfer function of the system

$$\begin{cases} x'(t) = Fx(t) + Gu(t), & t \in \mathbb{R}, \\ y(t) = Hx(t) + Du(t), \end{cases}$$

where

$$F = \begin{pmatrix} A & 0 \\ 0 & -A^* \end{pmatrix}, \quad G = \begin{pmatrix} -C^*J \\ B \end{pmatrix} D, \quad H = (C \quad -JB^*).$$

In the latter form the basic problem described in the previous paragraph has a natural time-varying analogue, which goes on as follows. Let $\omega = (C, A; Z, B; \Gamma)$ be a quintet of matrix functions of appropriate sizes with entries in $L_\infty(\mathbb{R})$. More precisely, let

(0.1) $C \in L_\infty^{p \times n}(\mathbb{R}), \quad A \in L_\infty^{n \times n}(\mathbb{R}), \quad Z \in L_\infty^{k \times k}(\mathbb{R}), \quad B \in L_\infty^{k \times p}(\mathbb{R}), \quad \Gamma \in L_\infty^{k \times n}(\mathbb{R}).$

Denote by $U_A(t)$ and $U_Z(t)$ the fundamental matrices (normalized to an identity matrix at $t = 0$) of the respective differential equations

$$x'(t) = A(t)x(t), \quad x'(t) = Z(t)x(t), \quad t \in \mathbb{R},$$

and assume that the following conditions hold:

(α) there exist constants $M \geq 0$, $0 < a < 1$, such that

$$\|U_A(t)U_A(s)^{-1}\| \leq Ma^{s-t}, \quad \|U_Z(t)U_Z(s)^{-1}\| \leq Ma^{s-t}, \quad s \geq t;$$

(β) there exists $\varepsilon > 0$ such that

$$\int_{-\infty}^t (U_A(t)^{-1})^* U_A(s)^* C(s)^* C(s) U_A(s) U_A(t)^{-1}\, ds \geq \varepsilon I_{\mathbb{C}^n}, \quad t \in \mathbb{R},$$

$$\int_t^\infty U_Z(t)U_Z(s)^{-1}B(s)B(s)^*(U_Z(s)^{-1})^*U_Z(t)^*\,ds \geq \varepsilon I_{\mathbb{C}^k}, \quad t \in \mathbb{R};$$

(γ) $\Gamma(\cdot)$ is absolutely continuous on finite intervals and

$$\frac{d\Gamma}{dt}(t) = Z(t)\Gamma(t) - \Gamma(t)A(t) + B(t)C(t), \quad \text{a.e.,} \ t \in \mathbb{R}.$$

For such a quintet of matrix functions our aim is to construct a time-varying system Σ,

$$\Sigma \begin{cases} x'(t) = F(t)x(t) + G(t)u(t), \\ y(t) = H(t)x(t) + D(t)u(t), \end{cases}$$

such that the input-output operator T of Σ is a J-unitary operator on $L_2^p(\mathbb{R})$ and

$$T\mathcal{LS}_2^{p\times 1} = \mathcal{S}_\omega.$$

Here $\mathcal{LS}_2^{p\times 1}$ denotes the set of all Hilbert-Schmidt integral operators from $L_2(\mathbb{R})$ into $L_2^p(\mathbb{R})$ (the Hilbert space direct sum of p copies of $L_2(\mathbb{R})$) that are lower triangular, and $\mathcal{S}_\omega$ is the subset of $\mathcal{S}_2^{p\times 1}$ consisting of all operators of the form

(0.2) $$C\left(\frac{d}{dt} - A\right)^{-1}g + K,$$

where $g \in L_2^{n\times 1}(\mathbb{R})$ and $K \in \mathcal{LS}_2^{p\times 1}$ are such that

$$\int_t^\infty U_Z(t)U_Z(s)^{-1}B(s)k(s,t)d\alpha = \Gamma(t)D(t), \quad t \in \mathbb{R}.$$

Here $k(s,t)$ is the kernel function of the Hilbert-Schmidt integral operator K, and in (0.2) the operators C and A are the operators of multiplication induced by the matrix functions $C(\cdot)$ and $A(\cdot)$. Notice that the conditions (α), (β) and (γ) on the matrix functions (0.1) are the time-dependent analogues of the conditions (a), (b) and (c) in the first paragraph of this introduction. In Section 3 we shall interpret the two inequalities in (β) as observability and controllability conditions.

In this paper the problem stated in the previous paragraph is solved. In a future publication the result will be used to obtain the solution of the time-varying version of the Nudelman interpolation problem.

The paper consists of six sections (not counting the present introduction). The first three sections have a preliminary character. In Section 1 we review the properties of dichotomy which are used in the present paper. Section 2 develops further the residue calculus for integral operators started in [BGK1]; here the emphasis is on Hilbert-Schmidt integral operators. Section 3 concerns time-varying systems and their input-output operators. In Section 4 we introduce for input-output operators of time-varying systems the analogue of a RHP-null-pole triple, and we define the notion of an admissible time-varying Sylvester data set. J-unitary input-output operators are studied in Section 5. The last section contains the main theorem and its proof.

1. PRELIMINARIES ABOUT DICHTOMY

In the sequel Δ denotes the operator of differentiation $\frac{d}{dt}$. The domain $\mathcal{D}(\Delta)$ of $\Delta = \frac{d}{dt}$ is the set of all functions $f \in L_2^p(\mathbb{R})$ (where p may be any positive integer) such that f is absolutely continuous on finite intervals and $\Delta f = f' \in L_2^p(\mathbb{R})$. Here $L_2^p(\mathbb{R})$ stands for the space of all square (Lebesgue) integrable vector functions with values in $\mathbb{C}^p$.

Let $A \in L_\infty^{p \times p}(\mathbb{R})$, that is, A is a $p \times p$ matrix with entries in $L_\infty(\mathbb{R})$. We shall use the symbol A also for the operator of multiplication on $L_2^p(\mathbb{R})$ by A. Thus $Af = A(\cdot)f$. It follows that $\Delta - A$ is a well-defined unbounded operator on $L_2^p(\mathbb{R})$ with domain equal to the domain of Δ. By $U_A(t)$ we denote the *fundamental matrix* (normalized to $I_{\mathbb{C}^p}$ at $t = 0$) associated with the differential equation

$$(1.1) \qquad x'(t) = A(t)x(t), \qquad -\infty < t < \infty.$$

It is known (see [BenG]) that the operator $\Delta - A$ is boundedly invertible if and only if the equation (1.1) has an (*exponential*) *dichotomy* P_A, that is, P_A is a projection of $\mathbb{C}^p$ and there exist constants $M \geq 0$, $0 < a < 1$, such that

$$(1.2a) \qquad \|U_A(t)P_A U_A(s)^{-1}\| \leq Ma^{t-s}, \quad t \geq s,$$

$$(1.2b) \qquad \|U_A(t)(I - P_A)U_A(s)^{-1}\| \leq Ma^{s-t}, \quad s \geq t.$$

Furthermore, in this case $(\Delta - A)^{-1}$ is an integral operator,

$$(1.3) \qquad ((\Delta - A)^{-1}\varphi)(t) = \int_{-\infty}^{\infty} \gamma_A(t, s)\varphi(s)\, ds,$$

whose kernel function γ_A is given by

(1.4) $$\gamma_A(t,s) = \begin{cases} U_A(t)P_AU_A(s)^{-1}, & t > s, \\ -U_A(t)(I-P_A)U_A(s)^{-1}, & t < s. \end{cases}$$

A dichotomy of (1.1), assuming it exists, is uniquely determined (see [Co], pp. 16, 17). In the time-invariant case when $A(t) = A$ for each t, the equation (1.1) admits a dichotomy if and only if the (constant) matrix A has no eigenvalues on the imaginary axis, and in this case the dichotomy P_A is the spectral projection of A corresponding to the eigenvalues in the open left half plane. There is a rich literature on the subject of dichotomy; see, e.g., the books [MS], [Co], [DK], and the papers [GKvS] and [BenG]. In what follows we review some facts about dichotomy that will be used later.

For simplicity, we call $A \in L_\infty^{p \times p}(\mathbb{R})$ *dichotomous* if the corresponding differential equation (1.1) has a dichotomy or, equivalently, if the corresponding operator $\Delta - A$ is boundedly invertible.

We say that $A \in L_\infty^{p \times p}(\mathbb{R})$ is *backward stable* if the associated differential equation (1.1) has dichotomy $P_A = 0$. In this case we also say that (1.1) is *backward stable*. Backward stability implies that the general solution of (1.1) decays exponentially in backward time. In the time-invariant case this property corresponds to a constant matrix which has all its eigenvalues in the open right half plane. If $A \in L_\infty^{p \times p}(\mathbb{R})$ is backward stable, then $(\Delta - A)^{-1}$ is an upper triangular integral operator. Indeed in this case

(1.5) $$\gamma_A(t,s) = \begin{cases} -U_A(t)U_A(s)^{-1}, & t < s, \\ 0, & t > s. \end{cases}$$

Similarly, $A \in L_\infty^{p \times p}(\mathbb{R})$ or the associated differential equation (1.1) is said to be *forward stable* if (1.1) has dichotomy $P_A = I_{\mathbb{C}^p}$. In this case $(\Delta - A)^{-1}$ is lower triangular.

It will be convenient to consider two types of operations on systems of the form (1.1). The first is connected with a time-varying basis transformation. Consider

(1.6) $$\tilde{x}(t) = \tilde{A}(t)\tilde{x}(t), \quad -\infty < x < \infty,$$

where $\tilde{A} \in L_\infty^{\tilde{p} \times \tilde{p}}(\mathbb{R})$. The systems (1.1) and (1.6) are said to be *kinematically similar* (cf., [Co], [DK]) if $p = \tilde{p}$ and there exists a $p \times p$ matrix function $E(\cdot)$ on $\mathbb{R}$, which is absolutely continuous on each finite interval, satisfies

$$(1.7) \qquad E'(t) = \tilde{A}(t)E(t) - E(t)A(t), \quad \text{a.e.,} \quad t \in \mathbb{R},$$

and is such that

$$(1.8) \qquad \sup_{t \in \mathbb{R}} \{ \|E(t)\|, \|E(t)^{-1}\| \} < \infty.$$

For such a matrix function $E(\cdot)$ the basis transformation $\tilde{x}(t) = E(t)x(t)$ transforms the equation (1.1) into (1.6). If the equations (1.1) and (1.6) are kinematically similar, then A is dichotomous if and only if $\tilde{A}$ is dichotomous, and in this case $P_{\tilde{A}} = E(0)P_A E(0)^{-1}$.

The second operation is that of forming direct sums. By definition, the *direct sum* of the systems (1.1) and (1.6) is the system

$$(1.9) \qquad z'(t) = \begin{pmatrix} A(t) & 0 \\ 0 & \tilde{A}(t) \end{pmatrix} z(t), \quad -\infty < t < \infty.$$

If both (1.1) and (1.6) have a dichotomy, then the same holds true for the direct sum (1.9). In fact, in this case the dichotomy Π of (1.9) is given by

$$\Pi = \begin{pmatrix} P_A & 0 \\ 0 & P_{\tilde{A}} \end{pmatrix}.$$

The proof of the next theorem may be found in [DK], Chapter IV.

THEOREM 1.1. *In order that the system (1.1) admits a dichotomy it is necessary and sufficient that (1.1) is kinematically similar to a direct sum of a forward stable and a backward stable system.*

Put $A^*(t) = A(t)^*$, and consider the equation

$$(1.10) \qquad x'(t) = -A(t)^* x(t), \quad -\infty < t < \infty.$$

The fundamental matrix of (1.10) is equal to $U_A(t)^{-*}$. Here T^{-*} stands for $(T^{-1})^*$. It follows that (1.1) admits a dichotomy if and only if the same holds true for (1.10), and in this case (see [BenG])

$$(1.11) \qquad P_{-A^*} = (I - P_A)^*,$$

$$(1.12) \qquad (\Delta + A^*)^{-1} = -(\Delta - A)^{-*}.$$

Let $A \in L_\infty^{p \times p}(\mathbb{R})$, and assume that $\Delta - A$ is boundedly invertible. Let X and Y be matrix functions of sizes $m \times p$ and $p \times r$, respectively, and with entries in $L_\infty(\mathbb{R})$. Then $X(\Delta - A)^{-1}Y$ is a well-defined operator from $L_2^r(\mathbb{R})$ into $L_2^m(\mathbb{R})$ whose action is given by

$$(1.13) \qquad (X(\Delta - A)^{-1}Y\varphi)(t) = \int_{-\infty}^{\infty} X(t)\gamma_A(t,s)Y(s)\varphi(s)\,ds, \quad t \in \mathbb{R},$$

where γ_A is given by (1.4). Formula (1.13) makes also sense if one of the matrix functions X and Y has entries in $L_2(\mathbb{R})$. In fact, in the latter case the kernel function

$$k(t,s) = X(t)\gamma_A(t,s)Y(s)$$

is square integrable on $\mathbb{R} \times \mathbb{R}$. For instance, if X has entries in $L_2(\mathbb{R})$ and Y in $L_\infty(\mathbb{R})$, then because of (1.2a) and (1.2b)

$$\int_{-\infty}^{\infty} \int_{-\infty}^{\infty} \|k(t,s)\|^2 \, dt\, ds \leq M^2 \|X\|_2^2 \|Y\|_\infty^2 \left(\int_{-\infty}^{\infty} a^{2|\alpha|} d\alpha \right) < \infty.$$

We conclude that $X(\Delta - A)^{-1}Y$ is a Hilbert-Schmidt integral operator if one of the functions X and Y has entries in $L_2(\mathbb{R})$ and the other has entries in $L_\infty(\mathbb{R})$.

2. RESIDUE CALCULUS FOR HILBERT-SCHMIDT OPERATORS

In [BGK1] a residue calculus was developed for operators from the nonstationary Wiener-class. Here we develop such a calculus from first principles for Hilbert-Schmidt operators, i.e., for operators $F : L_2^r(\mathbb{R}) \to L_2^m(\mathbb{R})$ that admit a representation of the form

$$(2.1) \qquad\qquad (F\varphi)(t) = \int_{-\infty}^{\infty} f(t,s)\varphi(s)\,ds, \quad t \in \mathbb{R},$$

where f is an $m \times r$ matrix function whose entries are in $L_2(\mathbb{R} \times \mathbb{R})$. This class of operators will be denoted by $S_2^{m \times r}$. An operator $F \in S_2^{m \times r}$ is called *lower triangular* (notation: $F \in \mathcal{LS}_2^{m \times r}$) if its kernel function f is zero a.e. on $s \geq t$. In this case

$$(2.2) \qquad\qquad (F\varphi)(t) = \int_{-\infty}^{t} f(t,s)\varphi(s)\,ds, \quad t \in \mathbb{R}.$$

By $\mathcal{U}S_2^{m \times r}$ we denote the *upper triangular* F *'s in* $S_2^{m \times r}$.

Now, let F be a Hilbert-Schmidt operator, $F \in S_2^{m \times r}$, and let f be its kernel function. In analogy with the stationary case it is natural (cf., [BGK1], Section 3) to consider the following limits

$$(2.3) \qquad \mathcal{R}es_{2,R}^+(F)(t) = \lim_{\varepsilon \downarrow 0} -\frac{1}{\varepsilon} \int_0^\varepsilon f(t, t + \alpha)\, d\alpha,$$

$$(2.4) \qquad \mathcal{R}es_{2,L}^+(F)(t) = \lim_{\varepsilon \downarrow 0} -\frac{1}{\varepsilon} \int_0^\varepsilon f(t - \alpha, t)\, d\alpha.$$

For fixed $\varepsilon > 0$ the integrals in (2.3) and (2.4) define $m \times r$ matrix functions with entries in $L_2(\mathbb{R})$. The limits in (2.3) and (2.4) have to be understood in the weak topology of $L_2^{m \times r}(\mathbb{R})$ and, if they exist, they do not depend on the particular choice of the kernel function f. Note that (2.3) and (2.4) involve only the upper triangular part of f.

Let $F \in S_2^{m \times r}$ and $\rho \in L_2^{m \times r}(\mathbb{R})$. We say that the *right nonstationary total RHP-residue* of F is equal to ρ (notation: $\mathcal{R}es_{2,R}^+(F) = \rho$) if the limit in the right hand side of (2.3) exists in the weak topology of $L_2^{m \times r}(\mathbb{R})$ and is equal to ρ. Similarly, $\mathcal{R}es_{2,L}^+(F) = \rho$ means that the limit in the right hand side of (2.4) exists in the weak topology of $L_2^{m \times r}(\mathbb{R})$ and is equal to ρ, and in this case we refer to ρ as the *left nonstationary total RHP-residue* of F. If both $\mathcal{R}es_{2,R}^+(F)$ and $\mathcal{R}es_{2,L}^+(F)$ exist and are equal, then we drop the subscripts R and L, and we call $\mathcal{R}es_2^+(F)$ the *(nonstationary) total RHP-residue* of F. The letters RHP stand for right half plane.

LEMMA 2.1. *Let* $F \in S_2^{m \times r}$ *have kernel function* f, *and assume that there exists a continuous function* ℓ *on* $[0, \infty)$ *such that*

$$(2.5) \qquad \int_{-\infty}^\infty \|f(t, t + \alpha)\|^2 dt \leq \ell(\alpha), \qquad \alpha \geq 0.$$

Then $\mathcal{R}es_{2,R}^+(F)$ *exists if and only if* $\mathcal{R}es_{2,L}^+(F)$ *exists, and in this case both residues are equal.*

PROOF. For $\varepsilon > 0$ put

$$g_\varepsilon(t) = \frac{1}{\varepsilon} \int_0^\varepsilon f(t - \alpha, t)\, d\alpha - \frac{1}{\varepsilon} \int_0^\varepsilon f(t, t + \alpha)\, d\alpha.$$

We have to show that g_ε converges to zero in the weak topology of $L_2^{m \times r}(\mathbb{R})$ whenever $\varepsilon \downarrow 0$. Without loss of generality we may assume that $m = r = 1$. Take $h \in L_2(\mathbb{R})$. Then

$$|\langle g_\varepsilon, h \rangle| \le \frac{1}{\varepsilon} \int_0^\varepsilon \left(\int_{-\infty}^\infty |f(t, t+\alpha)||h(t+\alpha) - h(t)| \, dt \right) d\alpha$$

$$\le \frac{1}{\varepsilon} \int_0^\varepsilon \left(\int_{-\infty}^\infty |f(t, t+\alpha)|^2 \, dt \right)^{\frac{1}{2}} \|h_\alpha - h\|_2 \, d\alpha.$$

Here $h_\alpha(t) = h(t+\alpha)$. So according to our assumption (2.5) we have

(2.6) $$|\langle g_\varepsilon, h \rangle| \le \frac{1}{\varepsilon} \int_0^\varepsilon \sqrt{\ell(\alpha)} \|h_\alpha - h\|_2 \, d\alpha.$$

Since $\|h_\alpha - h\|_2$ depends continuously on α, we conclude that the integrand in (2.6) is continuous in α. It follows that

$$\limsup_{\varepsilon \downarrow 0} |\langle g_\varepsilon, f \rangle| = \sqrt{l(0)} \|h_0 - h\| = 0,$$

which proves the lemma. □

PROPOSITION 2.2. *Let* $F \in \mathcal{LS}_2^{m \times r}$, *and let* $A \in L_\infty^{r \times r}(\mathbb{R})$ *and* $Z \in L_\infty^{m \times m}(\mathbb{R})$ *be dichotomous. Then* $\mathcal{R}es_2^+\{F(\Delta - A)^{-1}\} = X$ *and* $\mathcal{R}es_2^+\{(\Delta - Z)^{-1}F\} = Y$, *where*

(2.7) $$X(t) = \int_{-\infty}^t f(t, \alpha) U_A(\alpha)(I - P_A) U_A(t)^{-1} \, d\alpha, \quad \text{a.e.,} \quad t \in \mathbb{R},$$

(2.8) $$Y(t) = \int_t^\infty U_Z(t)(I - P_Z) U_Z(\alpha)^{-1} f(\alpha, t) \, d\alpha, \quad \text{a.e.,} \quad t \in \mathbb{R}.$$

PROOF. We shall prove the statement for $F(\Delta - A)^{-1}$; the result for $(\Delta - Z)^{-1}F$ is established in a similar way.

Let X be defined by (2.7). According to (1.2b),

$$\|X(t)\| \le \int_{-\infty}^t \|f(t, \alpha)\| M a^{t-\alpha} \, d\alpha \le \left(\int_0^\infty \|f(t, t-s)\|^2 \, ds \right)^{\frac{1}{2}} \left(\int_0^\infty M^2 a^{2s} \, ds \right)^{\frac{1}{2}}.$$

Put $C = \int_0^\infty M^2 a^{2s} \, ds$. Then

$$\int_{-\infty}^\infty \|X(t)\|^2 \, dt \le C \int_{-\infty}^\infty \int_0^\infty \|f(t, t-s)\|^2 \, ds \, dt < \infty,$$

and thus $X \in L_2^{m \times r}(\mathbb{R})$.

Let k be the kernel function of $F(\Delta - A)^{-1}$. For $t < s$ we have

$$k(t,s) = \int_{-\infty}^{\infty} f(t,\alpha)\gamma_A(\alpha,s)\,d\alpha = -\int_{-\infty}^{t} f(t,\alpha)U_A(\alpha)(I - P_A)U_A(s)^{-1}\,d\alpha.$$

Since $s > t \geq \alpha$ and P_A is a projection, we have

$$U_A(\alpha)(I - P_A)U_A(s)^{-1} = U_A(\alpha)(I - P_A)U_A(t)^{-1}U_A(t)(I - P_A)U_A(s)^{-1},$$

and therefore

(2.9) $$k(t,s) = -X(t)U_A(t)(I - P_A)U_A(s)^{-1}, \quad t < s.$$

Put $P_A(t) = U_A(t)P_A U_A(t)^{-1}$. Then $P_A(t)$ is a projection, and the integrand of (2.7) can be rewritten as

$$f(t,\alpha)U_A(\alpha)U_A(t)^{-1}(I - P_A(t)).$$

It follows that $X(t) = X(t)(I - P_A(t))$, and hence

(2.10) $$k(t,s) = -X(t)U_A(t)U_A(s)^{-1}, \quad t < s.$$

But then

(2.11)
$$X(t) + \frac{1}{\varepsilon}\int_0^\varepsilon k(t, t+\beta)\,d\beta = \frac{1}{\varepsilon}\int_0^\varepsilon \{X(t) - X(t)U_A(t)U_A(t+\beta)^{-1}\}\,d\beta$$
$$= X(t)\{\frac{1}{\varepsilon}\int_0^\varepsilon [I - U_A(t)U_A(t+\beta)^{-1}]\,d\beta.$$

From formula (2.5) in [BenG] we know that

$$\|I - U_A(t)U_A(t+\alpha)^{-1}\| \leq \varepsilon^{\alpha\|A\|_\infty} - 1, \quad \alpha \geq 0.$$

We conclude that

$$\frac{1}{\varepsilon}\int_0^\varepsilon [I - U_A(t)U_A(t+\beta)^{-1}]\,d\beta \to 0, \quad \varepsilon \downarrow 0,$$

and the convergence is uniform in $t \in \mathbb{R}$. By using the latter result in (2.11) we see that $\mathcal{R}es_{2,R}^+(F)(t) = X$.

Finally, from (2.9) and (1.2b) we obtain that

$$\int_{-\infty}^{\infty} \|k(t, t+\alpha)\|^2\,dt \leq \int_{-\infty}^{\infty} \|X(t)\|^2 M^2 a^{2\alpha}\,dt = M^2 a^{2\alpha}\|X\|_2,$$

for some constants $M \geq 0$ and $0 < a < 1$. Hence k satisfies the hypotheses of Lemma 2.1. Therefore, and by the result of the previous paragraph, $\mathcal{R}es^+_{2,L}(F) = X$.   □

PROPOSITION 2.3. *Let* $F \in \mathcal{L}S^{m \times r}_2$, *and let* $A \in L^{r \times r}_\infty(\mathbb{R})$ *and* $Z \in L^{m \times m}_\infty$ *be dichotomous. Put*

$$(2.12) \qquad P_A(t) = U_A(t)P_A U_A(t)^{-1}, \qquad P_Z(t) = U_Z(t)P_Z U_Z(t)^{-1}, \quad t \in \mathbb{R}.$$

Then $X := \mathcal{R}es^+_2\{F(\Delta - A)^{-1}\}$ *and* $Y := \mathcal{R}es^+_2\{(\Delta - Z)^{-1}F\}$ *are the unique elements in* $L^{m \times r}_2(\mathbb{R})$ *such that*

$$(2.13) \qquad F(\Delta - A)^{-1} - X(\Delta - A)^{-1} \in \mathcal{L}S^{m \times r}_2, \quad X(t)P_A(t) = 0 \text{ a.e. on } \mathbb{R};$$

$$(2.14) \qquad (\Delta - Z)^{-1}F - (\Delta - Z)^{-1}Y \in \mathcal{L}S^{m \times r}_2, \quad P_Z(t)Y(t) = 0 \text{ a.e. on } \mathbb{R};$$

PROOF. We shall prove the result for $F(\Delta - A)^{-1}$; the arguments for $(\Delta - Z)^{-1}F$ are similar. Let $\tilde{X}$ be an arbitrary element in $L^{m \times r}_2(\mathbb{R})$ such that $\tilde{X}(t)P_A(t) = 0$ a.e. on $\mathbb{R}$. From the remarks made in the last paragraph of the previous section we know that $\tilde{X}(\Delta - A)^{-1}$ is a Hilbert-Schmidt operator. Let h be its kernel function. Then (use (1.4) and (1.13))

$$h(t,s) = -\tilde{X}(t)U_A(t)(I - P_A)U_A(s)^{-1}, \quad t < s.$$

Also, $F(\Delta - A)^{-1}$ is a Hilbert-Schmidt operator, and from the proof of Proposition 2.2 (see (2.9)) we know that for $t < s$ its kernel function is given by

$$k(t,s) = -X(t)U_A(t)(I - P_A)U_A(s)^{-1}, \quad t < s.$$

We conclude that $F(\Delta - A)^{-1} - \tilde{X}(\Delta - A)^{-1}$ is lower triangular if and only if

$$(2.15) \qquad (\tilde{X}(t) - X(t))(I - P_A(t))U_A(t)U_A(s)^{-1} = 0, \quad t < s.$$

The proof of Proposition 2.2 also shows that $X(t)P_A(t) = 0$ a.e. on $\mathbb{R}$. By assumption the same holds true for $\tilde{X}$ in place of X. Thus (2.15) is equivalent to the requirement that $\tilde{X} = X$.   □

It is well-known that the space $S_2^{m \times r}$ is a Hilbert space with the inner product given by $\langle F, G \rangle = \operatorname{tr} G^* F$. In fact, $S_2^{m \times r}$ is isometrically isomorphic to $L_2^{m \times r}(\mathbb{R} \times \mathbb{R})$. Indeed, if F and G belong to $S_2^{m \times r}$ and have kernel functions f and g, respectively, then

$$\langle F, G \rangle = \int_{-\infty}^{\infty} \int_{-\infty}^{\infty} \operatorname{tr} \{g(t, s)^* f(t, s)\} \, dt \, ds.$$

It follows that the spaces $\mathcal{L} S_2^{m \times r}$ and $\mathcal{U} S_2^{m \times r}$ are mutually orthogonal, closed linear subspaces of $S_2^{m \times r}$. In the sequel we shall write $\mathbb{P}$ for the orthogonal projection of $S_2^{m \times r}$ onto $\mathcal{L} S_2^{m \times r}$ (along $\mathcal{U} S_2^{m \times r}$).

Fix $A \in L_\infty^{r \times r}(\mathbb{R})$ and $Z \in L_\infty^{m \times m}$, and assume that A and Z are dichotomous. Consider the maps

$$(2.16) \qquad \Lambda_R : \mathcal{L} S_2^{1 \times r} \to L_2^{1 \times r}(\mathbb{R}), \quad \Lambda_R(F) = \mathcal{R}es_2^+ \{F(\Delta - A)^{-1}\};$$

$$(2.17) \qquad \Lambda_L : \mathcal{L} S_2^{m \times 1} \to L_2^{m \times 1}(\mathbb{R}), \quad \Lambda_L(F) = \mathcal{R}es_2^+ \{(\Delta - Z)^{-1} F\}.$$

Both Λ_R and Λ_L act between Hilbert spaces. The next proposition describes the adjoints of Λ_R and Λ_L.

PROPOSITION 2.4. *Let Λ_R and Λ_L be defined by (2.16) and (2.17), respectively. Then*

(i) $\Lambda_R^*(h) = \mathbb{P}\{h(\Delta - A)^{-*}\}, \quad h \in L_2^{1 \times r}(\mathbb{R});$

(ii) $\Lambda_L^*(h) = \mathbb{P}\{(\Delta - Z)^{-*} h\}, \quad h \in L_2^{m \times 1}(\mathbb{R}).$

PROOF. We shall prove (i); the arguments for (ii) are similar. Take $F \in \mathcal{L} S_2^{1 \times r}$ and $h \in L_2^{1 \times r}(\mathbb{R})$. Then $K = h(\Delta - A)^{-*}$ is a Hilbert-Schmidt operator, $K \in S_2^{1 \times r}$ and its kernel function is given by

$$(2.18) \qquad k(t, s) = h(t)\gamma_A(s, t)^*, \quad t, s \in \mathbb{R}.$$

Let f be the kernel function of F. Then

$$\langle \Lambda_R(F), h \rangle = \int_{-\infty}^{\infty} \text{tr} \, \{h(t)^*(\Lambda_R(F))(t)\} \, dt = \int_{-\infty}^{\infty} (\Lambda_R(F))(t)h(t)^* \, dt$$

$$= \int_{-\infty}^{\infty} \left\{ \int_{-\infty}^{t} f(t,\alpha)U_A(\alpha)(I - P_A)U_A(t)^{-1} \, d\alpha \right\} h(t)^* \, dt$$

$$= -\int_{-\infty}^{\infty} \left\{ \int_{-\infty}^{t} f(t,\alpha)\gamma_A(\alpha,t)h(t)^* \, d\alpha \right\} dt$$

$$= -\int_{-\infty}^{\infty} \left\{ \int_{-\infty}^{t} f(t,\alpha)k(t,\alpha)^* \, d\alpha \right\} dt$$

$$= -\int_{-\infty}^{\infty} \left\{ \int_{-\infty}^{t} \text{tr} \, \{k(t,\alpha)^* f(t,\alpha)\} \, d\alpha \right\} dt$$

$$= \langle F, \mathbb{P}\{h(\Delta - A)^{-*}\} \rangle.$$

Here we use that $\mathbb{P}\{h(\Delta - A)^{-*}\}$ is a lower triangular integral operator whose kernel function coincides with $k(t,s)$ for $t > s$.   □

Let $A \in L_\infty^{r \times r}(\mathbb{R})$ and $Z \in L_\infty^{m \times m}$ be dichotomous, and let $F \in \mathcal{LS}_2^{m \times r}$. Then

(2.19) $(I - \mathbb{P})\{F(\Delta - A)^{-1}\} = X(\Delta - A)^{-1}$,

(2.20) $(I - \mathbb{P})\{(\Delta - Z)^{-1}F\} = (\Delta - Z)^{-1}Y$,

where $X = \mathcal{Res}_2^+\{F(\Delta - A)^{-1}\}$ and $Y = \mathcal{Res}_2^+\{(\Delta - Z)^{-1}F\}$. The identities (2.19) and (2.20) follow from (2.13) and (2.14), respectively. To see this one uses that $X(\cdot)P_A(\cdot) = 0$ and $P_Z(\cdot)Y(\cdot) = 0$ imply that both $X(\Delta - A)^{-1}$ and $(\Delta - Z)^{-1}Y$ are upper triangular.

3. SYSTEMS AND INPUT-OUTPUT OPERATORS

In the sequel we shall deal with input-output operators of time-varying systems of the following type:

(3.1) $$\Sigma \begin{cases} x'(t) = A(t)x(t) + B(t)u(t), & t \in \mathbb{R}, \\ y(t) = C(t)x(t) + D(t)u(t). \end{cases}$$

The main coefficient $A(\cdot)$, the input coefficient $B(\cdot)$, the output coefficient $C(\cdot)$ and the feedthrough coefficient $D(\cdot)$ are assumed to be matrix functions of sizes $N \times N$, $N \times r$, $m \times N$, and $m \times r$, respectively, which have their entries in $L_\infty(\mathbb{R})$. Furthermore, we shall assume that $A(\cdot)$ is dichotomous, i.e., the differential equation

(3.2) $x'(t) = A(t)x(t), \quad -\infty < t < \infty.$

has a dichotomy.

According to our hypotheses on Σ the differential operator $\Delta - A$ associated with the main coefficient of Σ is boundedly invertible on $L_2^N(\mathbb{R})$. It follows that for each $u \in L_2^r(\mathbb{R})$ the first equation has a unique solution in $L_2^N(\mathbb{R})$, namely $x = (\Delta - A)^{-1}Bu$. By inserting this solution into the second equation of (3.1), we obtain the output $y \in L_2^m(\mathbb{R})$ which is uniquely determined by the input u. The map which assigns to the input $u \in L_2^r(\mathbb{R})$ the corresponding output $y \in L_2^m(\mathbb{R})$ is called the *input-output operator* of Σ and will be denoted by T_Σ. From the above considerations we see that T_Σ acts as a bounded linear operator from $L_2^r(\mathbb{R})$ into $L_2^m(\mathbb{R})$, and

$$(3.3) \qquad\qquad T_\Sigma = D + C(\Delta - A)^{-1}B.$$

In the latter formula A, B, C and D stand for the operators of multiplication defined by the coefficients of Σ. Thus, more explicitly,

$$(3.4) \qquad (T_\Sigma\varphi)(t) = D(t)\varphi(t) + \int_{-\infty}^{\infty} C(t)\gamma_A(t,s)B(s)\varphi(s)\,ds, \quad t \in \mathbb{R}$$

where $\gamma_A(t,s)$ is defined by (1.4).

By $\mathcal{R}^{m \times r}$ we shall denote the class of all operators $T : L_2^r(\mathbb{R}) \to L_2^m(\mathbb{R})$ such that $T = T_\Sigma$ for some system Σ of the type considered in the first paragraph of this section. In this case we call Σ a *realization* of T.

Let

$$(3.5) \qquad \tilde{\Sigma} \begin{cases} \tilde{x}'(t) = \tilde{A}(t)\tilde{x}(t) + \tilde{B}(t)u(t), & t \in \mathbb{R}, \\ y(t) = \tilde{C}(t)\tilde{x}(t) + \tilde{D}(t)u(t), \end{cases}$$

be a system of the same type as Σ. We call $\tilde{\Sigma}$ and Σ (*kinematically*) *similar* if the order of $\tilde{A}(t)$ is equal to the order N of $A(t)$ and the coefficients of $\tilde{\Sigma}$ and Σ are related in the following way:

$$(3.6\text{a}) \qquad\qquad E'(t) = \tilde{A}(t)E(t) - E(t)A(t),$$

$$(3.6\text{b}) \qquad\qquad \tilde{B}(t) = E(t)B(t), \quad \tilde{C}(t) = C(t)E(t)^{-1},$$

$$(3.6\text{c}) \qquad\qquad \tilde{D}(t) = D(t),$$

almost everywhere on $\mathbb{R}$, where $E(\cdot)$ is an $N \times N$ matrix function which is absolutely continuous on each finite interval of $\mathbb{R}$ and satisfies

$$(3.7) \qquad \sup_{t \in \mathbb{R}}\{\|E(t)\|, \|E(t)^{-1}\|\} < \infty.$$

Similar systems have the same input-output operator.

By Theorem 1.1, given the system $\tilde{\Sigma}$ we can always find a system Σ similar to $\tilde{\Sigma}$ such that

$$(3.8a) \qquad A(t) = \begin{pmatrix} A_1(t) & 0 \\ 0 & A_2(t) \end{pmatrix}, \quad B(t) = \begin{pmatrix} B_1(t) \\ B_2(t) \end{pmatrix},$$

$$(3.8b) \qquad C(t) = (\, C_1(t) \quad C_2(t) \,), \quad D(t) = \tilde{D}(t),$$

with A_1 forward stable and A_2 backward stable. In this case the system Σ is called *decomposed*. Note that the partitionings of $B(t)$ and $C(t)$ in (3.8a) and (3.8b) are according to the partitioning of $A(t)$ in (3.8a).

Consider the system Σ (or the pair (A, B)). We define the *controllability operator at time* t to be the operator $\mathcal{C}_{(A,B)}(t) : L_2^r(\mathbb{R}) \to \mathbb{R}^N$ defined by $\mathcal{C}_{(A,B)}(t)u = x$, where x is the value of the state vector at time t if the input signal $u \in L_2^r(\mathbb{R})$ is fed into the system Σ, i.e.,

$$\mathcal{C}_{(A,B)}(t)(u) = \int_{-\infty}^{\infty} \gamma_A(t, s)B(s)u(s)ds.$$

The associated *controllability Gramian at time* t for the system Σ is defined to be $G_{(A,B)}(t) = \mathcal{C}_{(A,B)}(t)\mathcal{C}_{(A,B)}(t)^*$. We call the system Σ (or the pair (A, B)) to be *uniformly controllable* if $\mathcal{C}_{(A,B)}(t)$ is onto $\mathbb{R}^N$ uniformly in t, that is

$$(3.9) \qquad G_{(A,B)}(t) := \int_{-\infty}^{\infty} \gamma_A(t, s)B(s)B(s)^*\gamma_A(t, s)^* \, ds \geq \varepsilon I_{\mathbb{C}^N}, \quad t \in \mathbb{R}.$$

Similarly, we define the *observability operator* of the system Σ (or of the pair (C, A)) at time t to be the map $\mathcal{O}_{(C,A)}(t) : \mathbb{R}^N \to L_2^m(\mathbb{R})$ which yields the observation $y(\cdot) = C(\cdot)x(\cdot)$ resulting from the unique state trajectory $x(\cdot) \in L_2^N(\mathbb{R})$ with initial condition $x(t+) = P_A(t)x$, $x(t-) = -(I - P_A(t))x$; in other words,

$$\mathcal{O}_{(C,A)}(t) = C(\cdot)\gamma_A(\cdot, t)x.$$

We say that the system Σ (or the pair (C, A)) is *uniformly observable* if $\mathcal{O}_{(C,A)}(t)$ is uniformly one-to-one, i.e. if the *observability Gramian* $G_{(C,A)}(t) = \mathcal{O}_{(C,A)}(t)^*\mathcal{O}_{(C,A)}(t)$ at time t is strictly positive definite, that is

$$(3.10) \qquad G_{(C,A)}(t) := \int_{-\infty}^{\infty} \gamma_A(s,t)^*C(s)^*C(s)\gamma_A(s,t)\, ds \geq \varepsilon I_{\mathbb{C}^N}, \quad t \in \mathbb{R}.$$

Uniform controllability and uniform observability remain invariant under system similarity. In the case when there is no dichotomy ($P_A(t) = I$ or $P_A(t) = 0$), these are standard notions in system theorem (see, e.g., the recent text [CD]). Less standard is the following Hilbert-Schmidt operator version of uniform controllability/observability which will be useful for our purposes.

LEMMA 3.1. *Let $A \in L_\infty^{N \times N}(\mathbb{R})$ be dichotomous, and let $B \in L_\infty^{N \times r}(\mathbb{R})$. Define*

$$(3.11) \qquad \Xi : S_2^{r \times 1} \to L_2^{N \times 1}(\mathbb{R}), \quad (\Xi G)(t) = \int_{-\infty}^{\infty} \gamma_A(t,\alpha)B(\alpha)g(\alpha,t)\, d\alpha,$$

where g is the kernel function of G. Then the pair (A, B) is uniformly controllable if and only of $\Xi\Xi^*$ is positive definite.

PROOF. Let $G \in S_2^{r \times 1}$ with kernel function g, and take $h \in L_2^{N \times 1}(\mathbb{R})$. From (1.4) and the estimates (1.2a), (1.2b) it follows that $\Xi G \in L_2^{N \times 1}(\mathbb{R})$. Next,

$$\begin{aligned}
\langle \Xi G, h \rangle &= \int_{-\infty}^{\infty} h(s)^* \left(\int_{-\infty}^{\infty} \gamma_A(s,\alpha)B(\alpha)g(\alpha,s)\, d\alpha \right) ds \\
&= \int_{-\infty}^{\infty} \int_{-\infty}^{\infty} h(s)^* \gamma_A(s,\alpha)B(\alpha)g(\alpha,s)\, d\alpha\, ds \\
&= \int_{-\infty}^{\infty} \int_{-\infty}^{\infty} \{B(\alpha)^* \gamma_A(s,\alpha)^* h(s)\}^* g(\alpha,s)\, d\alpha\, ds \\
&= \langle G, R \rangle,
\end{aligned}$$

where R is the Hilbert-Schmidt integral operator with kernel function $B(t)^*\gamma_A(s,t)^*h(s)$, that is

$$(3.12) \qquad \Xi^* h = R = B^*(\Delta - A)^{-*}h.$$

Formula (3.10) implies that

$$(3.13) \qquad \langle \Xi\Xi^* h, h \rangle = \int_{-\infty}^{\infty} h(t)^* G_{(A,B)}(t)h(t)\, dt.$$

From (3.13) we see that $\Xi\Xi^*$ is positive definite whenever (A, B) is uniformly controllable. To prove the reverse implication, assume that $\Xi\Xi^* \geq \varepsilon I$. Fix $\tau \in \mathbb{R}$, and let

$$h(t) = \begin{cases} \frac{1}{\sqrt{\varepsilon}} x & \text{if } \tau \leq t \leq \tau + \varepsilon, \\ 0 & \text{otherwise.} \end{cases}$$

Here x is an arbitrary vector in $\mathbb{C}^N$. Then h is in $L_2^{N \times 1}(\mathbb{R})$ and $\|h\| = \|x\|$. Thus

$$\varepsilon \|x\|^2 = \varepsilon \|h\|^2 \leq \langle \Xi\Xi^* h, h \rangle = \int_{-\infty}^{\infty} h(t)^* G_{(A,B)}(t) h(t) \, dt$$

$$= \frac{1}{\varepsilon} \int_{\tau}^{\tau+\varepsilon} x^* G(A, B)(t) x \, dt \to x^* G_{(A,B)}(t) x, \quad \varepsilon \downarrow 0.$$

Here we use that $G_{(A,B)}$ is a continuous matrix function. The above calculation shows that $x^* G_{(A,B)} x \geq \varepsilon \|x\|^2$ for each $x \in \mathbb{C}^N$, and hence (A, B) is uniformly controllable. $\square$

LEMMA 3.2. *Let $A \in L_\infty^{N \times N}(\mathbb{R})$ be dichotomous, and let $C \in L_\infty^{m \times N}(\mathbb{R})$. Define*

(3.14) $$\Omega : L_2^{N \times 1}(\mathbb{R}) \to S_2^{m \times 1}, \quad \Omega h = C(\Delta - A)^{-1} h.$$

Then the pair (C, A) is uniformly observable if and only if $\Omega^* \Omega$ is positive definite.

PROOF. Let $G \in S_2^{m \times 1}$ have kernel function g. One computes that

(3.15) $$(\Omega^* G)(s) = \int_{-\infty}^{\infty} \gamma_A(t, s)^* C(t)^* g(t, s) \, dt, \quad \text{a.e., } s \in \mathbb{R}.$$

It follows that

(3.16) $$\langle \Omega^* \Omega h, h \rangle = \int_{-\infty}^{\infty} h(t)^* G_{(C,A)}(t) h(t) \, dt.$$

From (3.16) it follows (by using arguments similar to the ones used in the proof of Lemma 3.1) that $\Omega^* \Omega$ is positive definite if and only if (3.10) is fulfilled. $\square$

Assume that the system Σ is decomposed, with A, B and C as in (3.8a) and (3.8b). Then Σ is uniformly controllable if and only if the pairs (A_1, B_1) and (A_2, B_2) are uniformly controllable, and, similarly, Σ is uniformly observable if and only if the pairs (C_1, A_1) and (C_2, A_2) are uniformly observable. To see this one applies Lemmas 3.1 and

3.2, and one uses the fact that $(\Delta - A_1)^{-1}$ is lower triangular and $(\Delta - A_2)^{-1}$ is upper triangular. For example, let Ω be defined by (3.14), and for $\nu = 1, 2$ set

$$\Omega_\nu : L_2^{N_\nu \times 1}(\mathbb{R}) \rightarrow S_2^{m \times 1}, \quad \Omega_\nu h = C_\nu (\Delta - A_\nu)^{-1} h,$$

where N_ν is the order of $A_\nu(t)$. Then $\Omega = (\Omega_1 \quad \Omega_2)$. Since Ω_1 is lower triangular and Ω_2 is upper triangular, $\Omega_1^* \Omega_2 = 0$, and hence

$$\Omega^* \Omega = \begin{pmatrix} \Omega_1^* \Omega_1 & 0 \\ 0 & \Omega_2^* \Omega_2 \end{pmatrix},$$

and we can apply Lemma 3.2 to show that (C, A) is uniformly observable if and only if this property holds for (C_1, A_1) and (C_2, A_2).

We proceed with a result for inverse systems. Assume that the feedthrough coefficient $D(t)$ is identically equal to the $p \times p$ identity matrix (in particular, assume that $m = r = p$). In this case we may consider the inverse system:

$$(3.17) \qquad \Sigma^\times \begin{cases} x'(t) = A^\times(t)x(t) + B(t)u(t), & t \in \mathbb{R}, \\ y(t) = -C(t)x(t) + u(t), \end{cases}$$

where $A^\times(t) = A(t) - B(t)C(t)$ for $t \in \mathbb{R}$. From [BGK1], Theorem 2.1 we know that T_Σ is invertible if and only if $A^\times$ is dichotomous, and in this case $T_\Sigma^{-1} = T_{\Sigma^\times}$.

PROPOSITION 3.3. *Assume that the feedthrough coefficient of Σ is identically equal to the $p \times p$ identity matrix, and let T_Σ be invertible. Then $\Sigma^\times$ is uniformly controllable and uniformly observable if and only if Σ is uniformly controllable and uniformly observable.*

PROOF. First we show that

$$(3.18) \qquad T_\Sigma C(\Delta - A^\times)^{-1} = C(\Delta - A)^{-1},$$

$$(3.19) \qquad (\Delta - A^\times)^{-1} B T_\Sigma = (\Delta - A)^{-1} B.$$

We prove (3.18); the arguments for (3.19) are similar. Take $h \in L_2^N(\mathbb{R})$, and put $x = (\Delta - A^\times)^{-1} h$. Then x belongs to the domain of Δ, and hence

$$(3.20) \qquad BCx = (\Delta - A^\times)x - (\Delta - A)x.$$

It follows that

(3.21) $(\Delta - A)^{-1}BC(\Delta - A^{\times})^{-1} = (\Delta - A)^{-1} - (\Delta - A^{\times})^{-1}.$

Thus

$$T_{\Sigma}C(\Delta - A^{\times})^{-1} = \{I + C(\Delta - A)^{-1}B\}C(\Delta - A)^{\times})^{-1}$$
$$= C(\Delta - A^{\times})^{-1} + C(\Delta - A)^{-1}BC(\Delta - A^{\times})^{-1}$$
$$= C(\Delta - A)^{-1}.$$

Next, let $M_{\Sigma} : S_2^{p \times 1} \to S_2^{p \times 1}$ be the operator of multiplication by T_{Σ}. Then M_{Σ}^* is equal to the operator of multiplication by T_{Σ}^*. Indeed, if $F, G \in S_2^{p \times 1}$, then

$$\langle M_{\Sigma}F, G \rangle = \text{tr } \{G^* T_{\Sigma} F\} = \text{tr } \{(T_{\Sigma}^* G)^* F\} = \langle F, T_{\Sigma}^* G \rangle.$$

Since T_{Σ} is invertible by assumption, the operator M_{Σ} is also invertible. In fact $M_{\Sigma}^{-1} = M_{\Sigma^{\times}}$.

Now assume that $\Sigma^{\times}$ is uniformly observable. Consider the operators

$$\Omega : L_2^{N \times 1}(\mathbb{R}) \to S_2^{p \times 1}, \quad \Omega h = C(\Delta - A)^{-1}h,$$
$$\Omega^{\times} : L_2^{N \times 1}(\mathbb{R}) \to S_2^{p \times 1}, \quad \Omega^{\times}h = C(\Delta - A^{\times})^{-1}h.$$

Then (3.18) implies that $\Omega = M_{\Sigma}\Omega^{\times}$. Since M_{Σ} is invertible, $M_{\Sigma}^* M_{\Sigma} \geq \varepsilon I$ for some $\varepsilon > 0$. It follows that

$$\Omega^* \Omega = (\Omega^{\times})^* M_{\Sigma}^* M_{\Sigma} \Omega^{\times} \geq \varepsilon(\Omega^{\times})^* \Omega^{\times},$$

and we can apply Lemma 3.2 to show that Σ is uniformly observable.

In a similar way (using (3.19) and Lemmma 3.1) one can show that $\Sigma^{\times}$ is uniformly controllable implies that Σ is uniformly controllable.

Finally, note that $(\Sigma^{\times})^{\times} = \Sigma$ and $T_{\Sigma^{\times}}$ is invertible (because T_{Σ} is invertible). Thus we may apply the results proved so far to $\Sigma^{\times}$ instead of Σ. It follows that $\Sigma^{\times}$ is uniformly controllable and uniformly observable whenever Σ has these properties.   □

We conclude this section with a result about the uniqueness of a realization.

PROPOSITION 3.4. *Assume that Σ and $\tilde{\Sigma}$ are uniformly controllable and uniformly observable systems, and let Σ and $\tilde{\Sigma}$ have the same feedthrough coefficient.*

Then $T_\Sigma = T_{\tilde\Sigma}$ if and only if Σ and $\tilde\Sigma$ are similar, and in this case the similarity between Σ and $\tilde\Sigma$ is unique.

PROOF. We already know that similar systems have the same input-output operator. To prove the reverse implication, let Σ be given by (3.1) and $\tilde\Sigma$ by (3.5), with $D(t) = \tilde D(t)$ a.e. on $\mathbb{R}$, and assume that $T_\Sigma = T_{\tilde\Sigma}$. By Theorem 1.1 there exist kinematic similarities which transform the systems Σ and $\tilde\Sigma$ into decomposed ones. Under these transformations the systems remain uniformly controllable and uniformly observable. Hence without loss of generality we may assume that the coefficients A, B and C of Σ are as in (3.8a) and (3.8b). Note that

$$C(\Delta - A)^{-1}B = C_1(\Delta - A_1)^{-1}B_1 + C_2(\Delta - A_2)^{-1}B_2.$$

where $C_1(\Delta - A_1)^{-1}B_1$ is lower triangular and $C_2(\Delta - A_2)^{-1}B_2$ is upper triangular. From the remark made after the proof of Lemma 3.2 we know that for $\nu = 1, 2$ the system

$$\begin{cases} x_\nu'(t) = A_\nu(t)x_\nu(t) + B_\nu(t)u(t), & t \in \mathbb{R}, \\ y(t) = C_\nu(t)x_\nu(t), \end{cases}$$

is uniformly controllable and uniformly observable. Similarly, we may assume that

$$\tilde C(\Delta - \tilde A)^{-1}\tilde B = \tilde C_1(\Delta - \tilde A_1)^{-1}\tilde B_1 + \tilde C_2(\Delta - \tilde A_2)^{-1}\tilde B_2,$$

where $\tilde C_1(\Delta - \tilde A_1)^{-1}\tilde B_1$ is lower triangular, $\tilde C_2(\Delta - \tilde A_2)^{-1}\tilde B_2$ is upper triangular, and the corresponding systems are uniformly controllable and uniformly observable. Since $T_\Sigma = T_{\tilde\Sigma}$ and $D(\cdot) = \tilde D(\cdot)$, we have

$$C_\nu(\Delta - A_\nu)^{-1}B_\nu = \tilde C_\nu(\Delta - \tilde A_\nu)^{-1}\tilde B_\nu, \quad \nu = 1, 2.$$

It follows that without loss of generality we may assume that A and $\tilde A$ are both forward stable or both backward stable.

Let us assume that A and $\tilde A$ are both backward stable (the arguments for the case when A and $\tilde A$ are both forward stable are similar). Consider the operators Ξ and Ω defined by (3.11) and (3.12), respectively. Let $\tilde \Xi$ and $\tilde \Omega$ be the corresponding operators for $\tilde\Sigma$. Take $G \in \mathcal{LS}_2^{r \times 1}$ with kernel function g, and let k be the kernel function of $K := \Omega \Xi G$.

Since A is backward stable, K is upper triangular. Let us compute $k(t, s)$ for $t < s$. We have

$$k(t, s) = C(t)\gamma_A(t, s) \int_{-\infty}^{\infty} \gamma_A(s, \alpha) B(\alpha) g(\alpha, s) \, d\alpha$$

$$= C(t) U_A(t) U_A(s)^{-1} \int_{s}^{\infty} U_A(s) U_A(\alpha)^{-1} B(\alpha) g(\alpha, s) \, d\alpha$$

$$= \int_{s}^{\infty} C(t) U_A(t) U_A(\alpha)^{-1} B(\alpha) g(\alpha, s) \, d\alpha$$

$$= \int_{s}^{\infty} C(t) \gamma_A(t, \alpha) B(\alpha) g(\alpha, s) \, d\alpha$$

$$= \int_{-\infty}^{\infty} C(t) \gamma_A(t, \alpha) B(\alpha) g(\alpha, s) \, d\alpha, \quad t < s,$$

where the last equality is due to the fact that G is lower triangular. With G as above the kernel function h of $H := C(\Delta - A)^{-1} BG$ is given by

$$h(t, s) = \int_{-\infty}^{\infty} C(t) \gamma_A(t, \alpha) B(\alpha) g(\alpha, s) \, d\alpha.$$

It follows that for $G \in \mathcal{LS}_2^{r \times 1}$ the operator $K = \Omega \Xi G$ is uniquely determined by $C(\Delta - A)^{-1} BG$. Since A is backward stable, $\Xi G = 0$ if G is upper triangular. We conclude that the action of $\Omega \Xi$ is uniquely determined by $C(\Delta - A)^{-1} B$. Therefore, our hypotheses on Σ and $\tilde{\Sigma}$ imply that

$$(3.22) \qquad\qquad \Omega \Xi = \tilde{\Omega} \tilde{\Xi}.$$

According to our assumptions the systems Σ and $\tilde{\Sigma}$ are uniformly controllable and uniformly observable. So we may define

$$\Omega^+ = (\Omega^* \Omega)^{-1} \Omega^*, \qquad \tilde{\Omega}^+ = (\tilde{\Omega}^* \tilde{\Omega})^{-1} \tilde{\Omega}^*,$$

$$\Xi^+ = \Xi^* (\Xi \Xi^*)^{-1}, \qquad \tilde{\Xi}^+ = \tilde{\Xi}^* (\tilde{\Xi} \tilde{\Xi}^*)^{-1}.$$

Then Ω^+ and $\tilde{\Omega}^+$ are left inverses of Ω and $\tilde{\Omega}$, respectively, and Ξ^+ and $\tilde{\Xi}^+$ are right inverses of Ξ and $\tilde{\Xi}$, respectively. Put $E = \Omega^+ \tilde{\Omega} = \Xi \tilde{\Xi}^+$. From (3.22) it follows that Im $\Omega = $ Im $\tilde{\Omega}$ and Ker $\Xi = $ Ker $\tilde{\Xi}$. From these identities one derives that E is invertible with $E^{-1} = \tilde{\Omega}^+ \Omega = \tilde{\Xi} \Xi^+$. Next, observe that for each $h \in L_2^{\tilde{N} \times 1}(\mathbb{R})$, where $\tilde{N}$ is the order of $\tilde{A}(t)$, we have (because of (3.16))

$$(Eh)(t) = E(t) h(t), \quad \text{a.e., } t \in \mathbb{R},$$

with

(3.23) $\qquad E(t) = G_{(C,A)}(t)^{-1}\big(\int_{-\infty}^{\infty} \gamma_A(\alpha,t)^*C(\alpha)^*\tilde{C}(\alpha)\gamma_{\tilde{A}}(\alpha,t)\,d\alpha\big), \quad t \in \mathbb{R},$

(3.24) $\qquad E(t)^{-1} = \big(\int_{-\infty}^{\infty} \gamma_{\tilde{A}}(t,\alpha)\tilde{B}(\alpha)B(\alpha)^*\gamma_A(t,\alpha)^*\,d\alpha\big)G_{(A,B)}(t)^{-1}, \quad t \in \mathbb{R}.$

Since $C(\Delta - A)^{-1}B = \tilde{C}(\Delta - \tilde{A})^{-1}\tilde{B}$, we have $C(t)\gamma_A(t,s)B(s) = \tilde{C}(t)\gamma_{\tilde{A}}(t,s)\tilde{B}(s)$ almost everywhere on $\mathbb{R} \times \mathbb{R}$. Using the latter identity in (3.23) and (3.24), we obtain

(3.25) $\qquad E(t)\tilde{B}(t) = B(t), \qquad \tilde{C}(t)E(t)^{-1} = C(t), \text{ a.e.}, \quad t \in \mathbb{R}.$

A direct computation (using that A and $\tilde{A}$ are both backward stable) shows that

(3.26) $\dfrac{d}{dt}E(t) = A(t)E(t) - \tilde{A}(t)E(t) + G_{(C,A)}(t)^{-1}C(t)^*\{\tilde{C}(t) - C(t)E(t)\}, \text{ a.e.}, t \in \mathbb{R}.$

By the second part of (3.25) the term between brackets in the right hand side of (3.26) is zero almost everywhere on $\mathbb{R}$. We conclude that $E(\cdot)$ is a kinematic similarity transforming the system $\tilde{\Sigma}$ into Σ. The uniqueness of the similarity is a direct consequence of the fact that the observability operator Ω is injective. $\square$

4. GENERALIZED TIME-VARYING NULL-POLE TRIPLES

In this section $T \in \mathcal{R}^{p \times p}$, and our aim is to describe the set $T\mathcal{LS}_2^{p \times 1}$, which is the natural time-variant version of the right half plane null-pole subspace considered in [BGR]. Throughout this section we assume that T is the input-output operator of the system

(4.1) $\qquad \Sigma \begin{cases} x'(t) = A(t)x(t) + B(t)u(t), \quad t \in \mathbb{R}, \\ y(t) = C(t)x(t) + u(t), \end{cases}$

where $A \in L_\infty^{N \times N}(\mathbb{R})$ is dichotomous, $B \in L_\infty^{N \times p}(\mathbb{R})$, and $C \in L_\infty^{p \times N}(\mathbb{R})$. Since the feedthrough coefficient is identically equal to the $p \times p$ identity matrix, we may consider (see the previous section) the inverse system:

(4.2) $\qquad \Sigma^\times \begin{cases} x'(t) = A^\times(t)x(t) + B(t)u(t), \quad t \in \mathbb{R}, \\ y(t) = -C(t)x(t) + u(t), \end{cases}$

where $A^\times(t) = A(t) - B(t)C(t)$ for $t \in \mathbb{R}$.

We shall assume that T_Σ is invertible. In this case both A and $A^\times$ are dichotomous, and hence the matrix functions $U_A(t)P_A U_A(t)^{-1}$ and $U_{A^\times}(t)P_{A^\times}U_{A^\times}(t)^{-1}$ are uniformly bounded in t. In the sequel, $\mathcal{P}_A$ and $\mathcal{P}_{A^\times}$ denote the operators of multiplication on $L_2^N(\mathbb{R})$ induced by $U_A(\cdot)P_A U_A(\cdot)^{-1}$ and $U_{A^\times}(\cdot)P_{A^\times}U_{A^\times}(\cdot)^{-1}$, respectively. Note that both $\mathcal{P}_A$ and $\mathcal{P}_{A^\times}$ are projections on $L_2^N(\mathbb{R})$.

PROPOSITION 4.1. *Let T be the input-output operator of the system Σ in (4.1). Assume Σ is uniformly observable and uniformly controllable, and assume that T is invertible. Then $T\mathcal{LS}_2^{p\times 1}$ consists of all Hilbert-Schmidt operators of the form*

$$(4.3) \qquad\qquad C(\Delta - A)^{-1}h + K,$$

where $h \in \operatorname{Ker} \mathcal{P}_A$ and $K \in \mathcal{LS}_2^{p\times 1}$ are such that

$$(4.4) \qquad\qquad \mathcal{R}es_2^+\{(\Delta - A^\times)^{-1}BK\} = (I - \mathcal{P}_{A^\times})h.$$

Furthermore,

$$(4.5) \qquad\qquad \mathbb{Q}T\mathcal{LS}_2^{p\times 1} = \{C(\Delta - A)^{-1}h \mid h \in \operatorname{Ker} \mathcal{P}_A\}.$$

Here $\mathbb{Q}$ is the orthogonal projection of $\mathcal{S}_2^{p\times 1}$ onto $\mathcal{US}_2^{p\times 1}$ along $\mathcal{LS}_2^{p\times 1}$.

PROOF. Take $G \in \mathcal{LS}_2^{p\times 1}$, and put $h = \mathcal{R}es_2^+\{(\Delta - A)^{-1}BG\}$. According to Proposition 2.3 (in particular, formula (2.14)), we have $h \in \operatorname{Ker} \mathcal{P}_A$ and

$$(4.6) \qquad\qquad (\Delta - A)^{-1}BG - (\Delta - A)^{-1}h \in \mathcal{LS}_2^{N\times 1}.$$

Since $T = T_\Sigma = I + C(\Delta - A)^{-1}B$,

$$TG = G + C\{(\Delta - A)^{-1}BG - (\Delta - A)^{-1}h\} + C(\Delta - A)^{-1}h.$$

Apply the projection $\mathbb{Q}$ to both sides of the previous identity, and use (4.6) and the fact that G is lower triangular. This yields that $\mathbb{Q}TG = \mathbb{Q}C(\Delta - A)^{-1}h$. But $h \in \operatorname{Ker} \mathcal{P}_A$ implies that $C(\Delta - A)^{-1}h$ is upper triangular. Therefore $\mathbb{Q}TG = C(\Delta - A)^{-1}h$, and $\mathbb{Q}TG$ belongs to the set defined by the right hand side of (4.5).

Put $F_+ = \mathbb{P}TG$ and $F_- = \mathbb{Q}TG$. Then $G = T^{-1}F_+ + T^{-1}F_-$. Note that F_+ is lower triangular and $T^{-1} = T_{\Sigma^\times}$. So we may apply the result of the previous paragraph with T^{-1} in place of T and F_+ in place of G. This yields

$$(4.7) \qquad \mathbb{Q}T^{-1}F_+ = -C(\Delta - A^\times)^{-1}h^\times,$$

where $h^\times = \mathcal{R}es_2^+\{(\Delta - A^\times)^{-1}BF_+\}$ and $h^\times \in \mathrm{Ker}\,\mathcal{P}_{A^\times}$. Since

$$F_- = \mathbb{Q}TG = C(\Delta - A)^{-1}h,$$

we can apply (3.18) to show that $T^{-1}F_- = C(\Delta - A^\times)^{-1}h$. We conclude that

$$(4.8) \qquad \mathbb{Q}T^{-1}F_- = C(\Delta - A^\times)^{-1}(I - \mathcal{P}_{A^\times})h,$$

because $(\Delta - A^\times)^{-1}\mathcal{P}_{A^\times}$ is lower triangular.

Recall that $G = T^{-1}F_+ + T^{-1}F_-$ is lower triangular. Thus $\mathbb{Q}G = 0$, and hence (by (4.7) and (4.8))

$$(4.9) \qquad C(\Delta - A^\times)^{-1}\{(I - \mathcal{P}_{A^\times})h - h^\times\} = 0.$$

According to our hypotheses Σ is uniformly observable and uniformly controllable. Proposition 3.3 implies that the same holds true for $\Sigma^\times$. Thus the pair $(C, A^\times)$ is uniformly observable, and we can apply Lemma 3.2 to show that $h^\times = (I - \mathcal{P}_{A^\times})h$, because of (4.9). Thus we have proved that

$$TG = F_- + F_+ = C(\Delta - A)^{-1}h + F_+,$$

where $h \in \mathrm{Ker}\,\mathcal{P}_A$ and $F_+ \in \mathcal{LS}_2^{p\times 1}$ are such that

$$\mathcal{R}es_2^+\{(\Delta - A^\times)^{-1}BF_+\} = h^\times = (I - \mathcal{P}_{A^\times})h.$$

Next, we consider the reverse implications. Let F be equal to (4.3), where $h \in \mathrm{Ker}\,\mathcal{P}_A$ and $K \in \mathcal{LS}_2^{p\times 1}$ are such that (4.4) holds. Put $G = T^{-1}F$. We have to show that G is lower triangular. Note that $F_- = \mathbb{Q}F = C(\Delta - A)^{-1}h$, because $h \in \mathrm{Ker}\,\mathcal{P}_A$. So we can apply (3.18) to show that $T^{-1}F_- = C(\Delta - A^\times)^{-1}h$. We conclude that $\mathcal{Q}T^{-1}F_-$ is given by (4.8). Next, by Proposition 2.3 and (4.4),

$$\mathcal{Q}T^{-1}K = \mathcal{Q}K - \mathcal{Q}C(\Delta - A^\times)^{-1}BK = -C(\Delta - A^\times)^{-1}(I - \mathcal{P}_{A^\times})h.$$

Therefore, $\mathcal{Q}G = \mathcal{Q}T^{-1}F_- + \mathcal{Q}T^{-1}K = 0$, and hence G is lower triangular.

It remains to show that the right hand side of (4.5) is contained in $\mathcal{Q}T\mathcal{L}S_2^{p\times 1}$. So, let $F_- = C(\Delta - A)^{-1}h$ for some $h \in \operatorname{Ker} \mathcal{P}_A$. Since (A, B) is uniformly controllable, we know from Lemma 3.1 that there exists $K \in S_2^{p\times 1}$ with kernel function k such that

$$h(t) = \int_{-\infty}^{\infty} \gamma_A(t, \alpha)B(\alpha)k(\alpha, t)\,d\alpha, \ \text{a.e.,} \quad t \in \mathbb{R}.$$

Now, $h \in \operatorname{Ker} \mathcal{P}_A$. Thus $P_A(t)h(t) = 0$ a.e. on $\mathbb{R}$, and it follows that

$$h(t) = \int_{t}^{\infty} \gamma_A(t, \alpha)B(\alpha)k(\alpha, t)\,d\alpha, \ \text{a.e.,} \quad t \in \mathbb{R}.$$

Thus h is determined by the lower triangular part of k. In fact, $h = \mathcal{R}es_2^+\{(\Delta - A)^{-1}BG\}$ with $G = -\mathbb{P}K$. Note that $G \in \mathcal{L}S_2^{p\times 1}$. By the result proved in the first paragraph of the proof, we have $\mathcal{Q}TG = C(\Delta - A)^{-1}h$, and hence (4.5) holds.   □

Let T be the input-output operator of Σ. Assume Σ is uniformly controllable and uniformly observable, and assume that T is invertible. By Theorem 1.1 we can find kinematic similarities that transform Σ and the corresponding inverse system $\Sigma^\times$ into decomposed systems. In other words, there exist kinematic similarities $E(\cdot)$ and $E^\times(\cdot)$ such that almost everywhere on $\mathbb{R}$

(4.10) $\tilde{A}(t) := E(t)'E(t)^{-1} + E(t)A(t)E(t)^{-1} = \begin{pmatrix} A_1(t) & 0 \\ 0 & A_2(t) \end{pmatrix},$

(4.11) $\tilde{A}^\times(t) := E^\times(t)'E^\times(t)^{-1} + E^\times(t)A^\times(t)E^\times(t)^{-1} = \begin{pmatrix} A_1^\times(t) & 0 \\ 0 & A_2^\times(t) \end{pmatrix},$

where A_1 and $A_1^\times$ are forward stable, and A_2 and $A_2^\times$ are backward stable. Partition $C(t)E(t)^{-1}$ and $E^\times(t)B(t)$ according to the partitionings in (4.10) and (4.11), respectively, as follows:

(4.12) $C(t)E(t)^{-1} = (\,C_1(t) \quad C_2(t)\,), \qquad E^\times(t)B(t) = \begin{pmatrix} B_1^\times(t) \\ B_2^\times(t) \end{pmatrix}.$

Next, introduce

$$(4.13) \qquad \Gamma(t) = (0 \quad I_2^{\times}) E^{\times}(t) U_{A^{\times}}(t)(I - P_{A^{\times}}) U_{A^{\times}}(t)^{-1} E(t)^{-1} \begin{pmatrix} 0 \\ I_2 \end{pmatrix},$$

where I_2 and $I_2^{\times}$ are identity matrices of the same sizes as $A_2(t)$ and $A_2^{\times}(t)$, respectively. In the sequel we write n_2 for the order of $A_2(t)$ and $n_2^{\times}$ for the order of $A_2^{\times}(t)$. Now consider the triple

$$(4.14) \qquad \qquad \omega = (C_2, A_2, A_2^{\times}, B_2^{\times}; \Gamma).$$

Any triple ω constructed in this way is called a *generalized RHP-null-pole triple* for T. Using this notation, we have the following corollary to Proposition 4.1.

COROLLARY 4.2. *Let ω in (4.14) be a generalized RHP-null-pole triple for T. Then*

$$T\mathcal{LS}_2^{p \times 1} = \{ C_2(\Delta - A_2)^{-1} h + K \mid h \in L_2^{n_2}(\mathbb{R}), \ K \in \mathcal{LS}_2^{p \times 1} \text{are such that}$$

$$\mathcal{R}es_2^{+} \{ (\Delta - A_2^{\times})^{-1} B_2^{\times} K \} = \Gamma h \},$$

and

$$\mathcal{QT}\mathcal{LS}^{p \times 1} = \{ C_2(\Delta - A_2)^{-1} h_2 \mid h_2 \in L_2^{n_2}(\mathbb{R}).$$

Since two uniformly controllable and uniformly observable systems with the same transfer function and the same feedthrough coefficient are similar (by Proposition 3.4) the freedom in the construction of a generalized RHP-null-pole triple is limited. In fact, if

$$\tilde{\omega} = (\tilde{C}_2, \tilde{A}_2; \tilde{A}_2^{\times}, \tilde{B}_2^{\times}; \tilde{\Gamma})$$

is a second generalized RHP-null-pole triple, the ω and $\tilde{\omega}$ are *(kinematically)* equivalent, that is, there exist kinematic similarities $E(\cdot)$ and $F(\cdot)$ such that

$$\frac{d}{dt} E(t) = A_2(t) E(t) - E(t) \tilde{A}_2(t), \quad C_2(t) E(t) = \tilde{C}_2(t),$$

$$\frac{d}{dt} F(t) = A_2^{\times}(t) F(t) - F(t) \tilde{A}_2^{\times}(t), \quad F(t)^{-1} B_2^{\times}(t) = \tilde{B}_2^{\times}(t),$$

$$\tilde{\Gamma}(t) = F(t)^{-1} \Gamma(t) E(t),$$

almost everywhere on $\mathbb{R}$.

The definition of a generalized RHP-null-pole triple extends readily to input-output operators T of systems of which the feedthrough coefficient $D(t)$ satisfies

$$\sup_{t \in \mathbb{R}} \{\|D(t)\|, \|D(t)^{-1}\|\} < \infty.$$

In fact, in this case we define ω to a *generalized RHP-null-pole triple* for T if and only if ω is a generalized RHP-null-pole triple for TD^{-1}, where D^{-1} denotes the operator of multiplication by $D(\cdot)^{-1}$.

Let ω in (4.14) be a generalized RHP-null-pole triple for T. We already mentioned that A_2 and $A_2^\times$ are backward stable. Recall that in our construction of ω the initial system Σ is uniformly controllable and uniformly observable. By Proposition 3.3 the inverse system $\Sigma^\times$ has the same properties. Furthermore, uniform controllability and uniform observability is preserved under similarity of systems. It follows (see the paragraph after the proof of Lemma 3.2) that the pair (C_2, A_2) is uniformly observable and the pair $(A_2^\times, B_2^\times)$ is uniformly controllable. Using the properties of the kinematic similarities $E(\cdot)$ and $E^\times(\cdot)$, one deduces from (4.13) that $\Gamma(\cdot)$ is an $n_2^\times \times n_2$ matrix function whose entries are uniformly bounded on $\mathbb{R}$ and absolutely continuous on each finite interval. Furthermore, one computes that almost everywhere on $\mathbb{R}$

$$\frac{d}{dt}\{E^\times(t)U_{A^\times}(t)\} = \tilde{A}^\times(t)E^\times(t)U_{A^\times}(t),$$

$$\frac{d}{dt}\{U_{A^\times}(t)^{-1}E(t)^{-1}\} = U_{A^\times}(t)^{-1}B(t)C(t)E(t)^{-1} - U_{A^\times}(t)^{-1}E(t)^{-1}\tilde{A}(t),$$

$$(0 \quad I_2^\times)\,\tilde{A}^\times(t) = A_2^\times(t)\,(0 \quad I_2^\times), \qquad \tilde{A}(t)\begin{pmatrix} 0 \\ I_2 \end{pmatrix} = \begin{pmatrix} 0 \\ I_2 \end{pmatrix}A_2(t),$$

$$E^\times(t)U_{A^\times}(t)(I - P_{A^\times})U_{A^\times}(t)^{-1}E^\times(t)^{-1} = \begin{pmatrix} 0 & 0 \\ 0 & I_2^\times \end{pmatrix},$$

and hence

(4.15) $$\frac{d}{dt}\Gamma(t) = A_2^\times(t)\Gamma(t) - \Gamma(t)A_2(t) + B_2^\times(t)C_2(t), \quad \text{a.e.,} \quad t \in \mathbb{R}.$$

We summarize these properties in a definition.

Consider a quintet $\omega = (C, A; Z, B; \Gamma)$, where

(4.16) $\quad C \in L_\infty^{p \times n}(\mathbb{R}), \quad A \in L_\infty^{n \times n}(\mathbb{R}), \quad Z \in L_\infty^{k \times k}(\mathbb{R}), \quad B \in L_\infty^{k \times p}(\mathbb{R}), \quad \Gamma \in L_\infty^{k \times n}(\mathbb{R}).$

We call ω an *admissible time-varying Sylvester data set* if

   (i) A is a backward stable and (C, A) is uniformly observable;

   (ii) Z is backward stable and (Z, B) is uniformly controllable;

   (iii) Γ satisfies the Sylvester differential equation

(4.17) $$\frac{d}{dt}\Gamma(t) = Z(t)\Gamma(t) - \Gamma(t)A(t) + B(t)C(t), \text{ a.e, } t \in \mathbb{R}.$$

With an admissible time-varying Sylvester data set $\omega = (C, A; Z, B; \Gamma)$ we associate the following set of Hilbert-Schmidt operators:

(4.18)
$$S_\omega := \{C(\Delta - A)^{-1}h + K \mid h \in L_2^n(\mathbb{R}), \ K \in \mathcal{LS}_2^{p \times 1} \text{ are such that}$$
$$\mathcal{R}es_2^+\{(\Delta - Z)^{-1}BK\} = \Gamma h\}.$$

Obviously, $S_\omega \subset S_2^{p \times 1}$. Note that $T\mathcal{LS}_2^{p \times 1}$ is a right module over $\mathcal{LS}_2^{1 \times 1}$ (i.e., $TH \in S_\omega$ whenever $T \in S_\omega$ and $H \in \mathcal{LS}_2^{1 \times 1}$). It can be shown that a subspace S_ω of $S_2^{p \times 1}$ of the form (4.18) is a right module over $\mathcal{LS}_2^{1 \times 1}$ if and only if Γ satisfies the time-varying Sylvester equation (4.17); see Theorem 5.2 in [BGK2] for the discrete time version. The proof for the continuous time version is more technical and will not be done here.

We call S_ω in (4.18) the *singular subspace* associated with ω. Corollary 4.2 states that $T\mathcal{LS}_2^{p \times 1}$ is precisely the singular subspace associated with the generalized RHP-null-pole triple (4.14) of the input-output operator T. The main result of this paper (see Section 6) concerns an inverse problem, namely given an admissible time-varying Sylvester data set ω, construct (if possible) a J-unitary input-output operator T such that $T\mathcal{LS}_2^{p \times 1} = S_\omega$.

5. J-UNITARY INPUT-OUTPUT OPERATORS

Throughout this section T is the input-output operator of the system

(5.1) $$\Sigma \begin{cases} x'(t) = A(t)x(t) + B(t)u(t), & t \in \mathbb{R}, \\ y(t) = C(t)x(t) + u(t), \end{cases}$$

where $A(\cdot)$, $B(\cdot)$, $C(\cdot)$ and $D(\cdot)$ are matrix functions of sizes $N \times N$, $N \times p$, $p \times N$ and $p \times p$, respectively, which have their entries in $L_\infty(\mathbb{R})$. Moreover, the main coefficient is assumed to be dichotomous.

In what follows J is a $p \times p$ signature matrix, that is, $J = J^* = J^{-1}$. The matrix J induces a bounded linear operator on $L_2^p(\mathbb{R})$ which is also denoted by J. Thus $(J\varphi)(t) = J\varphi(t)$. This induced operator J is both selfadjoint and unitary, and therefore J is also a signature operator on $L_2^p(\mathbb{R})$. We are interested in describing a class of input-output operators $T \in \mathcal{R}^{p \times p}$ which are also J-unitary, i.e., for which

$$(5.2) \qquad\qquad T^*JT = J, \qquad TJT^* = J.$$

Let Σ be as in the first paragraph of this section, and let

$$(5.3) \qquad\qquad E(t, x) = \langle H(t)x, x \rangle, \quad t \in \mathbb{R}, \quad x \in \mathbb{C}^N,$$

be an *energy* (or *storage*) *function* on the state space of Σ. The latter means (see [HM], [W]) that $H(\cdot)$ is a Hermitian matrix-valued $N \times N$ matrix function with entries in $L_\infty(\mathbb{R})$ and E satisfies the *energy balance equation*

$$(5.4) \qquad E(t_2, x(t_2)) - E(t_1, x(t_1)) = \int_{t_1}^{t_2} \langle Ju(t), u(t) \rangle \, dt - \int_{t_1}^{t_2} \langle Jy(t), y(t) \rangle \, dt.$$

Here $t_1 < t_2$ are arbitrary time instances, u is an arbitrary input function from $L_2^p(\mathbb{R})$, the vector $x(t)$ represents the state at time t which one computes by solving the first equation of (5.1) in $L_2^N(\mathbb{R})$, and $y \in L_2^p(\mathbb{R})$ is the corresponding output function which one obtains from x and u by using the second equation of (5.1). In this case we say that Σ is *J-lossless* with respect to the energy function (5.3).

THEOREM 5.1. *If Σ is J-lossless, then T_Σ is a J-isometry, i.e., $T_\Sigma^* J T_\Sigma = J$.*

PROOF. Let Σ be J-lossless with respect to the energy function (5.3). Take $u \in L_2^p(\mathbb{R})$. Set $x = (\Delta - A)^{-1}Bu$ and $y = T_\Sigma u$. Then u, x and y satisfy the equations in (5.1), and hence for u, x and y chosen in this way the identity (5.4) holds. Since u and y are square integrable, the functions $\langle Ju(\cdot), u(\cdot) \rangle$ and $\langle Jy(\cdot), y(\cdot) \rangle$ are integrable on $\mathbb{R}$, and hence the following limits exist:

$$\lim_{t \to \infty} \langle H(t)x(t), x(t) \rangle = a, \quad \lim_{t \to -\infty} \langle H(t)x(t), x(t) \rangle = b.$$

The fact that $H(\cdot)$ is uniformly bounded and $x \in L_2^N(\mathbb{R})$ implies that $\langle H(\cdot)x(\cdot), x(\cdot) \rangle$ is integrable on $\mathbb{R}$. Thus both a and b must be zero. But then we see from (5.4) that

$$\int_{-\infty}^{\infty} \langle Ju(t), u(t) \rangle \, dt = \int_{-\infty}^{\infty} \langle Jy(t), y(t) \rangle \, dt.$$

In other words, $\langle Ju, u \rangle = \langle JT_\Sigma u, T_\Sigma u \rangle$. This holds for each $u \in L_2^p(\mathbb{R})$. Hence T_Σ is a J-isometry. $\square$

THEOREM 5.2. Let Σ be as in (5.1), and let J be $p \times p$ signature matrix. Assume $H(\cdot)$ is a Hermitian matrix valued $N \times N$ matrix function with entries in $L_\infty(\mathbb{R})$ such that $H(\cdot)$ is absolutely continuous on compact intervals and for almost all $t \in \mathbb{R}$ the following three identities hold:

$$(5.5a) \qquad \frac{d}{dt}H(t) + H(t)A(t) + A(t)^*H(t) = -C(t)^*JC(t),$$

$$(5.5b) \qquad H(t)B(t) = -C(t)^*JD(t),$$

$$(5.5c) \qquad D(t)^*JD(t) = J.$$

Furthermore, assume that $H(t)^{-1}$ exists for each t and is uniformly bounded on $\mathbb{R}$. Then the input-output operator T_Σ is J-unitary.

PROOF. The proof is divided into two parts. First we apply Theorem 5.1 to show that T_Σ is a J-isometry, and next we apply this result to the adjoint system.

Part (a). Consider the function (5.3). We claim that E satisfies the energy balance equation (5.4) if $H(\cdot)$ satisfies (5.5a)–(5.5c). To see this, we first note that (5.4) may be rewritten in the following equivalent form:

$$(5.6) \qquad \frac{d}{dt}E(t, x(t)) = \langle Ju(t), u(t) \rangle - \langle Jy(t), y(t) \rangle, \quad \text{a.e.,} \quad t \in \mathbb{R}.$$

Since the vector functions u, x and y are related as in (5.1). we have

$$\frac{d}{dt}E(t, x(t)) = \langle H'(t)x(t), x(t) \rangle + \langle H(t)A(t)x(t) + H(t)B(t)u(t), x(t) \rangle +$$
$$+ \langle H(t)x(t), A(t)x(t) + B(t)u(t) \rangle,$$

and

$$\langle Jy(t), y(t) \rangle = \langle JC(t)x(t) + JD(t)u(t), C(t)x(t) + D(t)u(t) \rangle,$$

almost everywhere on $\mathbb{R}$. It follows that

$$(5.7) \qquad \frac{d}{dt}E(t, x(t)) - \{\langle Ju(t), u(t) \rangle - \langle Jy(t), y(t) \rangle\} =$$
$$= (x(t)^* u(t)^*) \begin{pmatrix} M_{11}(t) & M_{12}(t) \\ M_{21}(t) & M_{22}(t) \end{pmatrix} \begin{pmatrix} x(t) \\ u(t) \end{pmatrix}, \quad \text{a.e.,} \quad t \in \mathbb{R}.$$

Here z^* denotes the conjugate transpose of the column vector z, and

$$M_{11}(t) = H'(t) + H(t)A(t) + A(t)^*H(t) + C(t)^*JC(t),$$

$$M_{12}(t) = H(t)B(t) + C(t)^*JD(t), \quad M_{21}(t) = M_{12}(t)^*,$$

$$M_{22}(t) = D(t)^*JD(t) - J.$$

Thus, if $H(\cdot)$ satisfies (5.5a)–(5.5c), then the energy function (5.3) satisfies (5.4), and we can apply Theorem 5.1 to show that $T_\Sigma^* JT_\Sigma = J$.

Part (b). In this part we use that the additional assumption that $H(t)^{-1}$ exists and is uniformly bounded. From (5.5c) we deduce that $D(t)^{-1} = JD(t)^*J$, and hence

$$(5.8) \qquad\qquad D(t)JD(t)^* = D(t)D(t)^{-1}J^{-1} = J^{-1} = J.$$

Furthermore, using (5.8) we see that

$$(5.9) \qquad\qquad H(t)^{-1}C(t)^* = -B(t)D(t)^{-1}J = -B(t)JD(t)^*.$$

Note that $\frac{d}{dt}H(t)^{-1} = -H(t)^{-1}H'(t)H(t)^{-1}$, and hence by multiplying (5.5a) on the left by $-H(t)^{-1}$ and on the right by $H(t)^{-1}$ we obtain

$$(5.10) \qquad \begin{aligned} \frac{d}{dt}H(t)^{-1} - A(t)H(t)^{-1} - A(t)^*H(t)^{-1} &= H(t)^{-1}C(t)^*JC(t)H(t)^{-1} = \\ &= B(t)JD(t)^*JD(t)B(t)^* = B(t)JB(t)^*. \end{aligned}$$

From (1.12) and (3.3) we deduce that

$$T_\Sigma^* = D^* - B^*(\Delta + A^*)^{-1}C^*,$$

and hence T_Σ^* is the input-output operator of the system

$$\Sigma_* \begin{cases} x'(t) = A_*(t)x(t) + B_*(t)u(t), \\ y(t) = C_*(t)x(t) + D_*(t)u(t), \end{cases}$$

where $A_*(t) = -A(t)^*$, $B_*(t) = C(t)^*$, $C_*(t) = -B(t)^*$, and $D_*(t) = D(t)^*$. Formulas (5.8)–(5.10) imply that (5.5a)–(5.5c) hold for $\Sigma_* = (A_*, B_*, C_*, D_*)$ in place of $\Sigma = (A, B, C, D)$, with $H(t)$ replaced by $H(t)^{-1}$, and with J replaced by $-J$. But then we can use the result of Part (a) to show that T_Σ^* is a J-isometry. Hence both T_Σ and T_Σ^* are J-isometries, and thus T_Σ is J-unitary. $\square$

Under certain additional minimality conditions the converse of Theorem 5.2 is also true. For instance, if Σ is uniformly controllable and uniformly observable and if its feedthrough coefficient is identically equal to the $p \times p$ identity matrix, then T_Σ is J-unitary implies the existence of a Hermitian matrix valued $N \times N$ matrix function satisfying all the conditions mentioned in Theorem 5.2. To prove this result one uses Propositions 3.3 and 3.4. Since the arguments are analogous to those used in the corresponding time-invariant case (see [BGR], Section 6), we omit the details.

6. MAIN THEOREM

Throughout this section

$$(6.1) \qquad\qquad \omega = (C, A; Z, B; \Gamma)$$

is an admissible time-varying Sylvester data set. Thus C, A, Z, B and Γ are matrix functions of sizes $p \times n$, $n \times n$, $k \times k$, $k \times p$ and $k \times n$, respectively, which have their entries in $L_\infty(\mathbb{R})$, and which satisfy the conditions (i)–(iii) listed in the one but last paragraph of Section 4. The singular subspace associated with ω is denoted by S_ω (see formula (4.18)).

THEOREM 6.1. *Let ω in (6.1) be an admissible time-varying Sylvester data set, and let J be a $p \times p$ signature matrix. Consider the matrix function*

$$(6.2) \qquad\qquad \Lambda(t) = \begin{pmatrix} \Lambda_1(t) & \Gamma(t)^* \\ \Gamma(t) & \Lambda_2(t) \end{pmatrix}, \quad t \in \mathbb{R},$$

where

$$\Lambda_1(t) = -\int_{-\infty}^t U_A(t)^{-*} U_A(s)^* C(s)^* J C(s) U_A(s) U_A(t)^{-1}\, ds,$$

$$\Lambda_2(t) = \int_t^\infty U_Z(t) U_Z(s)^{-1} B(s) J B(s)^* U_Z(s)^{-*} U_Z(t)^*\, ds.$$

If

$$(6.3) \qquad\qquad \sup_{t \in \mathbb{R}}\{\|\Lambda(t)\|,\ \|\Lambda(t)^{-1}\|\} < \infty,$$

then there exists a J-unitary input-output operator Θ such that ω is a generalized RHP-null-pole triple for Θ, and hence $S_\omega = \Theta \mathcal{L} S_2^{p \times 1}$. In fact, one such Θ is given by

$$(6.4) \qquad \Theta = D + (C \quad -JB^*) \begin{pmatrix} (\Delta - A)^{-1} & 0 \\ 0 & (\Delta + Z^*)^{-1} \end{pmatrix} \Lambda^{-1} \begin{pmatrix} -C^*J \\ B \end{pmatrix} D,$$

with

$$(6.5) \qquad \Theta^{-1} = D^* - D^* (C \quad -JB^*) \Lambda^{-1} \begin{pmatrix} (\Delta + A^*)^{-1} & 0 \\ 0 & (\Delta - Z)^{-1} \end{pmatrix} \begin{pmatrix} -C^*J \\ B \end{pmatrix},$$

for any $D \in L_\infty^{p \times p}(\mathbb{R})$ satisfying $D(t)^* J D(t) = J$ almost everywhere on $\mathbb{R}$. Conversely, let Θ be a J-unitary input-output operator of a uniformly controllable and uniformly observable system whose feedthrough coefficient $D(t)$ is J-unitary for almost each $t \in \mathbb{R}$, and assume that ω is a generalized RHP-null-pole triple for Θ. Then (6.3) holds and Θ is given by (6.4).

PROOF. Assume (6.3) is fulfilled. From the definitions of Λ_1 and Λ_2 it follows that

$$(6.6) \qquad \frac{d}{dt} \Lambda_1(t) + \Lambda_1(t) A(t) + A(t)^* \Lambda_1(t) = -C(t)^* J C(t),$$

$$(6.7) \qquad \frac{d}{dt} \Lambda_2(t) - \Lambda_2(t) Z(t)^* - Z(t) \Lambda_2(t) = -B(t) J B(t)^*.$$

By combining this with the Sylvester equation (4.17) we see that

$$(6.8) \qquad \begin{aligned} \frac{d}{dt} \Lambda(t) + \Lambda(t) \begin{pmatrix} A(t) & 0 \\ 0 & -Z(t)^* \end{pmatrix} &+ \begin{pmatrix} A(t)^* & 0 \\ 0 & -Z(t) \end{pmatrix} \Lambda(t) = \\ &= - \begin{pmatrix} C(t)^* \\ -B(t)J \end{pmatrix} J (C(t) \quad -JB(t)^*) \quad \text{a.e.,} \quad t \in \mathbb{R}. \end{aligned}$$

Obviously, $\Lambda(t)$ is Hermitian for each t. Since (6.3) holds, we can apply Theorem 5.2 (with $\Lambda(t)$ in place of $H(t)$) to show that Θ in (6.4) is unitary. The latter implies that $\Theta^{-1} = J \Theta^* J$, and hence Θ^{-1} is given by (6.5).

Next, we prove that ω is a generalized RHP-null-pole triple for Θ. For this purpose we apply the results of Section 4. Without loss of generality we may assume that $D(t)$ is identically equal to the $p \times p$ identity matrix. Thus Θ is the input-output operator of the system

$$(6.9) \qquad \Sigma \begin{cases} x'(t) = \begin{pmatrix} A(t) & 0 \\ 0 & -Z(t)^* \end{pmatrix} x(t) + \Lambda(t)^{-1} \begin{pmatrix} -C(t)^*J \\ B(t) \end{pmatrix} u(t), \quad t \in \mathbb{R}, \\ y(t) = (C(t) \quad -JB(t)^*) x(t) + u(t). \end{cases}$$

Note that Σ is decomposed, because A is backward stable and $-Z^*$ is forward stable. From (1.11) and (1.12) it follows that

$$G_{(-A^*,C^*)} = G_{(C,A)}, \qquad G_{(B^*,-Z^*)} = G_{(Z,B)},$$

and hence the conditions (i) and (ii) of an admissible time-varying Sylvester data set imply (cf., the remark made after the proof of Lemma 3.2) that the system Σ is uniformly observable. Next, observe that (6.8) may be rewritten in the following equivalent form:

$$\begin{pmatrix} -A(t)^* & 0 \\ 0 & Z(t) \end{pmatrix} = \Lambda'(t)\Lambda(t)^{-1} +$$
$$+ \Lambda(t) \left\{ \begin{pmatrix} A(t) & 0 \\ 0 & -Z(t)^* \end{pmatrix} - \Lambda(t)^{-1} \begin{pmatrix} -C(t)^*J \\ B(t) \end{pmatrix} (C(t) \quad -JB(t)^*) \right\} \Lambda(t)^{-1}.$$

Since (6.3) holds, it follows that $\Lambda(\cdot)$ is a kinematic similarity which transforms the inverse system $\Sigma^\times$ associated with Σ into the system:

$$\Sigma_* \begin{cases} x'(t) = \begin{pmatrix} -A(t)^* & 0 \\ 0 & Z(t) \end{pmatrix} x(t) + \begin{pmatrix} -C(t)^*J \\ B(t) \end{pmatrix} u(t), \quad t \in \mathbb{R}, \\ y(t) = - (C(t) \quad -JB(t)^*) \Lambda(t)^{-1}x(t) + u(t). \end{cases}$$

Furthermore,

$$\Gamma(t) = (0 \quad I) \Lambda(t) \begin{pmatrix} I \\ 0 \end{pmatrix}, \quad t \in \mathbb{R}.$$

The system Σ_* is uniformly controllable for the same reason as Σ is uniformly observable. Since kinematic similarity preserves uniform controllability, we conclude that $\Sigma^\times$ is uniformly controllable. It follows (cf., the proof of Proposition 3.3) that Σ is uniformly controllable. From the above remarks we see that $\omega = (C, A; Z, B; \Gamma)$ is a generalized RHP-null-pole triple for $\Theta = T_\Sigma$, and hence we can use Corollary 4.2 to show that $\Theta \mathcal{L}S_2^{p \times 1} = S_\omega$.

It remains to prove the converse statement. Let Σ_0 be a uniformly controllable and uniformly observable system whose feedthrough coefficient $D(t)$ is J-unitary for almost each $t \in \mathbb{R}$. Put $\Theta = T_{\Sigma_0}$, and assume that ω is a generalized RHP-null-pole triple for Θ. We want to prove that (6.3) holds and that Θ is given by (6.4). Without loss of generality we may assume that $D(t)$ is identically equal to the $p \times p$ identity matrix. By Theorem 1.1 there exists a kinematic similarity $E(\cdot)$ which transforms Σ_0 into a system

$\tilde{\Sigma}_0$ which is decomposed, i.e.,

$$\tilde{\Sigma}_0 \begin{cases} x'(t) = \begin{pmatrix} A_1(t) & 0 \\ 0 & A_2(t) \end{pmatrix} x(t) + \begin{pmatrix} B_1(t) \\ B_2(t) \end{pmatrix} u(t), \quad t \in \mathbb{R}, \\ y(t) = (C_1(t) \quad C_2(t)) x(t) + u(t), \end{cases}$$

where A_1 is backward stable and A_2 is forward stable. The system $\tilde{\Sigma}_0$ is also uniformly controllable and uniformly observable, and its input-output operator is Θ. Thus, by the remark made at the end of Section 5, we can find a kinematic similarity $H(\cdot)$ such that

(6.10)
$$\frac{d}{dt} H(t) + H(t) \begin{pmatrix} A_1(t) & 0 \\ 0 & A_2(t) \end{pmatrix} + \begin{pmatrix} A_1(t)^* & 0 \\ 0 & A_2(t)^* \end{pmatrix} H(t) =$$
$$= - \begin{pmatrix} C_1(t)^* \\ C_2(t)^* \end{pmatrix} J (C_1(t) \quad C_2(t)),$$

(6.11)
$$H(t) \begin{pmatrix} B_1(t) \\ B_2(t) \end{pmatrix} = - \begin{pmatrix} C_1(t)^* \\ C_2(t)^* \end{pmatrix} J.$$

Using (6.11) we may rewrite (6.10) in the following equivalent form

$$\begin{pmatrix} -A_1(t)^* & 0 \\ 0 & -A_2(t)^* \end{pmatrix} = H'(t) H(t)^{-1} +$$
$$+ H(t) \left\{ \begin{pmatrix} A_1(t) & 0 \\ 0 & A_2(t) \end{pmatrix} - \begin{pmatrix} B_1(t) \\ B_2(t) \end{pmatrix} (C_1(t) \quad C_2(t)) \right\} H(t)^{-1}.$$

It follows that H transforms the inverse system $\Sigma_0^\times$ associated with Σ_0 into the system

$$\Sigma_{0,*} \begin{cases} x'(t) = \begin{pmatrix} -A_1(t)^* & 0 \\ 0 & -A_2(t)^* \end{pmatrix} x(t) + \begin{pmatrix} -C_1(t)^* J \\ -C_2(t)^* J \end{pmatrix} u(t), \quad t \in \mathbb{R}, \\ y(t) = (J B_1(t)^* \quad J B_2(t)^*) x(t) + u(t), \end{cases}$$

Put

$$\Gamma_0(t) = (0 \quad I) H(t) \begin{pmatrix} I \\ 0 \end{pmatrix}, \quad t \in \mathbb{R}.$$

Then

$$\omega_0 = \{ C_1(\cdot), A_1(\cdot); -A_2(\cdot)^*, -C_2(\cdot)^* J; \Gamma_0(\cdot) \}$$

is a generalized RHP-null-pole-triple for Θ. Since such triples are unique up to equivalence, ω_0 and ω are equivalent, and hence we can find kinematic similarities $E_1(\cdot)$ and $E_2(\cdot)$ such that

$$\frac{d}{dt} E_1(t) = A(t) E_1(t) - E_1(t) A_1(t), \quad C(t) E_1(t) = C_1(t),$$
$$\frac{d}{dt} E_2(t) = Z(t) E_2(t) + E_2(t) A_2(t)^*, \quad E_2(t)^{-1} B(t) = -C_2(t)^* J,$$
$$\Gamma_0(t) = E_2(t)^{-1} \Gamma(t) E_1(t),$$

for almost each $t \in \mathbb{R}$. Now, put

$$F(t) = \begin{pmatrix} E_1(t) & 0 \\ 0 & E_2(t)^{-*} \end{pmatrix}, \qquad t \in \mathbb{R}.$$

Then F is a kinematic similarity, and one computes that for almost each $t \in \mathbb{R}$

$$(6.12) \qquad \frac{d}{dt}F(t) = \begin{pmatrix} A(t) & 0 \\ 0 & -Z(t)^* \end{pmatrix} F(t) - F(t) \begin{pmatrix} A_1(t) & 0 \\ 0 & A_2(t) \end{pmatrix},$$

$$(6.13) \qquad (C(t) \quad -JB(t)^*) F(t) = (C_1(t) \quad C_2(t)),$$

$$(6.14) \qquad F(t)^* \Lambda(t) F(t) = H(t).$$

To obtain (6.14) one partitions $H(t)$ as a 2×2 block matrix function according to the partitionings appearing in (6.10), and one uses the fact that the $(1,1)$-entry and the $(2,2)$-entry in this partitioning of $H(t)$ are uniquely determined by (6.10). From (6.14) we conclude that $\Lambda(t)^{-1}$ exists and is uniformly bounded. Thus (6.3) holds. It follows that the system Σ in (6.9) is well-defined. The identities (6.12)–(6.14) imply that $\tilde{\Sigma}_0$ and Σ are similar. Hence $\Theta = T_\Sigma$, and therefore Θ is given by (6.4). $\qquad \square$

REFERENCES

[BGK1] J.A. Ball, I. Gohberg and M.A. Kaashoek, Nevanlinna-Pick interpolation for time-varying input-output maps: the continuous case, in: *Time-variant systems and interpolation* (Ed. I. Gohberg), OT 56, Birkhäuser Verlag, Basel, 1992, pp. 52-89.

[BGK2] J.A. Ball, I. Gohberg and M.A. Kaashoek, Bitangential interpolation for input-output operators of time-varying systems: the discrete time case, *New aspects in interpolation and completion theories* (Ed. I. Gohberg), OT 64, Birkhäuser Verlag, Basel, 1993, pp. 33-72.

[BGR] J.A. Ball, I. Gohberg and L. Rodman, *Interpolation of rational matrix functions*, OT 45, Birkhäuser Verlag, Basel, 1990.

[BenG] A. Ben-Artzi and I. Gohberg, Dichotomy of systems and invertibility of linear ordinary differential operators, in: *Time-variant systems and interpolation* (Ed. I. Gohberg), OT 56, Birkhäuser Verlag, Basel, 1992, pp. 90-120.

[CD] F.M. Callier and C.A. Desoer, *Linear System Theory*, Springer-Verlag, New York, 1991.

[Co] W.A. Coppel, *Dichotomies in stability theory*, Lecture notes in Mathematics 629, Springer-Verlag, Berlin, 1978.

[DK] Ju.L. Daleckii and M.G. Kreĭn, *Stability of solutions of differential equations in Banach space*, Transl. Math. Monographs 43, Amer. Math. Soc., Providence, Rhode Island, 1974.

[DKV] P. Dewilde, M.A. Kaashoek and M. Verhaegen (Eds.), *Challenges of a generalized
       system theory*, Koninklijke Nederlandse Akademie van Wetenschappen, Verhan-
       delingen, Afd. Natuurkunde, Eerste reeks, deel 40, North-Holland Publ. Co.,
       Amsterdam, 1993.

  [G] I. Gohberg (Ed.), *Time-variant systems and interpolation*, OT 56, Birkhäuser
       Verlag, Basel, 1992.

[GKvS] I. Gohberg, M.A. Kaashoek and F. van Schagen, Non-compact integral operators
       with semi-separable kernels and their discrete analogues: inversion and Fredholm
       properties, *Integral Equations and Operator Theory* 7 (1984), 642-703.

 [HM] D.J. Hill and P.J.Moylan, Dissipative dynamical systems: basic input-output
       properties, *J. Franklin Inst.* 309 (1980), 327-357.

 [MS] J.L. Massera and J.J. Schäffer, *Linear differential equations and function spaces*,
       Academic Press, New York, 1966.

  [W] J.C. Willems, Dissipative dynamical systems, Part I: General Theory, *Arch. Rat.
       Mech. Anal.* 45 (1972), 321–351.

J.A. Ball
Department of Mathematics, Virginia Tech
Blacksburg, VA 24061, U.S.A.

I. Gohberg
Raymond and Beverley Sackler Faculty of Exact Sciences
School of Mathematical Sciences, Tel-Aviv University
Ramat-Aviv, Israel.

M.A. Kaashoek
Faculteit Wiskunde en Informatica, Vrije Universiteit
De Boelelaan 1081a
1081 HV Amsterdam, The Netherlands.

Operator Theory:
Advances and Applications, Vol. 75
© 1995 Birkhäuser Verlag Basel/Switzerland

OPTIMIZATION WITHOUT COMPACTNESS, AND ITS APPLICATIONS

Alexander V. Bukhvalov

Dedicated to Professor A.C. Zaanen on the occasion of his eightieth birthday

In this paper I give a survey of the "near-compactness" properties of norm-bounded convex subsets of Banach Function Spaces which are closed in measure. This originated in the work of G.Ya.Lozanovskiĭ and the author [BL1], and has subsequently been generalized in various directions, including the vector-valued setting. Numerous applications are discussed in the following areas: optimal control, minimax theorems and best approximation in Banach Function Spaces (L^1 being the main example), G.Godefroy's proof of the weak sequential completeness of L^1/H_0^1, geometry of Banach lattices.

INTRODUCTION

Many results about compact convex subsets of locally convex and Banach spaces have numerous extremely useful applications—especially in problems concerning the existence of optimal solutions, minimax theorems, best approximation. We here present an approach which shows that the same circle of results holds true in the case of norm-bounded convex subsets of "good" spaces of measurable functions which are closed in measure. In Section 1 we deal with the classical space L^1; the case of general Function Spaces is considered in Section 2 and finally generalization to the vector-valued case is discussed in Section 3. Other Sections are devoted to applications. In Section 4 some straightforward applications of the general theory to optimal control, best approximation and mathematical economics are given. Sections 5 and 6 are devoted to much more sophisticated and unexpected applications. The first one is G.Godefroy's proof (see [Go1]) of the well-known Mooney–Khavin theorem about the weak sequential completeness of the space L^1/H_0^1. This proof is based on very modest analytical tools but it is easier than previous ones and works in the case of many complex variables without any changes. The second application is due to V.Caselles (see [C1,C2]) and deals with the geometrical properties of Banach lattices. In Section 6 we also present a discussion of the wonderful G.Ya.Lozanovskii's identities: $L^2 = E^{\frac{1}{2}}(E')^{\frac{1}{2}}$ and $L^1 = EE'$.

To begin with, we shall explain the general idea of our approach. In fact, we reduce the situation to consideration of weak*-compact sets in the second dual, via the following procedure. Let V be a norm-bounded convex set in some Banach space X. We shall fix a linear subspace Y of X^* which is 1-norming for X. In this case, the canonical imbedding $\pi\colon X \to Y^*$ is norm-preserving. We assume that there is a projection P from Y^* onto $\pi(X)$. Let W be the weak*-closure of $\pi(V)$ in Y^*. This set is weak*-compact. In general, we only have the inclusion $\pi(V) \subset P(W)$. Let us consider a class $\mathcal{K}$ of sets V in X for which the following equality holds

$$P(W) = \pi(V). \tag{$*$}$$

It is easy to see that the class $\mathcal{K}$ inherits many of the properties of the $(Y^*, \sigma(Y^*, Y))$-compact subsets of Y^*. It is not difficult to give general results in terms of this abstract scheme, but we prefer to restrict ourselves to some concrete situations below. It is clear that the main difficulty lies in finding suitable examples for spaces X and Y, which admit a projector P, and which single out an interesting class $\mathcal{K}$ of sets. Certainly, it is desirable to have an intrinsic description of the class $\mathcal{K}$ as subsets of X.

It happens that in most Banach Function Spaces, including L^1 as the main example, we can take for $\mathcal{K}$ in the abstract scheme above the class of all norm-bounded convex sets which are closed in measure. In this case we can choose the space of integral functionals for Y and a band projection for P. This band projection is given by the generalized form of the Yosida–Hewitt theorem (see [YH, KA, Z2, BBY]).

The theory presented gives one of the few contributions of the theory of vector lattices which has proved its importance in a variety of concrete situations and areas. At this place I would like to tell some words concerning the general historical background of its appearance.

SOME HISTORY

The theory of vector lattices has its origins in the works of several mathematicians written about 1935. An impetus was given by a talk of F.Riesz at the Mathematical Congress in Bologna in 1928. The formal priority, in publication of the final version of the set of axioms defining a vector lattice, belongs to L.Kantorovich from Leningrad. His article is separated by a few months from the publication of a Dutch mathematician H.Freudenthal. From this time the three main schools, devoted themselves to the investigation of ordered spaces in the framework of functional analysis, have been working in

Russia (Leningrad, Voronezh, Novosibirsk), the Netherlands (expanding afterwards to the USA, Australia and South Africa) and Japan. Later the wonderful Japanese school had declined, but its place has been occupied by the powerful German school.

In 1950s the notion of Banach Function Space was introduced and deeply investigated. A.C.Zaanen and his collaborators were among the main contributors to this area (see [Z1]). But, in a certain extent, this theory was a parallel to that of vector lattices (and some results were actually known in another framework) though some deeper properties from the measure theory have been used (including the possibility of investigation of the spaces of Banach-space-valued functions). On the other hand, many concrete spaces of measurable functions (Orlicz, Lorentz, Marcinkiewicz spaces) have been treated from the point of view of general Banach Function Spaces.

In the middle of 1960s a historical series of papers by W.A.J.Luxemburg and A.C.Zaanen, entitled 'Notes on Banach Function Spaces' (Proc. Acad. Sci. Amsterdam, Ser. A, 1963–65), opened new research horizons generating a synthesis of the theory of Banach Function Spaces and that of vector lattices. This gave a lucky opportunity for functional analysis, algebra and measure theory to work together.

Independently, about the same time, the similar ideas of synthesis came to G.Ya.Lozanovskiĭ, a young mathematician from Leningrad. So, he became one of the first and most careful readers of the series by W.A.J.Luxemburg and A.C.Zaanen. The whole chapters of his optional courses lectured in the end of 60s were devoted to the exposition of the highlights of this series. The generalized Yosida–Hewitt theorem, which is an important tool for the results of this paper, was published by W.A.J.Luxemburg within this series (1965). By this time G.Ya.Lozanovskiĭ also had a (quite different) proof of this theorem (analogous to the proof given in exercises in [Z2]; for discussion and generalizations see [BBY]). He has not published this result because of the paper by W.A.J.Luxemburg. At the late 60s and 70s G.Ya.Lozanovskiĭ became a recognized leader of the group working in the field of Banach lattices in Leningrad (he died aged 38 in 1976).

Now let us return back to the optimization on the sets which are closed with respect to convergence in measure. The very idea of this theory belongs to G.Ya.Lozanovskiĭ. The history of the appearance of the main results is the following. In 1973 G.Ya.Lozanovskiĭ presented a version of the theorems from Sections 1 and 2, for the case of Banach Function Spaces on σ-finite measure spaces, at the Seminar of Prof. B.Z.Vulikh (Leningrad University). In the week following the talk the author was able to give a simplified version of the proof of the main theorem and a generalization to the case of an (almost) arbitrary measure and to more general classes of Function Spaces. These results were announced, without proofs, in the joint paper [BL1] and the proofs were given in

§1 of [BL2]. Some important generalizations to the vector-valued case were obtained by the author and presented in §4 of [BL2]. G.Ya.Lozanovskiĭ searched for the most general classes of spaces and sets for which the theory could be applied. In §2 of [BL2] he developed a version of the theory for Banach Function Spaces E with a projection from E^{**} onto E. Finally, he introduced a general scheme, in terms of abstract topological vector spaces and Banach spaces (see [Lo3]). All known implementations of the theory may be derived as particular cases of this scheme via an appropriate version of the Yosida–Hewitt theorem. It should be mentioned that the paper [BL2] was submitted before the work [Lo3]. Some additional bibliographical information will be presented below.

1. SETS CLOSED IN MEASURE IN THE SPACE L^1

In this Section we present the main results in the special but important case of the space L^1. We restrict ourselves to the case of a σ-finite measure space (T, Σ, μ). Let π denote the canonical imbedding of L^1 into $(L^1)^{**} = (L^\infty)^*$, and let P denote the canonical projection of $(L^\infty)^*$ onto $\pi(L^1)$ which is given by the Yosida–Hewitt theorem [YH, KA]. The term 'convergence in measure' means the convergence in measure on every set of finite measure. We say a subset V of a space E of measurable functions is closed in measure if it is closed with respect to the relative topology on E induced by the topology of the convergence in measure. In the σ-finite case it is equivalent to the closedness with respect to almost everywhere convergence of sequences in E, i.e. $(\{e_n\} \subset V, e \in E, e_n \to e \text{ a.e.}) \implies (e \in V)$.

A realization of the scheme presented in the Introduction is given by the following main theorem.

THEOREM 1.1. *Let V be a nonempty convex subset of L^1, and let W be the $\sigma((L^\infty)^*, L^\infty)$-closure of the set $\pi(V)$ in $(L^\infty)^*$. Then the following assertions are true:*

(i) *If V is closed in measure in L^1, then the equality*

$$P(W) = \pi(V) \tag{$*$}$$

   is valid;

(ii) *If V is norm-bounded in L^1 and satisfies condition $(*)$, then V is closed in measure in L^1.*

One can find the detailed proof of Theorem 1.1 in [BL2, Theorem 1.1′], [KA, §X.5].

SKETCH OF THE PROOF. (i) Take $\varphi \in W$. Then there is a net $\{e_\alpha\}$ in L^1 such that $\pi(e_\alpha) \to \varphi$ in the weak*-topology. There exists $e \in L^1$ satisfying $\pi(e) = P\varphi$. It is sufficient to show that $e \in V$.

Let G be an order dense ideal in L^∞ such that $(\varphi - P\varphi)(G) = \{0\}$. The existence of such an ideal comes from the order-theoretic description of the singular component of a functional in the Yosida–Hewitt decomposition. An easy computation shows that $e_\alpha \to e$ in the weak topology $\sigma(L^1, G)$. Since without any loss of generality we may assume that G is norm closed, there is a strictly positive function $g \in G$. Hence, $e_\alpha \to e$ in the weak topology of weighted L^1-space. Now, by the Mazur lemma there is a net of some convex combinations of e_α (which belong to V because of its convexity) converging to e in the norm topology of weighted L^1-space and, hence, also in measure. Since V is closed in measure then $e \in V$.

(ii) As it was mentioned above it is sufficient to prove that $(\{e_n\} \subset V,\ e \in E,\ e_n \to e$ a.e.$) \implies (e \in V)$. Using relative weak*-compactness of $\{\pi(e_n)\}$ we can choose a subnet $\{e_\alpha\}$ of the given sequence such that $\pi(e_\alpha) \to \varphi \in W$ in the weak*-topology. Hence, due to the condition $(*)$, there is an element $f \in V$ such that $\pi(f) = P\varphi$. Now, arguing exactly as in the proof of (i), we see that $e_\alpha \to f$ in the weak topology $\sigma(L^1, G)$ for an order dense ideal G in L^∞.

It is sufficient to prove that $e = f$ since we know that $f \in V$. For this purpose we notice that any a.e. convergent sequence is order bounded in L^0, the space of all (finite) measurable functions. Since $e_n \to e$ a.e., and hence also in measure, the subnet $e_\alpha \to e$ in measure. Passing to a partition of the measure space (to get a summable majorant) and using the Lebesgue dominated convergence theorem for nets, we easily prove our claim. $\square$

All the theorems below are easy consequences of the main theorem and for this reason may not merit the name 'theorem'. But, on the other hand, they are far from obvious without this result, and they have numerous applications.

THEOREM 1.2. *Let V_1 and V_2 be nonempty convex nonintersecting sets in L^1 which are closed in measure, and suppose further that V_1 is norm-bounded. Then V_1 and V_2 can be strictly separated by a continuous functional.*

We give a very simple proof of this result here as an illustration of the method.

PROOF. Let W_i be the weak*-closure of the sets $\pi(V_i)$ (i=1,2). Then by $(*)$ we have $W_1 \cap W_2 = \varnothing$, and if, say, V_1 is norm-bounded, then W_1 is weak*-compact. Now we can apply the usual separation theorem in the locally convex space $(L^\infty)^*$ with its weak* topology. $\square$

THEOREM 1.3. *If a collection of convex norm-bounded subsets of L^1 which are closed in measure has the finite intersection property, then it has a non-empty intersection.*

This assertion implies that the unit ball of L^1 has the following surrogate of compactness.

THEOREM 1.4. *If* $\{e_n\}_1^\infty$ *is a norm-bounded sequence in* L^1, *then there exist a sequence* $1 = n_1 < n_2 < \ldots$, *numbers* $\{\lambda_i\}_1^\infty$ *and a function* $e \in L^1$ *such that*

$$\lambda_i \geq 0, \quad \sum_{i=n_k}^{n_{k+1}-1} \lambda_i = 1 \; (\forall k) \quad \text{and} \quad g_k(t) = \sum_{i=n_k}^{n_{k+1}-1} \lambda_i e_i(t) \to e(t)$$

for almost all $t \in T$.

PROOF. It is an immediate corollary of Theorem 1.3. $\square$

The following two results describe permanence properties of the class of sets which are closed in measure.

THEOREM 1.5. *Let* V_1 *and* V_2 *be convex, norm-bounded sets in* L^1 *which are closed in measure. Then* $V = \{v_1 + v_2 \colon v_1 \in V_1, v_2 \in V_2\}$ *is closed in measure.*

THEOREM 1.6. *Let* V *be a convex, norm-bounded subset of* L^1 *which is closed in measure. For each measurable* $A \in \Sigma$ *the set* $\{v\chi_A \colon v \in V\}$ *is closed in measure.*

The following result takes us into the field of best approximation problems.

THEOREM 1.7. *Let* V_1 *and* V_2 *be nonempty convex sets in* L^1 *which are closed in measure, and suppose that* V_1 *is norm-bounded. Then there exist* $v_1 \in V_1$ *and* $v_2 \in V_2$ *such that*

$$\|v_1 - v_2\| = \inf\{\,\|e_1 - e_2\| \colon e_1 \in V_1, e_2 \in V_2\}.$$

And finally, the theorem below leads us to the topic of "optimization without compactness".

THEOREM 1.8. *Any convex functional, which is lower semicontinuous with respect to convergence in measure, attains its minimum value on every convex, norm-bounded set* V *which is closed in measure.*

Theorems 1.1–1.7 were given in [BL1, BL2]. Theorem 1.8, which is an immediate consequence of Theorem 1.3 (or, Theorem 1.4—as you like), was derived in [Le1] from the results of [BL1] in a very complicated way by making use of subdifferential calculus. A simple proof was presented by the author in [KA, Theorem X.5.6]. As [BL1] contained no proofs, some of the proofs were given in [V].

It was noticed in [BL2] that Theorem 1.4 is a consequence of the Komlós theorem, which is stronger and deeper result in the theory of functions.

THEOREM 1.9 (Komlós). *If $\{e_n\}$ is a norm-bounded sequence in L^1, then there exist a function $f \in L^1$ and a subsequence $\{e_{n_k}\}$ such that any of its subsequences $\{g_m\}$ possesses the strong law of large numbers:*

$$\frac{1}{m}(g_1 + \ldots + g_m) \to e \quad a.e.$$

It appears to be impossible to derive Theorems 1.1, 1.2 and 1.3 (the last one in the uncountable case) from the Komlós theorem (at least in straightforward way). All of the other results can be obtained from the Komlós theorem. We should like to mention that the situation changes in the vector-valued case, as we will see later in Section 3.

Some interrelations between the Komlós theorem and Theorem 1.1 are discussed in [Len].

2. SETS CLOSED IN MEASURE IN FUNCTION SPACES

In fact, the results in [BL1, BL2] were presented not for the L^1 case, but for the more general case of Banach Function Spaces on general measure spaces. We shall briefly discuss some of the peculiarities of this case.

Let (T, Σ, μ) be a measure space with the direct sum property [KA, I.6.9]. In this case the topology of convergence in measure is not metrizable, and we have to use generalized sequences. Moreover, this topology is strictly stronger than the topology induced by almost everywhere convergence. Let L^0 denote the space of all a.e. finite real-valued functions on (T, Σ, μ) with the usual identification of equivalent functions. It is well-known that L^0 is an order complete vector lattice. An ideal space on (T, Σ, μ) is a linear subset E of L^0 such that

$$(e_1 \in L^0, \ e_2 \in E, \ |e_1| \le |e_2|) \implies (e_1 \in E).$$

We may assume that the support of E is equal to T, that is, E is an order dense ideal in L^0. A norm $\| \cdot \|$ on a ideal space E is called monotone if

$$(e_1, e_2 \in E, \ |e_1| \le |e_2|) \implies (\|e_1\| \le \|e_2\|).$$

A Banach Function Space is an ideal space E endowed with a monotone norm with respect to which E is a Banach space.

For any ideal space E we define the dual (ideal) space E' by

$$E' = \{e' \in L^0 : \int |ee'| \, d\mu < \infty\}.$$

The dual space E' can be identified with the space of integral functionals on E and hence with the space of order continuous functionals as well (see [KA]). We can introduce the second dual via the formula: $E'' = (E')'$. In general, we have $E \subset E''$. An ideal space E is called Nakano-reflexive if $E = E''$. If $E^\sim$ denote the order bounded dual of E, then by the generalized Yosida–Hewitt theorem there is a band projection from $(E')^\sim$ onto the canonical image of E. In the case of a Banach Function Space E we have $(E')^\sim = (E')^*$.

Versions of Theorems 1.1–1.6, 1.8 are true for Nakano-reflexive ideal spaces. We have to substitute there E for L^1, E' for L^∞, and boundedness in the weak topology $\sigma(E, E')$ for norm-boundedness. It is interesting to mention the following new feature of Theorem 1.2—we can separate sets by an integral functional (this is not trivial because, in general, there are continuous non-integral functionals).

A Banach Function Space E on a σ-finite measure space is called perfect if it has the strong Fatou property:

$$(\{e_n\} \subset E,\ 0 \le e_n \uparrow,\ \sup \|e_n\| < \infty \text{ and } \|e_n\| \uparrow \|e\|).$$

Almost all concrete examples of Banach Function Spaces possess this property (in the case of a non-σ-finite measure space one should consider the strong Fatou property with respect to generalized sequences (nets) rather than usual ones). It is well-known that every perfect Banach Function Space is Nakano-reflexive, and so the theory works in this case. Moreover, Theorem 1.7 is valid in this case. We shall refer to the analogue of Theorem 1.N as Theorem 2.N. It is easy to see that the conditions on the space E, introduced in this Section, are necessary in the class of ideal spaces.

Some non-trivial examples of spaces for which the results of this section can be applied are given by spaces with mixed norm such as $L^1[L^p]$, $L^p[L^1]$, $(1 \le p \le \infty)$. These spaces are not Banach conjugates, and for this reason one cannot use weak*-compactness arguments.

In the case of the spaces with mixed norm two topologies of convergence in measure appear, i.e. the usual convergence in two variables and the convergence in the sense of the theory of vector-valued functions (the latter is strictly stronger in any non-trivial situation). In [B1] their interrelations are investigated.

Numerous generalizations have been considered to spaces of measurable functions with a weaker form of ideal property (see [FG, Gi1, Gi2, Le2]).

At the Fifth Meeting on Real Analysis and Measure Theory (Capri, 1992) the author has set up a question whether it is possible to derive a non-commutative version of the present theory. In response to this O.E. Tikhonov from Kazan' has obtained (1993) an analogue for non-commutative L^1-spaces associated with a von Neumann algebra with a faithful normal finite trace.

3. SETS CLOSED IN MEASURE IN
THE VECTOR-VALUED CASE

The results of Sections 1,2 were extended to the vector-valued case by the author in [BL2] (and were announced at the All-Union Conference "Theory of Operators in Function Spaces" (Novosibirsk, 1975), see [BL3]).

Let E be an ideal space, X be a Banach space, and $E(X)$ denote the space of all measurable functions $\vec{f}\colon T \to X$ such that $|\vec{f}| = \|\vec{f}(\cdot)\|_X \in E$. If E is a Banach Function Space, then $E(X)$ is a Banach space with the norm $\|\vec{f}\| = \||\vec{f}|\|$. We say that a sequence $\{\vec{f}_n\}$ is convergent in measure to $\vec{f}$ in the vector-valued case provided $|\vec{f}_n - \vec{f}| \to 0$ in measure.

The results of Section 2 are valid for the spaces $E(X)$ if X is any reflexive Banach space (and with the same conditions on E). In the case of Banach Function Spaces the canonical imbedding of $E(X)$ into $(E'(X^*))^*$ is considered. These results and the appropriate version of the Yosida–Hewitt theorem were established in [BL2] (the latter is valid for any Banach space). We shall refer to the analogue of Theorem 1.N as Theorem 3.N. Obviously, the condition of reflexivity is necessary for the validity of any one of Theorems 3.1–3.8. Theorem 3.4 was explicitly stated in [B1] where some of its applications were given. This theorem was repeated later in [Le2].

As we have mentioned in Section 1, the vector-valued results cannot be reduced to the vector-valued version of the Komlós theorem. To make this clear, we recall the following definitions. A Banach space X is said to have the Banach–Saks property ($X \in$ (BS)) provided any bounded sequence $\{x_n\}$ in X has a subsequence $\{x_{n_k}\}$ such that

$$\frac{1}{k}\left(x_{n_1} + \ldots + x_{n_k}\right) \to x$$

in norm. A Banach space X is said to have the Komlós property ($X \in$ (K)) provided the Komlós theorem is valid for any norm-bounded sequence in $L^1(X)$ (with a.e. norm convergence in X).

It is evident that $X \in$ (K) implies $X \in$ (BS). An example of J.Bourgain (see also [B2]) shows that the converse is not true (see [DU]). It is known (see [Ga, Kob]) that uniform convexity (more general, superreflexivity) is a sufficient condition for $X \in$ (K). But what is essential for us is that there is a reflexive Banach space X without the Banach–Saks property [Ba], and, hence, without the Komlós property. The vector-valued version of our theory works in such spaces, but the Komlós theorem is not valid there. So in this case we cannot avoid the use of Theorem 3.1 in the proofs of its corollaries (Theorems 3.2–3.8).

One may find some extensions of the vector-valued setting in [FG, Gi1, Gi2, Le1]. It is interesting that there is no pathology, connected with the geometry of Banach spaces, when we derive an analogue of the Komlós theorem for weak* convergence rather than norm convergence [Bd].

4. OPTIMIZATION AND APPROXIMATION

Theorems 1.8, 2.8 and 3.8 have lead to the investigation of "optimization without compactness" in [CA, Fe, FG, FP, Gi1, Gi2, Le1, Le2, Lu, V]. This reference list is far from complete. Of course, some versions of minimax theorems may be derived almost immediately from Theorems 1.1, 2.1, 3.1 (see [Fe, Le2]). We present one result from [Le2] in a slightly weaker form.

THEOREM 4.1. *Let E be a perfect Banach Function Space, X be a reflexive space. Let Q be a convex set in a vector space M, and let B denote a convex, norm-bounded set in $E(X)$ which is closed in measure. Suppose that the function $\Phi\colon B \times Q \to R^1 \cup \{\infty\}$ is convex and lower semicontinuous in measure in the first variable and concave in the second variable. Then functionals $\Phi(\cdot, z)$, $z \in Q$ and $\sup_{z \in Q} \Phi(\cdot, z)$ attain their minimum value on B and the following minimax equality holds:*

$$\sup_{z \in Q} \min_{x \in B} \Phi(x, z) = \min_{x \in B} \sup_{z \in Q} \Phi(x, z).$$

This minimax theorem gives a way to develop a version of the Nikishin–Maurey factorization theory "without compactness" (cf. [M, Lemma 3]); we will return to this theme later in Section 6. We hope that this approach will provide us with new results in the field of existence problems for weighted norm inequalities (cf. [RF]).

Some applications to mathematical economics are given in [Ca, Le2].

Certainly, some theorems on the existence of fixed points can be derived from the theory presented (see [Be] and [J]; the latter work is hardly the only one which uses the general scheme from [Lo3]). I hope that this and minimax Theorem 4.1 will lead to some new results in the general equilibrium theory and in game theory.

In contrast with Theorem 1.8, the best approximation Theorem 1.7 has not attracted the experts' attention. But, in fact, some recent results are immediate corollaries of it (and it was even rediscovered in [Lu, Theorem 3.1] in a very special case).

A subspace W of a Banach space X is said to be proximinal in X if to each $x \in X$ there corresponds a closest point in W. As an immediate corollary of Theorem 3.4 we have

THEOREM 4.2. *If E is a perfect Banach Function Space and Y is a reflexive subspace of a Banach space X, then $E(Y)$ is proximinal in $E(X)$.*

This result was published in [Kh1] for $E = L^1$. It is easy to prove the main result from [HLS] about the proximinality of sums of some tensor products in $L^1(T \times S)$, combining the results of Sections 1 and 3 for two different types of convergence in measure.

Another result of R.Khalil [Kh2] is a particular case of the following theorem, which is an immediate consequence of Theorem 3.7.

THEOREM 4.3. *Let Σ_0 be a σ-subalgebra in Σ. Let E be a perfect Banach Function Space and X be a reflexive space, and $L^0(\Sigma_0, X)$ denote the space of X-valued Σ_0-measurable functions. The subspace $E(X) \cap L^0(\Sigma_0, X)$ is proximinal in $E(X)$.*

I think that there is some interest in search for unusual applications concerned with the spaces of measures and operator spaces rather than function spaces. The main problem here is to describe an abstract analogue of the convergence in measure in an appropriate way (in intrinsic terms).

To begin with, note that the theory holds true in the case of arbitrary (abstract) Nakano-reflexive vector lattices, because any such space may be represented as a Banach Function Space on an appropriate measure space (maybe, not σ-finite but it does not matter [BL2]). This may be applied to many of spaces of operators. But in these cases it is difficult to describe the sense of the convergence in measure. If the corresponding measure space is σ-finite, then closedness in measure is equivalent to the closedness with respect to the order convergence in the maximal extension of the initial vector lattice. It is possible to describe it in the terms of truncations and, hence, in the intrinsic terms of the initial lattice. In the general case closedness in measure differs from the closedness with respect to order convergence. So the problem of adequate description of the convergence in measure arises.

I have in mind, first of all, the applications to operator spaces $B(L^1)$ and $B(L^\infty)$. It is well-known (and easily verified) that these are Nakano-reflexive. But only in that component, which corresponds to usual integral operators and which is canonically isomorphic to an appropriate space with mixed norm, we can describe the measure convergence—it is usual convergence in measure of the kernels. I see no straightforward approach for any use of abstract theory or pseudo-integral representation.

Nevertheless, it possible to derive some results for certain simpler classes of operators. For example, it is not difficult to prove, using Theorem 1.7 that the subspace of triangular operators in $B(L^1)$ is proximinal [AW, Theorem 2].

5. APPLICATIONS TO COMPLEX ANALYSIS

Let us consider L^1 on the circle with Lebesgue measure and the quotient space L^1/H_0^1, where H_0^1 is the space of boundary values of holomorphic functions on the unit disk, vanishing at 0 and belonging to the classical Hardy space H^1. This space is a natural member of the scale of Hardy spaces since $(L^1/H_0^1)^* = H^\infty$. Alike L^1 this space is not isomorphic to any Banach conjugate. The well-known Mooney–Khavin theorem claims the following.

THEOREM 5.1. *The Banach space L^1/H_0^1 is weakly sequentially complete.*

In [Go1] G.Godefroy has found a very elegant proof of this theorem, based on Theorem 1.1. The main advantage of G.Godefroy's proof is in the unified approach to the one-dimensional and multidimensional cases using very modest tools from the theory of functions. In [Go2] this result is generalized to weakly sequentially complete Banach Function Spaces in place of L^1. The proof is based on the following fact, which is derived from Theorem 1.1 or 2.1.

THEOREM 5.2. *If E is a weakly sequentially complete Banach Function Space and X is a subspace of E whose unit ball is closed in measure in E, then the quotient space E/X is weakly sequentially complete.*

Theorem 5.2 reduces the problem to verifying that in many cases the unit ball is closed in measure, and in many cases this is easy. K.B.Petrenko, a research student of the author, has proved the following result along the same lines in the vector-valued case.

THEOREM 5.3. *If X is a Banach space, then the space $L^1(X)/H_0^1(X)$ is weakly sequentially complete in the following two cases: (i) X is reflexive; (ii) X is a weakly sequentially complete Banach Lattice.*

Here, again, we can substitute an arbitrary weakly sequentially complete Banach Function Space for L^1. The statement of Theorem 5.3 is no longer true in the class of all weakly sequentially complete Banach spaces X. Namely, W.Hensgen [He] has noticed that one counterexample by G.Pisier [Pi1], which had been built up for another purpose, is suitable in this case.

In [Ki, Go3, GoLP] other interrelations are considered between problems in the theory of functions and closedness in measure. In particular, connections between the results of Section 1 and the famous Kadec–Pełczyński splitting lemma [KP] are established. The approach of G.Godefroy has been developed in [Li1, Li2, GoLi].

6. APPLICATIONS TO BANACH LATTICES

The first applications of the main theorem to questions about the geometry of Banach spaces and Banach lattices were given by G.Ya.Lozanovskiĭ [BL2, Theorems 1.7, 2.2, 2.3] and the author [B1].

Some very interesting and unexpected applications were derived by V.Caselles in [Ca1, Ca2]. There he proved the following two theorems.

THEOREM 6.1. *There exists a weakly sequentially complete Banach lattice (Banach Function Space) without the Radon–Nikodym property and such that all operators from L^1 into E are Dunford–Pettis.*

THEOREM 6.2. *Let E be a perfect Banach Function Space. Then, (i) E is a dual Banach lattice if and only if (ii) E has the lattice Radon–Nikodym property, i.e. each operator from L^1 into E possesses an integral representation.*

One of the main steps in the proofs consists of establishing the following result, which is derived from Theorem 1.1.

THEOREM 6.3. *Let $L^1 = L^1(0,1)$, U be the unit ball of L^1 and $U_+ = U \cap L^1_+$.*

(i) *Let C be a convex norm-bounded subset of L^1, which is closed in measure. Denote $H^+(C) = \{f \in L^1_+ : 0 \le f \le g \text{ for some } g \in C\}$. Let $0 \le T : L^1 \to L^1$ be such that $T(U_+) \subset H^+(C)$. Then there exists a positive operator $\hat{T} : L^1 \to L^1$ such that $0 \le T \le \hat{T}$ and $\{\hat{T}(\chi_A)/\mu(A) : \mu(A) > 0\} \subset C$.*

(ii) *Let C_1 and C_2 be solid, convex, norm-bounded subsets of L^1, which are closed in measure. Let $T : L^1 \to L^1$ be such that $T(U) \subset C_1 + C_2$. Then there exist some bounded operators $T_1, T_2 : L^1 \to L^1$ such that $T = T_1 + T_2$ and $T_1(U) \subset C_1$, $T_2(U) \subset C_2$.*

Theorem 1.2 has been used by G.Ya.Lozanovskiĭ when investigating the dual of the spaces $E_0^{1-\theta}E_1^\theta$ and $\varphi(E_0, E_1)$ (paper [Lo5] contains the complete proofs and details, while in [Lo4] the principle results of [Lo5] are announced). In fact, some deep and technically sophisticated results on this duality have been derived by G.Ya. Lozanovskiĭ earlier (including the identities (1)–(4) below; see [Lo1, Lo2]), but just thinking of this lead him to the main idea of the theory presented here.

Both of spaces mentioned above are important for interpolation theory: the first one arises naturally in the methods of complex interpolation by A.Calderon and the second one is connected with the φ-interpolation methods by V.I.Ovchinnikov (and Gustavsson–Peetre, see [Ni]). The first applications to the second method by A.Calderon were derived by V.Shestakov, a student by G.Ya.Lozanovskiĭ. Theorem 2.4 was applied in [B2, B3] to the complex method of interpolation in the vector-valued case.

These investigations are connected with the following wonderful identities by G.Ya.Lozanovskiĭ:

$$L^2 = E^{\frac{1}{2}}(E')^{\frac{1}{2}} \tag{1}$$

and

$$L^1 = EE',\qquad(2)$$

which hold true for any Banach Function Space E.

Identity (2) is an easy consequence of identity (1). Both identities claim also the possibility of factorization with the equality of norms. Namely, say, (2) states that for every $x \in L^1$ and any $\varepsilon > 0$ there exist $e \in E, e' \in E'$ such that

$$x = ee' \quad \text{and} \quad \|x\|_{L^1} \geq (1 - \varepsilon)\|e\|_E\|e'\|_{E'}.\qquad(3)$$

In the case of a perfect Banach Function Space E we can get the equality in (3):

$$\|x\|_{L^1} = \|e\|_E\|e'\|_{E'}.\qquad(4)$$

Factorization (2)–(4) is obvious for $L^1 = L^p L^{p'}$ $(1/p + 1/p' = 1)$ but this is practically the only such a case (try it for Orlicz spaces!). Formulas (1)–(4) were announced by G.Ya.Lozanovskiĭ in 1967 [Lo1] and published with the detailed proofs in [Lo2]. It is interesting that the result was new even in finite-dimensional setting (we can treat a finite-dimensional space E as a space with an unconditional basis and E' as E with the biorthogonal system; in this case multiplication means simply coordinate-wise multiplication) and for the Banach spaces with unconditional basis (in the latter case the possibility of factorization was reopened in [JR]). In [Gill] T.A.Gillespie was able to find a beautiful and quite different proof for (1)–(4) having been inspired by some issues from the theory of reflexive algebras of operators in a Hilbert space. In [Lo6] an extension of the equality (2) to the spaces of measures was considered.

It is interesting to investigate some equalities of more general type than (2), i.e. for an arbitrary Banach Function Space F and another Banach Function Space E, which is in a certain sense 'smaller' than F, to establish the validity of the following formula:

$$F = E\,M(E, F),\qquad(5)$$

where $M(E, F)$ stands for the space of multipliers from E to F $(M(E, L^1) = E')$. This topic is closely connected with the Nikishin–Maurey factorization theory for operators in spaces of measurable functions. For example, the case of Orlicz spaces was under consideration in the article [M] by B.Maurey. He has shown that if φ and ψ are the Young functions, related to each other in a suitable way, then a natural formula defines the third Young function θ such that

$$L_\varphi = L_\psi L_\theta$$

provided L_θ is reflexive (Theorem 109). Using the minimax theorem 4.1 (or directly Theorem 1.3) we can drop out the artificially looking reflexivity condition. See [Pi2, Re] for some related material on factorization.

ACKNOWLEDGEMENT. The main part of this work was written in the autumn of 1990 when the author was a visitor in the University of Cambridge on an invitation of the Royal Society. The author wishes to express his gratitude to Dr. D.J.H. Garling for making this visit both enjoyable and rewarding, and for many helpful comments concerning this manuscript. Many copies of that preliminary version have been sent to various universities.

This work was upgraded in the autumn of 1991 when the author was a visiting professor in the University of Franche–Comté (Besançon). The popularity of this theory in France and contacts with colleagues were a major impetus for that. The author would like to thank Prof. W. Arendt for his friendly influence.

At last, some additions and improvements have been done on the occasion of this anniversary publication.

Finally, the author would like to thank Prof. A.C.Zaanen for his encouraging recognition and support which have done the author's results available to the Western auditory in the difficult times of the late 70s.

REFERENCES

[AW]   K.T. Andrews, J.D. Ward, *Proximinality in operator algebras on L_1*, J. Operator Theory **17** (1987), No.2, 213–221.

[Ba]   A. Baernstein, *On reflexivity and summability*, Studia Math. **42** (1972), 91–94.

[Bd]   E.J. Balder, *New sequential compactness results for spaces of scalarly integrable functions*, J. Math. Anal. Appl. **151** (1990), No.1, 1–16.

[BBY]   A.A. Basile, A.V. Bukhvalov, M.Ya. Yakubson, *The generalized Yosida–Hewitt theorem*, Math. Proc. Cambr. Phil. Soc. (submitted).

[Be]   M. Besbes, *Points fixes des contractions définies sur un convexe L^0-fermé de L^1*, C. R. Acad. Sci. Paris, Sér I, **311** (1990), No.5, 243–246.

[B1]   A.V.Bukhvalov, *Geometrical properties of Banach spaces of measurable vector-valued functions*, Dokl. Akad. Nauk SSSR **239** (1978), No.6, 1279–1282; English transl.: Soviet Math. Dokl. **19** (1978), No.2, 501–505.

[B2]   _____, *Kernel operators and spaces of measurable vector-valued functions*, Dissertation for the degree of doctor of physical and mathematical sciences, LOMI, Leningrad, 1984. (Russian)

[B3]   _____, *Interpolation of linear operators in spaces of vector-valued functions with mixed norm*, Sibirsk. Mat. Zh. **28** (1987), No.1, 37–51; English transl.: Siberian Math. J. **28** (1987), 24-36.

[BL1]   A.V. Bukhvalov, G.Ya. Lozanovskiĭ, *On sets closed with respect to convergence in measure in spaces of measurable functions*, Dokl. Akad. Nauk SSSR **212** (1973), 1273–1275; English transl.: Soviet Math. Dokl. **14** (1973), 1563–1565.

[BL2]   _____, *On sets closed in measure in spaces of measurable functions*, Trudy
            Moskov. Mat. Ob-va **34** (1977), 129–150; English transl.: Trans. Moscow Math.
            Soc. (1978), issue 2, 127–148.

[BL3]   _____, *Representation of linear functionals and operators on vector lattices and
            certain applications of these representations*, in: Theory of Operators in Func-
            tion Spaces, (Proc. School, Novosibirsk, 1975), "Nauka", Novosibirsk (1977),
            71–98. (Russian)

[C1]   V. Caselles, *Dunford-Pettis operators and the Radon-Nikodym property*, Arch.
            Math. **50** (1988), No.2, 183–188.

[C2]   _____, *A characterization of dual Banach lattices*, Proc. Nederl. Akad. Weten-
            sch. **A92** (1989), No.1, 35–47.

[Ca]   C. Castaing, *Sur la décomposition de Slaby. Applications aux problèmes de con-
            vergence en probabilités, economie mathématique, théorie du contrôle. Minimi-
            sation*, Séminaire d'Analyse Convexe, Montpellier **19** (1989), exp.3.

[DU]   J. Diestel, J.J. Uhl Jr., *Progress in vector measures: 1977-83*, Lect. Notes Math.
            1033 (1983), 144–192.

[Fe]   R. Fennich, *Application de la méthode de relaxation-projection aux problèmes
            min-max*, Séminaire d'Analyse Convexe, Montpellier (1980), exp.8.

[FG]   A. Fougères, E. Giner, *Applications de la décomposition du dual d'un espace
            d'Orlicz engendré L_φ: polarité et minimisation "sous compacité", φ-équicontinuité
            et orthogonalité*, C. R. Acad. Sci. Paris **A284** (1977), A299–A302.

[FP]   A. Fougères, J.-Cl. Peralba, *Application au calcul des variations de l'optimisation
            intégrale convexe*, Lecture Notes Math. **1978**, 70–99.

[Ga]   D.J.H. Garling, *Subsequence principles for vector-valued random variables*, Math.
            Proc. Cambr. Phil. Soc. **86** (1979), No.2, 301–311.

[Gill]   T.A. Gillespie, *Factorization in Banach function spaces*, Proc. Nederl. Akad.
            Wetensch. **A84** (1981), No.3, 287–300.

[Gi1]   E. Giner, *Topologies de dualité sur les espaces intégraux de type Orlicz. Applica-
            tions à l'optimisation. I, II*, Séminaire d'Analyse Convexe, Montpellier (1977),
            exp.17,18.

[Gi2]   _____, *Topologies de dualité sur les espaces intégraux de type Orlicz. Applica-
            tions à l'optimisation*, C. R. Acad. Sci. Paris **287** (1978), A425–A428.

[Go1]   G. Godefroy, *Sous-espaces bien disposés de L^1 - applications*, Trans. Amer.
            Math. Soc. **286** (1984), 227–249.

[Go2]   _____, *Nicely placed subspaces of Banach lattices*, Semesterbericht Funktion-
            alanalysis, Sommersemester (1984), 205–218.

[Go3]   _____, *On Riesz subsets of Abelian discrete groups*, Israel J. Math. **61** (1988),
            No.3, 301–331.

[GoLi]   _____, D. Li, *Some natural families of M-ideals*, Math. Scand. **66** (1990),
            249–263.

[GoLP]   _____, F. Lust-Piquard, *Some applications of geometry of Banach spaces to
            harmonic analysis*, Colloq. Math. **60/61** (1990), 443–456.

[He]   W. Hensgen, *Contributions to the Geometry of Vector-Valued H^∞ and L^1/H_0^1 Spaces*, Habilitation Thesis, Regensburg, 1992.

[HLS]   S.M. Holland, W.A. Light, L.J. Sulley, *On proximinality in $L_1(T \times S)$*, Proc. Amer. Math. Soc. **86** (1982), No.2, 279 – 282.

[JR]   R.E. Jamison, W.H. Ruckle, *Factoring absolutely convergent series*, Math. Ann. **224** (1976), 143–148.

[J]   L.P. Janovskiĭ, *Some theorems of nonlinear analysis in non-reflexive Banach spaces*, in: Proc. XV-th School on the Theory of Operators in Function Spaces, Part II, Ul'yanovsk (1990), 136. (Russian)

[KP]   M.I. Kadec, A. Pełczyński, *Bases, lacunary sequences and complemented subspaces in the spaces L^p*, Studia Math. **21** (1961/62), No.1, 161–176.

[KA]   L.V. Kantorovich, G.P. Akilov, *Functional Analysis*, 2nd rev. ed, "Nauka", Moscow, 1977 (Russian); English transl.: Pergamon Press, Oxford, 1982.

[Kh1]   R. Khalil, *Best approximation in $L^p(X)$*, Math. Proc. Cambr. Phil. Soc. **94** (1983), No.2, 277–279.

[Kh2]   _____, *Best approximation in L^1*, Numer. Funct. Anal. and Optim. **9** (1987), No.9-10, 1031–1037.

[Ki]   S.V. Kislyakov, *More on free interpolation by functions which are regular outside a prescribed set*, Zap. Nauchn. Sem. LOMI **107** (1982), 71–88; English transl.: J. Soviet Math. **36** (1987), No.3, 342–352.

[Kob]   V.N. Kobzev, *On the strong law of large numbers and $S_X(p,r), \tilde{S}_X(p,r)$-systems*, Soobtch. AN Gruz. SSR (Bull. Acad. Sci. Georgian SSR) **86** (1977), No.1, 53–55. (Russian, English summary)

[Kom]   J. Komlós, *A generalization of a problem of Steinhaus*, Acta Math. Acad. Sci. Hungar. **18** (1967), No.1-2, 217–229.

[Len]   C. Lennard, *On Komlós and convergence in measure compact, convex subsets of L_1* (1990), Preprint.

[Le1]   V.L. Levin, *Extremal problems with convex functionals that are lower semicontinuous with respect to convergence in measure*, Dokl. Akad. Nauk SSSR **224** (1975), No.6, 1256–1259; English transl.: Soviet Math. Dokl. **16** (1976), No.5, 1384–1388.

[Le2]   _____, *Convex Analysis in the Spaces of Measurable Functions, and its Applications in Mathematics and Economics*, "Nauka", Moscow, 1985. (Russian)

[Li1]   D. Li, *Espaces L-facteurs de leurs biduaux: bonne disposition, meilleure approximation et propriete de Radon–Nikodym*, Quart. J. Math. Oxford (2) **38** (1987), 229–243.

[Li2]   _____, *Lifting properties for some quotients of L^1-spaces and other spaces L-summand in their bidual*, Math. Z. **199** (1988), 321–329.

[Lo1]   G.Ya. Lozanovskiĭ, *On Banach lattices of Calderón*, Dokl. Akad. Nauk SSSR **172** (1967), No.5, 1018–1020; English transl.: Soviet Math. Dokl. **8** (1967).

[Lo2]   _____, *On some Banach lattices*. I, Sibirsk. Mat. Zh. **10** (1969), No.3, 584–599; English transl.: Siberian Math. J. **10** (1969), 419–431.

[Lo3]   _____, *Concerning one class of linear operators and its applications to the theory of spaces of measurable functions*, Sibirsk. Mat. Zh. **16** (1975), No.4, 755–760; English transl.: Siberian Math. J. **16** (1975), 577–581.

[Lo4]   _____ , *Mappings of Banach lattices of measurable functions*, Izv. Vyssh. Uchebn. Zaved. Matematika (1978), No.5, 84–86; English transl.: Soviet Math. (Iz. VUZ) **22** (1978), No.5, 61–63.

[Lo5]   _____ , *Transformations of ideal Banach spaces by means of concave functions*, Qualitative and Approximate Methods of the Investigation of Operator Equations, Jaroslavl' **3** (1978), 122–147. (Russian)

[Lo6]   _____ , *On the conjugate space of a Banach lattice*, Theory of Functions, Funct. Anal., and Their Applications (Khar'kov) (1978), No.30, 85–90. (Russian)

[Lu]   D.Q. Luu, *A short proof of biting lemma*, Séminaire d'Analyse Convexe, Montpellier **19** (1989), exp.1.

[M]   B. Maurey, *Théorèmes de factorisation pour les opérateurs linéaires à valeurs dans les espaces L_p*, Astérisque (1974), No.11.

[Ni]   P. Nilsson, *Interpolation of Banach lattices*, Studia Math. **82** (1985), 135–154.

[Pi1]   G. Pisier, *Counterexamples to a conjecture of Grothendieck*, Acta Math. **151** (1983), 181–208.

[Pi2]   G. Pisier, *Some applications of the complex interpolation method to Banach lattices*, J. D'Analyse Math. **35** (1979), 264–280.

[Re]   S. Reisner, *A factorization theorem in Banach lattices and its application to Lorentz spaces*, Ann. Inst. Fourier, Grenoble **31** (1981), 239–255.

[RF]   J. Rubio de Francia, *Weighted norm inequalities and vector valued inequalities*, Lect. Notes Math. **908** (1982), 86–101.

[V]   M. Valadier, *Convergence en mesure et optimisation*, Travaux du séminaire d'analyse convexe, Univ. Languedoc **6** (1976), exp.14.

[YH]   K. Yosida, E. Hewitt, *Finitely additive measures*, Trans. Amer. Math. Soc. **72** (1952), 46–66.

[Z1]   A.C. Zaanen, *Integration*, North-Holland, Amsterdam, 1967.

[Z2]   A.C. Zaanen, *Riesz Spaces. II*, North-Holland, Amsterdam, 1983.

1991 *Mathematics Subject Classification.* Primary 49A27; Secondary 46E30, 46E15, 46E40, 46B20.

Department of Mathematics
St.Petersburg University of Economics and Finance
Sadovaya street 21
191023 St.Petersburg, Russia

Operator Theory:
Advances and Applications, Vol. 75
© 1995 Birkhäuser Verlag Basel/Switzerland

ON A SUBMAJORIZATION INEQUALITY OF T. ANDO

Peter G. Dodds and Theresa K. Dodds *

Dedicated to Professor A.C. Zaanen on the occasion of his 80-th birthday

A submajorization inequality of T.Ando for operator monotone functions is extended to the setting of measurable operators affiliated with a semi-finite von Neumann algebra. The general form yields certain norm inequalities for the absolute value in symmetric operator spaces which were previously known in the setting of trace ideals.

0. INTRODUCTION

Over a period of many years, a significant contribution to the study of Banach function spaces has been made by A.C.Zaanen and his students. In its various aspects,this study has ranged from problems in concrete function spaces such as Orlicz spaces and normed Köthe spaces [Za1,2] to the more abstract viewpoint given by the theory of vector lattices [Za3] and the associated operator theory. Somewhat more recently, the present authors, together with Ben de Pagter, have sought to widen the perspective by extending the general theory of rearrangement-invariant Banach function spaces to the more general setting of Banach spaces of measurable operators affiliated with a semi-finite von Neumann algebra. Such an aproach is based on the theory of non-commutative integration created by Segal [Se] and unifies the classical theory of (commutative) symmetric function spaces (see[KPS]) with the theory of trace ideals of compact operators in Hilbert space developed by Schatten (see [GK1]).a central role in this study is played by techniques based on classical rearrangement inequalitiesrelated to the notion of submajorization, as shown by Ovcinnikov [Ov1,2,3]. It is the purpose of this note to place in general setting a submajorization inequality for operator monotone functions due to T.Ando [An] for the

---

* Research partially supported by A.R.C.

case of finite matrices and to indicate some applications for a wise class of symmetric operator norms. We proceed to recall the necessary terminology.

Let M be a semifinite von Neumann algebra with a normal faithful semifinite trace τ, acting in the Hilbert space H. Throughout, we denote by 1 the identity of M. If x is a (densely defined) self-adjoint operator in H and if $x = \int_{(-\infty,\infty)} s \, de_s^x$ is its spectral decomposition then, for any Borel subset $B \subseteq \mathbb{R}$, we denote by $X_B(x)$ the corresponding spectral projection $\int_{(-\infty,\infty)} X_B(s) de_s^x$. A closed densely defined linear operator x affiliated with M is called τ-measurable if and only if, there exists a number $s \geq 0$ such that $\tau(X_{(s,\infty)}(|x|)) < \infty$. We denote by $\widetilde{M}$ the set of all τ-measurable operators. Sum and product in $\widetilde{M}$ are defined as the respective closures of the algebraic sum and product. For $x \in \widetilde{M}$, the *generalized singular value function* (or *decreasing rearrangement*) $\mu_{.}(x)$ of x is defined by $\mu_t(x) = \inf\{s \geq 0 : \tau(X_{(s,\infty)}(|x|)) \leq t\}$, $t \geq 0$. It follows that $\mu_{.}(x)$ is a decreasing, right-continuous function on the half line $[0,\infty)$. For basic properties of decreasing rearrangements, we refer to [FK]. The topology defined by the translation invariant metric d on $\widetilde{M}$ obtained by setting

$$d(x,y) = \inf\{t \geq 0 : \mu_t(x-y) \leq t\}, \qquad \text{for } x,y \in \widetilde{M},$$

is called the *measure topology*. It is shown in [Ne] and [Te] that $\widetilde{M}$ equipped with the measure topology is a complete, Hausdorff, topological *-algebra in which M is dense.

We remark that if M is the algebra $\mathcal{L}(H)$ of all bounded linear operators on H ond if τ is the standard trace, then $\widetilde{M} = \mathcal{L}(H)$ and $z \in \mathcal{L}(H)$is compact if and only if $\mu_t(z) \to 0$ as $t \to \infty$. In this case, for each $n = 0,1,2,\ldots$,

$$\mu(z) = \mu_t(z), \quad t \in [n, n+1),$$

and the sequence $\{\mu_n(z)\}_{n=0}^{\infty}$ is just the usual singular value sequence of z counted according to multiplicity [GK].

1. ANDO'S INEQUALITY

The continuous real function F on the half-line $\mathbb{R}^+$ is said to be *operator monotone* if and only if, for all $n = 1,2,\ldots$, whenever x,y are $n \times n$ matrices which satisfy

$0 \leq x \leq y$, it follows that $F(x) \leq F(y)$. It is well known [RR] that F is operator monotone if and only if F admits a (unique) integral representation

$$F(\lambda) = \alpha + \beta\lambda + \int_{(0,\infty)} \frac{\lambda s}{\lambda + s} \nu(ds), \quad \lambda \geq 0, \tag{I}$$

where $\alpha \in \mathbb{R}$ and $\beta \geq 0$ are constants and $\nu(\cdot)$ is a positive Borel measure on $(0,\infty)$ such that $\int_{(0,\infty)} s(1+s)^{-1}\nu(ds) < \infty$.

We remark that if $0 < p < 1$, then it follows from the well-known formulae

$$\lambda^p = \frac{\sin(p\pi)}{\pi} \int_{(0,\infty)} \frac{\lambda s}{\lambda + s} s^{p-2} ds, \quad \lambda \geq 0,$$

and

$$\ln(1+\lambda) = \int_{[1,\infty)} \frac{\lambda s}{\lambda + s} s^{-2} ds, \quad \lambda \geq 0.$$

that each of the functions $\lambda \to \lambda^p$ for $0 < p \leq 1$, and $\lambda \to \ln(1+\lambda)$ are operator monotone.

For ease of notation, if $x \in \widetilde{\mathcal{M}}$ and if $\alpha > 0$, let us set

$$\Phi_\alpha(x) = \int_0^\alpha \mu_t(x)dt.$$

It is shown in [FK] Lemma 4.1 that if $\mathcal{M}$ has no minimal projections and if $x \in \widetilde{\mathcal{M}}$, then

$$\Phi_\alpha(x) = \sup\{\tau(e|x|e) : e \text{ is a projection in } \mathcal{M} \text{ with } \tau(e) \leq \alpha\}.$$

In the more general case that $\mathcal{M}$ has minimal projections, we consider the tensor product $\mathcal{N} = \mathcal{M} \otimes L^\infty[0,1]$ equipped with the trace η given by the tensor product of τ with the trace

$$f \to \int_0^1 f(s)ds, \quad f \in L^\infty[0,1].$$

It is clear that $\mathcal{N}$ has no atoms and it is not difficult to see that the embedding $x \to x \otimes 1$, $x \in \mathcal{M}$ extends in the obvious way to a rearrangement-preserving map of $\widetilde{\mathcal{M}}$ onto the $\star$-subalgebra $\widetilde{\mathcal{M}} \otimes 1 \subseteq \widetilde{\mathcal{N}}$. For details, see for example [St]. We remark that if $x \in \widetilde{\mathcal{M}}$ is self -adjoint, and if f is any Borel function on $\mathbb{R}$, then $f(x \otimes 1) = f(x) \otimes 1$. This follows readily, for example from [St], Theorem 8.2.

The submajorisation inequality which follows is due to Ando [An] in the matrix setting. The present formulation is due to H. Kosaki in connection with a general version of the Powers-Stormer inequality and is given in the appendix to [HN].

THEOREM 1.1 *If the function F is non-negative and operator monotone and if $0 \leq x, y \in \widetilde{\mathcal{M}}$, then for all $\alpha > 0$,*

$$\Phi_\alpha \left(F(x) - F(y) \right) \leq \Phi_\alpha \left(F(|x - y|) \right).$$

We remark that Kosaki's proof of Theorem 1.1 preceding is based essentially on Ando's method, and uses a result of Tychonov [Ty] concerning continuity for the measure topology of a wide class of Borel functions. It is the purpose of the next paragraph to indicate that Ando's original proof in the matrix case carries over directly to the more general setting without recourse to the theorem of Tychonov. As noted by Kosaki, some additional remarks are necessary.

We recall [DDP] that if $0 \leq x, y \in \widetilde{\mathcal{M}}$ then $0 \leq x \leq y$ holds if and only if

$$\mathcal{D}(y^{\frac{1}{2}}) \subseteq \mathcal{D}(x^{\frac{1}{2}}) \quad \text{and} \quad \|x^{\frac{1}{2}}\xi\| \leq \|y^{\frac{1}{2}}\xi\|, \quad \xi \in \mathcal{D}(y^{\frac{1}{2}}).$$

Moreover, the ordered vector space $(\widetilde{\mathcal{M}}, \leq)$ is order complete and if $0 \leq x_\sigma \uparrow_\sigma \leq x$ holds in $\widetilde{\mathcal{M}}$, then $x = \sup x_\sigma$ if and only if

$$\mathcal{D}(x^{\frac{1}{2}}) = \{\xi : \sup_\sigma \|x_\sigma^{\frac{1}{2}}\xi\| < \infty\} \quad \text{and} \quad \|x^{\frac{1}{2}}\xi\| = \sup_\sigma \|x_\sigma^{\frac{1}{2}}\xi\|, \quad \xi \in \mathcal{D}(x^{\frac{1}{2}}).$$

If $0 \leq x, y \in \widetilde{\mathcal{M}}$, and if $0 \leq x \leq y$, observe that

$$(x+1)^{-1} - (y+1)^{-1} = (x+1)^{-\frac{1}{2}} \left(1 - \left[1 + (x+1)^{-\frac{1}{2}}(y-x)(x+1)^{-\frac{1}{2}} \right]^{-1} \right) (x+1)^{-\frac{1}{2}}.$$

$$\text{(II)}$$

Now since $0 \leq w \in \widetilde{\mathcal{M}}$ implies $0 \leq (w+1)^{-1} \leq 1$, and since

$$sx(x+s)^{-1} = s - s^2(x+s)^{-1} \tag{III}$$

for all $0 < s \in \mathbb{R}$ and $0 \leq x \in \widetilde{\mathcal{M}}$, it follows from (II) and (III) that the inequality

$$0 \leq sx(x+s)^{-1} \leq sy(y+s)^{-1} \tag{IV}$$

holds for all $0 < s \in \mathbb{R}$ and $0 \leq x, y \in \widetilde{\mathcal{M}}$ with $0 \leq x \leq y$.

Suppose now that $\nu(\cdot)$ is a positive Borel measure on $(0, \infty)$ such that

$$\int_{(0,\infty)} s(1+s)^{-1} \nu(ds) < \infty.$$

We set

$$f(\lambda) = \int_{(0,\infty)} \frac{\lambda s}{\lambda + s} \nu(ds), \quad f_I(\lambda) = \int_I \frac{\lambda s}{\lambda + s} \nu(ds), \quad \lambda \in \mathbb{R}^+.$$

for any Borel set $I \subseteq (0, \infty)$. As usual, if $0 \leq x \in \widetilde{\mathcal{M}}$ then the operators $f(x), f_I(x)$ are defined by setting

$$f(x) = \int_{\mathbb{R}+} f(\lambda) de_\lambda^x, \quad f_I(x) = \int_{\mathbb{R}+} f_I(\lambda) de_\lambda^x.$$

It follows from Fubini's theorem that $\xi \in \mathcal{D}(f_I^{\frac{1}{2}}(x))$ if and only if

$$\int_I \langle sx(x+s)^{-1}\xi, \xi \rangle \nu(ds) < \infty$$

in which case

$$\|f_I^{\frac{1}{2}}(x)\xi\|^2 = \int_{\mathbb{R}+} f_I(\lambda) d\langle e_\lambda^x \xi, \xi \rangle = \int_I \langle sx(x+s)^{-1}\xi, \xi \rangle \nu(ds).$$

It now follows from (IV) above that if $0 \leq x, y \in \widetilde{\mathcal{M}}$ and if $0 \leq x \leq y$, then $0 \leq f_I(x) \leq f_I(y)$. In particular, if F is given by (I) then it follows also that $0 \leq F(x) \leq F(y)$. We summarise these remarks in the following result, due to Heinz [H].

PROPOSITION 1.2   If F is operator monotone, then $0 \leq F(x) \leq F(y)$ for all $x, y \in \widetilde{\mathcal{M}}$ with $0 \leq x \leq y$.

We shall need the following.

LEMMA 1.3   If $0 \leq x \in \widetilde{\mathcal{M}}$, if $[a, b] \subseteq (0, \infty)$ is a compact subinterval, and if $e \in \mathcal{M}^+$ is a projection with $\tau(e) < \infty$, then

$$\tau \left(e \int_{(a,b]} sx(x+s)^{-1} \nu(ds) e \right) = \int_{(a,b]} \tau \left(esx(x+s)^{-1} e \right) \nu(ds).$$

PROOF. Let $[a, b] \subseteq (0, \infty)$ be a compact subinterval. For each finite partition $P : a = a_0 < a_1 < \ldots < a_n = b$, and $0 \leq x \in \widetilde{\mathcal{M}}$, define

$$l(x; P) = \sum_{i=0}^{n-1} a_i x(x + a_i)^{-1} \chi_{(a_i, a_{i+1}]}.$$

It is clear that

$$0 \leq l(x; P)(s) \leq sx(x + s)^{-1} \chi_{(a, b]}(s), \quad s \in (0, \infty)$$

and so also

$$0 \leq \int_{(a, b]} l(x; P)(s) \nu(ds) = \sum_{i=0}^{n-1} a_i x(x + a_i)^{-1} \nu\left((a_i, a_{i+1}]\right)$$

$$\leq \int_{(a, b]} sx(x + s)^{-1} \nu(ds).$$

If $s, t \in [a, b]$, and if $\xi \in \mathcal{H}$ then it follows from the inequality

$$|xs(x + s)^{-1} - xt(x + t)^{-1}| \leq |s - t| x^2 (x + a)^{-2} \leq |s - t| 1,$$

that the mapping $s \mapsto \langle sx(x+s)^{-1}\xi, \xi \rangle$, $s \in [a, b]$, is uniformly continuous. This implies that, given $\xi \in \mathcal{H}$ and $\epsilon > 0$, there exists a partition P such that

$$0 \leq \int_{(a, b]} \langle sx(x + s)^{-1}\xi, \xi \rangle \nu(ds) - \int_{(a, b]} \langle l(x; P)(s)\xi, \xi \rangle \nu(ds) < \epsilon.$$

This implies that

$$\int_{(a, b]} l(x; P)(s) \nu(ds) \uparrow_P \int_{(a, b]} sx(x + s)^{-1} \nu(ds).$$

Note that this holds even in $\mathcal{M}^+$, since

$$\int_{(a, b]} sx(x + s)^{-1} \nu(ds) \leq bx(b + x)^{-1} \nu\left((a, b]\right).$$

Now suppose that $e \in \mathcal{M}^+$ is a projection with $\tau(e) < \infty$. We observe that

$$\tau\left(e \int_{(a, b]} sx(x + s)^{-1} \nu(ds) e\right) = \tau\left(\int_{(a, b]} esx(x + s)^{-1} e\nu(ds)\right)$$

$$= \sup_P \tau\left(\int_{(a, b]} el(x; P)(s) e\nu(ds)\right)$$

$$= \sup_P \int_{(a, b]} \tau\left(el(x; P)(s)e\right) \nu(ds)$$

$$= \int_{(a, b]} \tau\left(esx(x + s)^{-1}e\right) \nu(ds),$$

where the last step follows from the uniform continuity of the map

$$s \mapsto \tau(esx(x+s)^{-1}e), \quad s \in [a,b],$$

and this completes the proof of the Lemma.

LEMMA 1.4   If $0 \le x \le y \in \widetilde{\mathcal{M}}$ and if $\alpha > 0$, then

$$\Phi_\alpha \left(\int_{(0,\infty)} \left(\frac{sy}{y+s} - \frac{sx}{x+s} \right) \nu(ds) \right) \le \int_{(0,\infty)} \Phi_\alpha \left(\frac{sy}{y+s} - \frac{sx}{x+s} \right) \nu(ds).$$

PROOF.   We suppose that $0 \le x \le y \in \widetilde{\mathcal{M}}$, and that $\alpha > 0$. Let e be a projection in $\mathcal{M}^+$ with $\tau(e) \le \alpha$. For simplicity of notation, we set

$$g(x,y;s) = sy(y+s)^{-1} - sx(x+s)^{-1}, \quad s \in (0,\infty).$$

From Lemma 1.3, it now follows that

$$\tau \left(e \int_{(a,b]} g(x,y;s)\nu(ds)e \right) = \int_{(a,b]} \tau\left(eg(x,y;s)e \right) \nu(ds)$$
$$\le \int_{(0,\infty)} \nu(ds) \int_0^{\tau(e)} \mu_t\left(g(x,y;s) \right) dt.$$

If $\xi \in \mathcal{H}$, observe that

$$\int_{(0,\infty)} \chi_{(\frac{1}{n},n]}(s) \langle g(x,y;s)\xi, \xi \rangle \nu(ds) \uparrow_n \int_{(0,\infty)} \langle g(x,y;s)\xi, \xi \rangle \nu(ds),$$

and this implies that

$$\int_{(\frac{1}{n},n]} g(x,y;s)\nu(ds) \uparrow_n \int_{(0,\infty)} g(x,y;s)\nu(ds)$$

holds in $\widetilde{\mathcal{M}}^+$. Consequently

$$\tau \left(e \int_{(\frac{1}{n},n]} g(x,y;s)\nu(ds)e \right) \uparrow_n \tau \left(e \int_{(0,\infty)} g(x,y;s)\nu(ds)e \right)$$

and so,

$$\tau \left(e \int_{(0,\infty)} g(x,y;s)\nu(ds)e \right) \le \int_{(0,\infty)} \nu(ds) \int_0^{\tau(e)} \mu_t\left(g(x,y;s) \right) dt.$$

If $\mathcal{M}$ is non-atomic, this implies that

$$\int_0^{\tau(e)} \mu_t \left(\int_{(0,\infty)} g(x,y;s)\nu(ds) \right) dt \le \int_{(0,\infty)} \nu(ds) \int_0^{\tau(e)} \mu_t \left(g(x,y;s) \right) dt.$$

The assumption that $\mathcal{M}$ has no atoms may now be removed by embedding $\mathcal{M}$ into the tensor product $\mathcal{M} \otimes L^\infty[0,1]$ and using the remarks concerning this embedding preceding the statement of Theorem 1.1. The details are omitted, and this suffices to complete the proof of the Lemma.

Before proceeding, we recall the following basic properties of rearrangements ([FK], Lemma 2.5).

LEMMA 1.5   (i) If ϕ is any continuous increasing function on $[0,\infty)$ with $\phi(0) \ge 0$, then $\phi(\mu(z)) = \mu(\phi(|z|))$, for all $z \in \widetilde{\mathcal{M}}$.

(ii) For all $z \in \widetilde{\mathcal{M}}$, $\mu(z^*z) = \mu(zz^*)$.

Let us make the following remark concerning Lemma 1.4. If $0 \le x \in L^1(\mathcal{M}) + \mathcal{M}$ and if $\alpha > 0$ then the map $s \to sx(s+x)^{-1}$, $s > 0$ is continuous from $(0,\infty)$ to the Banach space $L^1(\mathcal{M}) + \alpha\mathcal{M}$, equipped with the norm $\Phi_\alpha(\cdot)$. Via Lemma 1.5 (i), we observe that

$$\int_{(0,\infty)} \Phi_\alpha \left(sx(s+x)^{-1} \right) \nu(ds) = \int_{(0,\infty)} \nu(ds) \int_{(0,\alpha)} \frac{s\mu_t(x)}{s+\mu_t(x)} dt$$

$$\le \left(\int_0^\alpha \max\{1, \mu_t(x)\} dt \right) \left(\int_{(0,\infty)} \frac{s}{1+s} \nu(ds) \right),$$

it follows that $\int_{(0,\infty)} sx(s+x)^{-1}\nu(ds) \in L^1(\mathcal{M}) + \alpha\mathcal{M}$ exists as a Bochner integral. Consequently, if $0 \le x, y \in L^1(\mathcal{M}) + \mathcal{M}$, then the assertion of Lemma 1.4 follows directly from properties of the Bochner integral.

The calculation which follows is due to Ando in the matrix case; that its validity in the more general setting follows from Lemma 1.5 has already been observed by Kosaki.

LEMMA 1.6   If $0 \le x \le y \in \widetilde{\mathcal{M}}$ and if $s > 0$, then

$$\mu \left(y(y+s)^{-1} - x(x+s)^{-1} \right) \le \mu \left((y-x)(y-x+s)^{-1} \right).$$

PROOF   Replacing x, y by sx, sy respectively, it may be assumed that $s = 1$. Further, since

$$y(y+1)^{-1} - x(x+1)^{-1} = (x+1)^{-1} - (y+1)^{-1},$$

it follows from the identity (II) that

$$\mu\left(y(y+1)^{-1} - x(x+1)^{-1}\right) \leq \mu\left(1 - \left[1 + (x+1)^{-\frac{1}{2}}(y-x)(x+1)^{-\frac{1}{2}}\right]^{-1}\right).$$

Lemma 1.5 now yields that

$$\mu\left(1 - \left[1 + (x+1)^{-\frac{1}{2}}(y-x)(x+1)^{-\frac{1}{2}}\right]^{-1}\right)$$

$$= 1 - \left[1 + \mu\left((x+1)^{-\frac{1}{2}}(y-x)(x+1)^{-\frac{1}{2}}\right)\right]^{-1}$$

$$= 1 - \left[1 + \mu\left((y-x)^{\frac{1}{2}}(x+1)^{-1}(y-x)^{\frac{1}{2}}\right)\right]^{-1}$$

$$= \mu\left(1 - \left[1 + (y-x)^{\frac{1}{2}}(x+1)^{-1}(y-x)^{\frac{1}{2}}\right]^{-1}\right)$$

$$\leq \mu\left(1 - (1+y-x)^{-1}\right)$$

$$= \mu\left((y-x)(y-x+1)^{-1}\right),$$

where the last inequality follows by noting that

$$0 \leq (y-x)^{\frac{1}{2}}(x+1)^{-1}(y-x)^{\frac{1}{2}} \leq y-x.$$

The assertion of Theorem 1.1 now follows for the special case that $0 \leq x \leq y$. The general case now follows by Ando's original argument based on the following submajorization inequalities.

LEMMA 1.7   *If $x, y \in \widetilde{M}$ are self-adjoint and if $x \leq y$, then $\Phi_\alpha(x^+) \leq \Phi_\alpha(y^+)$ for all $\alpha > 0$.*

PROOF.   As in the proof of Lemma 1.4, it may be assumed that M has no atoms. If $\alpha > 0$ is given, let e be any projection in M with $\tau(e) \leq \alpha$. If e^+ is the carrier projection of x^+, then

$$\tau(x^+e) = \tau(ee^+xe^+e) \leq \tau(ee^+ye^+e)$$

$$\leq \tau(ee^+y^+e^+e) \leq \int_0^{\tau(e)} \mu_t(e^+y^+e^+)dt \leq \Phi_\alpha(y^+).$$

The statement of the Lemma now follows from [FK] Lemma 4.1.

LEMMA 1.8　*If $0 \leq x, y \in \mathcal{M}$, if $xy = 0$ and if $\alpha > 0$ then*

$$\Phi_\alpha(x + y) = \sup_{0 \leq \beta \leq \alpha} \left(\Phi_\beta(x) + \Phi_{\alpha - \beta}(y) \right).$$

PROOF.　Let $\alpha > 0$ and let $E_1, E_2 \subseteq \mathbb{R}^+$ be measurable subsets with $E_1 \cap E_2 = \emptyset$ and $|E_1| = |E_2| = \infty$. Let $\phi_i : E_i \mapsto \mathbb{R}^+$, $i = 1, 2$, be measure preserving bijections. It is not difficult to see that

$$\mu(x + y) = \mu \left(\mu(x) \circ \phi_1 + \mu(y) \circ \phi_2 \right)$$

so that

$$\begin{aligned}
\Phi_\alpha(x + y) &= \int_0^\alpha \mu_t \left(\mu(x) \circ \phi_1 + \mu(y) \circ \phi_2 \right) dt \\
&= \sup_{|E| = \alpha} \left(\int_E \mu(x) \circ \phi_1(t) dt + \int_E \mu(y) \circ \phi_2(t) dt \right) \\
&= \sup_{|E| = \alpha} \left(\int_{E \cap E_1} \mu(x) \circ \phi_1(t) dt + \int_{E \cap E_2} \mu(y) \circ \phi_2(t) dt \right) \\
&\leq \sup_{|E| = \alpha} \left(\int_0^{|E \cap E_1|} \mu_t(x) dt + \int_0^{|E \cap E_2|} \mu_t(y) dt \right) \\
&\leq \sup_{0 \leq \beta \leq \alpha} \left(\Phi_\beta(x) + \Phi_{\alpha - \beta}(y) \right).
\end{aligned}$$

The reverse inequality follows by the argument of [Fr] Lemma 16, and by this the proof of the Lemma is complete.

We may now complete the proof of Theorem 1.1 via Ando's original argument. We include the details for the sake of completeness. Suppose then that $0 \leq x, y \in \mathcal{M}$, and that F is given by (I). Observing that $0 \leq x \leq y + (x - y)^+$, and using Proposition 1.2, it follows that

$$F(x) - F(y) \leq F(y + (x - y)^+) - F(y).$$

Consequently, using Lemma 1.7 and the special case of Theorem 1.1 already established, it follows, for every $\alpha > 0$, that

$$\Phi_\alpha \left((F(x) - F(y))^+ \right) \leq \Phi_\alpha \left(F(y + (x - y)^+) - F(y) \right) \leq \Phi_\alpha \left(F((x - y)^+) \right).$$

Interchanging x, y we obtain, for all $\alpha > 0$,

$$\Phi_\alpha\left((F(y) - F(x))^+\right) \le \Phi_\alpha\left(F\left((y-x)^+\right)\right).$$

We now observe that the equality $(x-y)^+(y-x)^+ = 0$ implies that

$$F\left((x-y)^+\right) F\left((y-x)^+\right) = 0.$$

On the other hand, it is trivial that

$$(F(x) - F(y))^+ (F(y) - F(x))^+ = 0.$$

Using Lemma 1.8, it now follows, for all $\alpha > 0$, that

$$\begin{aligned}
\Phi_\alpha\left(F(x) - F(y)\right) &= \sup_{0 \le \beta \le \alpha}\left(\Phi_\beta\left((F(x) - F(y))^+\right) + \Phi_{\alpha - \beta}\left((F(y) - F(x))^+\right)\right) \\
&\le \sup_{0 \le \beta \le \alpha}\left(\Phi_\beta\left(F\left((x-y)^+\right)\right) + \Phi_{\alpha - \beta}\left(F\left((y-x)^+\right)\right)\right) \\
&= \Phi_\alpha\left(F\left((x-y)^+\right) + F\left((y-x)^+\right)\right) \\
&= \Phi_\alpha\left(F(|x-y|)\right),
\end{aligned}$$

and this completes the proof of Theorem 1.1.

THEOREM 1.9   *Let G be a continuous, increasing function on $[0, \infty)$ with $G(0) = 0$ and $G(\infty) = \infty$. If the function F inverse to G is operator monotone, then*

$$\Phi_\alpha\left(G(|x-y|)\right) \le \Phi_\alpha\left(G(x) - G(y)\right)$$

for all $\alpha > 0$ and $0 \le x, y \in \widetilde{\mathcal{M}}$.

PROOF   Via Theorem 1.1, the proof is identical to that of [An], Theorem 2. We include the details for sake of completeness. If $\alpha > 0$, then it follows from Theorem 1.1 that

$$\Phi_\alpha(x - y) = \Phi_\alpha\left(F(G(x)) - F(G(y))\right) \le \Phi_\alpha\left(F(|G(x) - G(y)|)\right).$$

Since F is operator monotone, it follows that G is convex, and consequently (see, for example, the proof of [Ch] Theorem 2.1), it follows that

$$\Phi_\alpha\left(G(x-y)\right) \le \Phi_\alpha\left(G \circ F\left(|G(x) - G(y)|\right)\right) = \Phi_\alpha\left(|G(x) - G(y)|\right),$$

for every $\alpha > 0$, and this completes the proof of the theorem.

When applied to each of the functions

$$\lambda \to e^\lambda - 1; \quad \lambda \to \lambda^p, \ p \geq 1; \quad \lambda \to \lambda^p . \log(\lambda + 1), \ p \geq 1,$$

the preceding theorem yields extensions of the norm inequalities given in [An], Corollary 4. We omit the details.

THEOREM 1.10   *Let F be non-negative and operator monotone. If $0 \leq x, y \in L^1(\mathcal{M}) + \mathcal{M}$, if $0 \leq a, b$ are numbers such that $0 \leq a1 \leq x, 0 \leq b1 \leq y$ and if $z \in \mathcal{M}$, then*

$$\Phi_\alpha\left(F(x)z - zF(y)\right) \leq C(a, b)\Phi_\alpha(xz - zx)$$

for all $\alpha > 0$ where

$$C(a, b) = \begin{cases} \frac{F(a) - F(b)}{a - b}, & \text{if } a \neq b; \\ F'(a) & \text{if } a = b. \end{cases}$$

The inequality of the preceding theorem is due to Kittaneh and Kosaki [KK] for the case of trace ideals and goes back to an earlier inequality of Ando and van Hemmen [AvH]. Via remarks similar to those immediately following Lemma 1.4, the proof is a direct consequence of the representation (I) for operator monotone functions. The details are identical to those given in in the proof of [KK], Theorem 3.1, and are accordingly omitted.

2. CONCERNING THE ABSOLUTE VALUE

Let E be a complex quasi-Banach lattice. If $0 < \gamma \leq \infty$ then E is said to be γ-*convex* if there exists a constant $C > 0$ such that for all finite sequences $\{x_n\}$ in E,

$$\left\|\left(\sum |x_n|^\gamma\right)^{\frac{1}{\gamma}}\right\|_E \leq C\left(\sum \|x_n\|_E^\gamma\right)^{\frac{1}{\gamma}}, \quad (\gamma < \infty);$$

$$\left\|\sup_{1 \leq i \leq n} |x_i|\right\|_E \leq C \sup_{1 \leq i \leq n} \|x_i\|_E, \quad (\gamma = \infty).$$

The least such constant C is called the γ-*convexity constant* of E and is denoted by $M^{(\gamma)}(E)$. For $0 < p < \infty$, $E^{(p)}$ will denote the quasi-Banach lattice defined by

$$E^{(p)} = \{x : |x|^p \in E\}$$

equipped with the quasi-norm

$$\|x\|_{E^{(p)}} = \||x|^p\|_E^{1/p}, \quad x \in E^{(p)}.$$

If E is γ−convex then $E^{(p)}$ is γp−convex with $M^{(\gamma p)}\left(E^{(p)}\right) \leq M^{(\gamma)}\left(E\right)^{\frac{1}{p}}$. Consequently, if E is γ−convex, with $0 \leq \gamma < \infty$ then $E^{(1/\gamma)}$ is 1−convex and so can be renormed as a Banach lattice (cf. [LT 2]).

Following [Xu1], by a symmetric quasi-Banach function space on the positive half-line $\mathbb{R}^+$ is meant a quasi-Banach lattice E of measurable functions with the following properties: (i) E is an order-ideal in the linear space $L^0(\mathbb{R}^+)$ of all finite almost-everywhere measurable functions on $\mathbb{R}^+$; (ii) E is rearrangement-invariant in the sense that whenever $x \in E$ and y is a measurable function with $\mu(y) = \mu(x)$, it follows that $y \in E$ and $\|y\|_E = \|x\|_E$; (iii) E contains all finitely-supported simple functions. The norm on E is said to be a $\sigma-Fatou$ norm if, for every sequence $\{x_n\}$ in E,

$$0 \leq x_n \uparrow x, \quad x \in E \text{ implies } \|x_n\|_E \uparrow \|x\|_E.$$

Now suppose that the symmetric quasi-Banach function space E is γ-convex for some $0 < \gamma < \infty$ and that $M^{(\gamma)}(E) = 1$. Let F be the Banach lattice $E^{(1/\gamma)}$. If the norm on E is a $\sigma-Fatou$ norm, then so is the norm on F and consequently the natural embedding of F into the second associate space F'' is an isometry. Consequently [KPS], there exists a family W of non-increasing functions on $\mathbb{R}^+$ such that

$$\|x\|_F = \sup\{\int_{[0,\infty)} \mu_t(x)w(t)dt \ : \ w \in W\} \tag{2.1}$$

for all $x \in F$, and so

$$\|x\|_E^\gamma = \sup\{\int_{[0,\infty)} \mu_t^\gamma(x)w(t)dt \ : \ w \in W\} \tag{2.2}$$

for all $x \in E$.

If E denotes a symmetric quasi-Banach function space on the positive half-line $\mathbb{R}^+$, we define the symmetric space $E(\mathcal{M})$ of measurable operators associated with E by setting

$$E(\mathcal{M}) = \{x \in \widetilde{\mathcal{M}} \ : \mu(x) \in E\}$$

and

$$\|x\|_{E(\mathcal{M})} = \|\mu(x)\|_E, \quad x \in E(\mathcal{M}).$$

It follows immediately that $\|x\|_{E(\mathcal{M})} = \||x|\|_{E(\mathcal{M})} = \|x^*\|_{E(\mathcal{M})}$ for all $x \in E(\mathcal{M})$. If E has σ–Fatou norm and is γ-convex for some $0 < \gamma < \infty$ with $M^{(\gamma)} = 1$, then it follows from [FK] Theorem 4.7 (i) that the functional $\|\cdot\|_{E(\mathcal{M})}$ is a norm if $\gamma \geq 1$ and a γ–norm if $0 < \gamma < 1$. This latter assertion means that if $x, y \in E(\mathcal{M})$, and if $0 \leq \gamma < 1$, then

$$\|x + y\|_{E(\mathcal{M})}^{\gamma} \leq \|x\|_{E(\mathcal{M})}^{\gamma} + \|y\|_{E(\mathcal{M})}^{\gamma}.$$

Equipped with this norm or γ–norm, the space $E(\mathcal{M})$ is complete. See, for example, [Xu], [Ov1,2], [DDP1,2].

If E is a symmetric Banach function space with σ–Fatou norm and if $x, y \in E$ satisfy $\Phi_\alpha(x) \leq \Phi_\alpha(y)$ for all $\alpha > 0$, then $\|x\|_E \leq \|y\|_E$; if E has order-continuous norm or is *maximal* [KPS] in the sense that the natural embedding of E into its second associate space is an isometric surjection, then E has the following property: if $y \in E$ and if $\Phi_\alpha(x) \leq \Phi_\alpha(y)$ for all $\alpha > 0$, then $x \in E$ and $\|x\|_E \leq \|y\|_E$. Any space E with this property is called *fully symmetric* in which case the norm $\|\cdot\|_E$ is called a *fully symmetric norm*. Suppose now that E is γ–convex with $M^{(\gamma)} = 1$ for some $\gamma \in (0, \infty)$, and that $0 < q < \infty$. If $x, y \in \widetilde{\mathcal{M}}$, then it follows from [FK] Theorem 4.2(iii), and the representation (2.2) that for $0 < q_0, q_1, q < \infty$ with $1/q = 1/q_0 + 1/q_1$,

$$\|xy\|_{E^{(q)}(\mathcal{M})} \leq \|x\|_{E^{(q_0)}(\mathcal{M})} \|y\|_{E^{(q_1)}(\mathcal{M})}. \tag{2.3}$$

It should be noted that the special case of (2.3) preceding obtained by taking $q_0 = q_1 = 1$ contains [Bh] Proposition 5. We now give an application of Ando's inequality to certain norm estimates for the absolute value. The basic ideas here go back to Kosaki [Ko1].

THEOREM 2.1 *Suppose that E is fully symmetric and γ-convex for some $\gamma \geq 1$, with γ-convexity constant 1. If $x, y \in E(\mathcal{M})$, and if $\beta = \min\{\gamma, 2\}$ then*

(i) $\quad \| |x| - |y| \|_{E(\mathcal{M})} \leq 2^{\frac{1}{\beta} - \frac{1}{2}} \|x + y\|_{E(\mathcal{M})}^{\frac{1}{2}} \|x - y\|_{E(\mathcal{M})}^{\frac{1}{2}}$

(ii) $\quad \| |x| - |y| \|_{E(\mathcal{M})} \leq \left(\|x\|_{E(\mathcal{M})}^{\frac{\beta}{2}} + \|y\|_{E(\mathcal{M})}^{\frac{\beta}{2}} \right)^{\frac{1}{\beta}} \|x - y\|_{E(\mathcal{M})}^{\frac{1}{2}}$

PROOF. (i) Observe that $E^{\left(\frac{1}{2}\right)}$ is $\gamma/2$-convex with $\gamma/2$-convexity constant 1 and so the norm on $E^{\left(\frac{1}{2}\right)}(\mathcal{M})$ is an $\gamma/2$-norm if $1 \leq \gamma \leq 2$ and a norm if $\gamma \geq 2$. Setting $\beta = \min\{\gamma, 2\}$, and using Ando's inequality, we obtain that

$$\| \, |x| - |y| \, \|_{E(\mathcal{M})}^{\beta} = \|\sqrt{x^*x} - \sqrt{y^*y}\|_{E(\mathcal{M})}^{\beta}$$

$$\leq \| \, |x^*x - y^*y|^{\frac{1}{2}} \, \|_{E(\mathcal{M})}^{\beta}$$

$$= \|x^*x - y^*y\|_{E^{\left(\frac{1}{2}\right)}(\mathcal{M})}^{\frac{\beta}{2}}$$

$$= \|\frac{1}{2}[(x+y)^*(x-y) + (x-y)^*(x+y)]\|_{E^{\left(\frac{1}{2}\right)}(\mathcal{M})}^{\frac{\beta}{2}}$$

$$\leq \frac{1}{2^{\beta/2}} \left(\|(x+y)^*(x-y)\|_{E^{\left(\frac{1}{2}\right)}(\mathcal{M})}^{\frac{\beta}{2}} + \|(x-y)^*(x+y)\|_{E^{\left(\frac{1}{2}\right)}(\mathcal{M})}^{\frac{\beta}{2}} \right)$$

$$\leq 2^{1-\frac{\beta}{2}} \|x+y\|_{E(\mathcal{M})}^{\frac{\beta}{2}} \|x-y\|_{E(\mathcal{M})}^{\frac{\beta}{2}},$$

where the last inequality follows by applying (2.3) in the special case that $q_0 = q_1 = 1$. Consequently

$$\| \, |x| - |y| \, \|_{E(\mathcal{M})} \leq 2^{\frac{1}{\beta} - \frac{1}{2}} \|x+y\|_{E(\mathcal{M})}^{\frac{1}{2}} \|x-y\|_{E(\mathcal{M})}^{\frac{1}{2}},$$

and by this the proof of (i) is complete.

The proof of (ii) is identical using the identity

$$x^*x - y^*y = x^*(x-y) + (x-y)^*y.$$

Since any norm is $1-$convex, the theorem applies to any fully symmetric norm and contains [Bh] Theorem 2 in the trace ideal setting. When specialised to the Schatten p-classes $\mathcal{C}_p$, the Theorem reduces to [Bh] Theorem 1 in the case that $1 \leq p \leq 2$, and to an estimate of Kittaneh and Kosaki [KK] in the case that $p \geq 2$. Finally, let us observe that if E is γ-convex for some $\gamma \geq 2$, then the equality

$$\|x\|_{E(\mathcal{M})} = \| \, |x|^2 \, \|_{E^{\left(\frac{1}{2}\right)}(\mathcal{M})}^{\frac{1}{2}}, \qquad x \in E(\mathcal{M}), \tag{2.4}$$

shows that $\| \cdot \|_{E(\mathcal{M})}$ is a $Q-$norm in the sense of [Bh].

We now assume that $\mathcal{M}$ is a general von Neumann algebra (not necessarily semi-finite). Let $\mathcal{U}$ be the crossed product of $\mathcal{M}$ by the modular automorphism group

$\{\sigma_t\}_{t\in\mathbf{R}}$ of a fixed weight on $\mathcal{M}$. It is proved in [Ta2] that $\mathcal{U}$ admits the dual action $\{\theta_s\}_{s\in\mathbf{R}}$ and the normal faithful semi-finite trace τ satisfying $\tau \circ \theta_s = e^{-s}\tau$, $s \in \mathbf{R}$. For $0 < p \le \infty$, the Haagerup L^p-space, again denoted $L^p(\mathcal{M})$, associated with $\mathcal{M}$ consists of those τ-measurable operators x affiliated with $\mathcal{U}$ which satisfy $\theta_s(x) = e^{-s/p}x$, $s \in \mathbf{R}$. For full details, see [Te]. If $p \ge 1$, the spaces $L^p(\mathcal{M})$ are Banach spaces, while the quasi-norm on $L^p(\mathcal{M})$ is a p-norm if $0 < p \le 1$. See, for example [FK], Theorem 4.9. Further, it is shown in [FK], Lemma 4.8, that if $x \in L^p(\mathcal{M})$, $0 < p < \infty$, then

$$\mu_t(x) = t^{-1/p}\|x\|_p, \quad t > 0, \tag{2.5}$$

where the decreasing rearrangement is taken relative to the canonical trace on $\mathcal{U}$. It now follows from (2.4) and Ando's inequality, that if $p > 1$ and if $x, y \in L^p(\mathcal{M})$ then

$$\left\| \sqrt{x^*x} - \sqrt{y^*y} \right\|_p \le \left\| |x^*x - y^*y|^{\frac{1}{2}} \right\|_p.$$

Further, if $p = 1$, then it has been shown by Kosaki ([Ko] Proposition 7) that the same inequality continues to hold, without appealing to Ando's inequality. We may now state the following result.

THEOREM 2.2.   *If* $p \ge 1$, *and if* $\beta = \min\{p, 2\}$ *then*

$$\text{(i)} \quad \left\| \, |x| - |y| \, \right\|_p \le 2^{\frac{1}{\beta} - \frac{1}{2}} \|x + y\|_p^{\frac{1}{2}} \|x - y\|_p^{\frac{1}{2}},$$

$$\text{(ii)} \quad \left\| \, |x| - |y| \, \right\|_p \le \left(\|x\|_p^{\frac{\beta}{2}} + \|y\|_p^{\frac{\beta}{2}} \right)^{\frac{1}{\beta}} \|x - y\|_p^{\frac{1}{2}},$$

for all $x, y \in L^p(\mathcal{M})$.

The proof of the theorem follows exactly the same steps as the proof of Theorem 2.1 but using [FK] Theorem 4.9(1) instead of (2.3), and by using the identity

$$\|x\|_p = \left\| \, |x|^2 \, \right\|_{\frac{p}{2}}^{\frac{1}{2}}, \quad x \in L^p(\mathcal{M}),$$

in place of (2.4).

By way of final remarks, we note that Theorem 2.1 implies Hölder continuity of the absolute value map when restricted to bounded subsets of an arbitrary fully symmetric operator space. Using methods based on a theorem of Macaev [GK2], it has been shown by Davies [Da] that the absolute value map is in fact Lipschitz continuous on each of the Schatten ideals $C_p, 1 < p < \infty$, and that this fails in each of the extreme cases $p = 1, \infty$. More recently, it has been shown by Kosaki [Ko1] that the absolute value map is Lipschitz continuous on a trace ideal C_Φ (in the sense of [GK1]) if and only if C_Φ is an interpolation space for some pair C_p, C_r, with $1 < p < r < \infty$.

REFERENCES

[An]   T.Ando, *Comparison of norms* $|||f(A) - f(B)|||$ *and* $|||f(|A - B|)|||$, Math.Zeit. **197** (1988), 403-409.

[AH]   T.Ando and J.H.van Hemmen, *An inequality for trace ideals*, Comm. Math. Phys. **76**(1980), 143-148.

[Bh]   R. Bhatia, *Perturbation inequalities for the absolute value map in norm ideals of operators*, J.Operator Theory **21** (1988),129-136.

[BKS]   M.S. Birman, L.S.Koplienko and M.Z.Solomyak, *Estimates for the spectrum of the difference between fractional powers of two self-adjoint operators*, Izv.Vyssh. Uchebn.Zaved.,Mat.**19**(1975) 3-10.

[Ch]   K-M Chong, *Some extensions of a theorem of Hardy, Littlewood and Pólya and their applications*, Can.J.Math **26** (1974), 1321-1340.

[Da]   E.B. Davies, *Lipschitz continuity of functions of operators in the Schatten classes*, J. London Math. Soc. **37** (1988), 148-157.

[DDP1]   P.G. Dodds, T.K. Dodds and B. de Pagter, *Non-commutative Banach function spaces*, Math. Z. **201** (1989), 583-597.

[DDP2]   P.G. Dodds, T.K. Dodds and B. de Pagter, *A general Markus inequality*, Proc. CMA (ANU), **24** (1989), 47-57.

[DDP3]   P.G. Dodds, T.K. Dodds and B. de Pagter, *Non-commutative Köthe duality*, Trans. Amer. Math. Soc. **339** (1993),717-750.

[Fa]   T. Fack, *Type and cotype inequalities for non-commutative L^p-spaces*, J. Operator Theory **17** (1987), 255-279.

[FK]   T. Fack and H. Kosaki, *Generalized s-numbers of τ-measurable operators*, Pacific J. Math. **123** (1986), 269-300.

[Fr]   D.H. Fremlin, *Stable subspaces of $L^1 + L^\infty$*, Math. Proc. Cambridge Philos. Soc. **64**(1968), 625-643.

[GK1]   I.C. Gohberg and M.G. Krein, *Introduction to the theory of non-selfadjoint operators* Translations of Mathematical Monographs, vol.18, AMS (1969).

[GK2]   I.C. Gohberg and M.G. Krein, *Theory and applications of Volterra operators on Hilbert space*, Translations of Mathematical Monographs, vol.24, AMS (1970).

[He]   E.Heinz, *Beiträge zur Störungstheorie der Spektralzerlegung*, Math. Annalen, **123** (1951), 415-438.

[HN]   F.Hiai and Y.Nakamura, *Distance between unitary orbits in von Neumann algebras*, Pacific J.Math. **138**(1989), 259-294.

[KK]   F.Kittaneh and H.Kosaki, *Inequalities for the Schatten $p-norm$ V*, Publ. RIMS, Kyoto Univ. **23**(1987), 433-443.

[Ko1]   H.Kosaki, *On the continuity of the map $\phi \rightarrow |\phi|$ from the predual of a W^*-algebra*, J.Funct. Anal.,**59** (1984),123-131.

[Ko2]   H.Kosaki, *Unitarily invariant norms under which the map $A \rightarrow |A|$ is continuous*, Preprint (1991).

[KPS]   S.G. Krein, Ju.I. Petunin and E.M. Semenov, *Interpolation of linear operators*, Translations of Mathematical Monographs, Amer. Math. Soc. **54** (1982).

[LT]   J. Lindenstrauss and L. Tzafriri, *Classical Banach Spaces II*, Springer-Verlag, 1979.

[LS1]   G.G. Lorentz and T. Shimogaki, *Interpolation theorems for operators in function spaces*, J. Functional Anal. **2** (1968), 31-51.

[Lu]   W.A.J. Luxemburg, *Rearrangement invariant Banach function spaces*, Queen's Papers in Pure and Applied Mathematics, No. **10** (1967), 83-144.

[Ne]   E. Nelson, *Notes on non-commutative integration*, J. Functional Anal. **15** (1974), 103-116.

[Ov1]   V.I. Ovčinnikov, *s-numbers of measurable operators*, Funktsional'nyi Analiz i Ego Prilozheniya **4** (1970), 78-85 (Russian).

[Ov2]   V.I. Ovčinnikov, *Symmetric spaces of measurable operators*, Dokl. Nauk SSSR **191**(1970), 769-771 (Russian), English Translation: Soviet Math. Dokl. **11** (1970), 448-451.

[Ov3]   V.I. Ovčinnikov, *Symmetric spaces of measurable operators*, Trudy inst. matem. VGU **3** (1971), 88-107.

[PS]   R.T. Powers and E. Stormer, *Free states of the canonical anticommutation relations* Commun. Math. Phys. **16**(1970), 1-33.

[RR]   M.Rosenblum and J.Rovnyak, *Hardy Classes and Operator Theory*. New york: Oxford University Press 1985.

[Se]   I.E. Segal, *A non-commutative extension of abstract integration*, Ann. Math. **57** (1953), 401-457.

[St]   W.F. Stinespring, *Integration theorems for gages and duality for unimodular groups*, Trans. Amer. Math. Soc.**90**(1959), 15-56.

[Ta1]   M. Takesaki, *Theory of Operator Algebras I*, Springer-Verlag, New York-Heidelberg Berlin, 1979.

[Ta2]   M.Takesaki, *Duality for crossed products and structure of von Neumann algebras of type III*, Acta Math., **131**(1973),249-310.

[Te]   M.Terp, *L^p-spaces associated with von Neumann algebras*,Notes, Copenhagen Univ. (1981).

[Ty]   O.E. Tychonov, *Continuity of operator functions in topologies connected with a trace on a von Neumann algebra*(Russian),Izv.Vyssh.Uchebn. Zaved.Mat., 1987,no.1, 77-79; translated in Soviet Math. (Iz.VUZ),**31** (1987), 110-114.

[Xu]   Q.Xu, *Analytic functions with values in lattices and symmetric spaces of measurable operators*, Math. Proc.Camb.Phil.Soc. **109** (1991), 541-563.

[Za1]   A.C.Zaanen, *Linear Analysis*, 2nd edition.,Amsterdam:North-Holland (1956).

[Za2]   A.C.Zaanen, *Integration* ,Amsterdam, cvNorth -Holland (1967).

[Za3]   A.C.Zaanen, *Riesz Spaces II*, 2nd edition.,Amsterdam:North -Holland (1983).

Peter G. Dodds and Theresa K. Dodds
School of Information Science & Technology
The Flinders University of South Australia
GPO Box 2100, Adelaide 5001
Australia

1991 *Mathematics Subject Classification.* Primary 46E30; Secondary 46L50, 47B55.

Operator Theory:
Advances and Applications, Vol. 75
© 1995 Birkhäuser Verlag Basel/Switzerland

SPECTRAL THEORY IN BANACH LATTICES

JJ Grobler*

Dedicated to Professor A.C. Zaanen on the occasion of his eightieth birthday.

We survey the recent results related to the Jentzsch-Perron and Frobenius theorems for a single positive operator on a Banach lattice. A number of new results are added for σ-order continuous band irreducible operators and also some results on the primitivity of an operator.

1. Introduction. Positive finite square matrices are known to have spectra with extraordinary beautiful properties. This was discovered by O. Perron [46], who proved the first results in this direction,

THEOREM 1.1 (O. Perron, 1907.) *Let $T = [t_{ij}]_{i,j=1}^{n}$ be a matrix with $t_{ij} > 0$ for all i, j. Then*

(i) *T has a strictly positive spectral radius $r(T)$;*

(ii) *$r(T)$ is a simple eigenvalue with strictly positive eigenvector;*

(iii) *T is primitive, i.e., $r(T)$ is the unique eigenvalue on the circle $|\lambda| = r(T)$.*

An $n \times n$-matrix P is called a *permutation matrix* if, for some permutation π of $\{1, 2, \ldots, n\}$, we have $Pe_i = e_{\pi(i)}$ where $e_i = (\delta_{ij})$. Perron's result was extended and

*This paper was written during the fall semester of 1993 while the author visited the Department of Mathematics of the University of Leiden as the guest of Dr. C.B. Huijsmans, to whom he expresses his gratitude for the hospitality offered to him.

generalized by G. Frobenius [23], [24] who called an $n \times n$-matrix T with non-negative entries *irreducible* if there exists no permutation matrix P such that

$$P^{-1}TP = \begin{pmatrix} T_1 & 0 \\ B & T_2 \end{pmatrix},$$

with T_i square of order m_i $(1 \leq m_i < n)$, and proved:

THEOREM 1.2. (G. Frobenius, 1909, 1912.) *Let $T = [t_{ij}]_{i,j=1}^n$ be a matrix with $t_{ij} \geq 0$ for all i,j. If T is irreducible, then*

(i) *T has a strictly positive spectral radius $r(T)$;*

(ii) *$r(T)$ is a simple eigenvalue with strictly positive eigenvector;*

(iii) *all eigenvalues on the circle $|\lambda| = r(T)$ are simple and if there are k of them, they are the roots of the equation $r(T)^k - \lambda^k = 0$;*

(iv) *the entire spectrum of T is invariant under a rotation of the complex plane by the angle $2\pi/k$.*

Generalizations of aspects of this theory to the infinite dimensional case came early on from R. Jentzsch [35] and E. Hopf [30] who investigated integral equations with positive kernels which were continuous or at least continuous in the mean. They were followed by A.C. Zaanen [59] who, following Jentzsch's method, considered integral operators with positive kernels on the Hilbert space $L_2(X,\mu)$.

A landmark in the history of the development of these ideas was the paper of M.G. Krein and M.A. Rutman [39] who considered operators on Banach spaces leaving some cone in the space invariant. In modern terms, positive operators on ordered Banach spaces. The applicability of their results was hampered by the fact that the cone had to have interior points, which in effect, restricted attention to spaces of continuous functions on compact Hausdorff spaces. However, this was the starting point of the abstract consideration of the problems involved which to this day are still studied. At present the theory has been developed to the point where the classical theorems of Perron-Frobenius are well understood, known to hold under very general hypotheses and even extended. The quest to obtain these results, it has been said, has been one of the motivating forces in developing the theory of Banach lattices and positive operators [51].

It is our aim to survey the modern theory starting with the result of T. Andô ([6], 1957). This is best done by analyzing and describing the four distinct circles of questions and ideas which comprise the classical theorems of Perron-Frobenius for a positive operator. These are questions about: the strict positivity of the spectral radius, the cyclicity of that part of the spectrum which intersects the spectral circle, the general structure of the spectrum and eigenspaces and the primitivity of the operator.

In section 3 results on the strict positivity of the spectral radius of a positive operator are surveyed, starting with the Andô-Krieger theorem. The main result in this section is the theorem of de Pagter (theorem 3.2) which contains essentially all information needed for the extension of the Andô-Krieger theorem in many directions. We also present what we consider the key to all extensions of the theorems from ideal irreducible operators to σ-order continuous band irreducible operators in Propositions 3.8 and 3.9 and Corollary 3.10.

Section 4 contains a description of the results known for the peripheral spectrum of a positive operator. Two results are central; firstly the theorem of H.P. Lotz on the cyclicity of the peripheral spectrum and secondly the result of F. Niiro and I. Sawashima on the existence of poles on the spectral circle. Using a result of E. Emilion here, we observe that a result of Lotz on uniformly ergodic positive operators can be slightly improved, and using the idea mentioned in section 3 on the extension of results on ideal irreducible operators to σ-order continuous band irreducible operators, we prove such an extension of the Niiro-Sawashima theorem.

The results in sections 3 and 4 are of such a nature, that if compactness in some form is added to the hypothesis on the positive operator, then the results of Perron and Frobenius follow easily. This is done in section 5. We draw attention to the fact that the Frobenius theorem is proved here in greater generality than was done before in the literature.

Section 6 contains the results on the primitivity of a positive operator. Here we prove an extension of a result of H.H. Schaefer (theorem 6.1) and two theorems relating the primitivity of the operator to the existence of a trace of some kind for the operator. The first of these theorems is due to the author and the second to B. de Pagter, but it appears here for the first time.

We draw the attention of the reader to some known open problems in this area in the final section 7.

In the next section we fix our notation and present the structural framework necessary to understand the subsequent theorems.

2. Preliminaries. Let E be a real Riesz space, i.e., a partially ordered real vector space in which the supremum and infimum of each pair of elements exist. As is usual, $|x| = x \vee (-x)$, $x^+ = x \vee 0$, $x^- = -x \vee 0$ and $E^+ = \{x \in E : x \geq 0\}$. A sublattice I of E is called an *ideal* if it is a solid subset of E, which means that if $x \in I$ and $0 \leq |y| \leq |x|$, then also $y \in I$. An ideal B is called a *band* in E if it is order closed. If E is normed and the norm satisfies $\|x\| \leq \|y\|$ whenever $|x| \leq |y|$, then E is called a *lattice normed space*. A lattice normed Banach space is called a *Banach lattice*. If E is a Banach lattice, every band in E is norm closed. An element $u > 0$ is called a *weak order unit* of E if the band generated by u is equal to E; it is called a *quasi-interior point* if the closed ideal E_u generated by it is E. A Riesz space is called *Dedekind complete* whenever every non-empty set with an upper bound has a least upper bound in the space. A Banach lattice is said to have an *order continuous norm* if $\|x_\alpha\| \downarrow 0$ for every downwards directed net x_α in E with 0 as infimum. The gauge-function of the interval $[-u, u]$ is a norm on the ideal E_u generated by u in the Riesz space E. E is called *uniformly complete* if E_u is norm complete with this norm for every $u \geq 0$. Every Banach lattice is uniformly complete. Every uniformly complete real Riesz space E has the property that an absolute value can be introduced in the *complexification* $E + iE$ of the vector space E by means of the formula

$$(1) \qquad |z| = |x + iy| = \sup\{x \cos \theta + y \sin \theta : 0 \leq \theta \leq 2\pi\}.$$

The existence of the supremum on the right hand side is shown in [12] using elementary means. It follows that every Banach lattice E has a complexification with this property and the norm of an element z is defined by $\|z\| := \||z|\|$. The linear subspace A of $E + iE$ is now called an ideal if $x \in A$ and $|y| \leq |x|$ implies $y \in A$. Every ideal A in $E + iE$ is of the form $A_r + iA_r$ with A_r an ideal in E. The ideal A in $E + iE$ is called a band if A_r is a band in E. If $F = E + iE$ is a complex Riesz space, we call $z \in F$ *real* if $z \in E$ and *positive* (writing $z \geq 0$) if $z \in E^+$.

In the sequel we shall need some concepts from the theory of f-algebras. A real Riesz space E is called a *lattice algebra* if it is endowed with an associative multiplication with the usual algebra properties and also satisfies $E^+E^+ \subset E^+$. The Riesz algebra E is called an f-*algebra* if it has the property that $u \wedge v = 0$ in E implies $uw \wedge v = 0 = wu \wedge v$ for all $0 \leq w \in E$. An f-algebra is commutative and satisfies $|xy| = |x||y|$ for all $x, y \in E$. If E is a uniformly complete f-algebra, the multiplication in E extends to $E + iE$ in the obvious way and $E + iE$ is again an f-algebra, i.e., if $x, y, z \in E + iE$ with $|x| \wedge |y| = 0$, then also $|zx| \wedge |y| = |xz| \wedge |y|$. In [12] there is an elementary proof of the fact that for $x = a + ib \in E + iE$ one has $|x| = (a^2 + b^2)^{\frac{1}{2}}$ and it easily then follows that $|xy| = |x||y|$ holds also in the complex case.

By an operator T on E we will always mean a linear operator $T \colon E \to E$. An operator T on a real Riesz space is said to be *positive* whenever $x \geq 0$ implies $Tx \geq 0$, *regular* if it is the difference of two positive operators and *order bounded* whenever it maps order intervals into order intervals. We denote by $L^r(E)$ the set of all regular operators and by $L^b(E)$ the set of all order bounded operators. The first of these sets is contained in the latter and if E is Dedekind complete, they are equal. In this case it is a Dedekind complete Riesz space with absolute value $|T|$ where

(2) $$|T|x = \sup\{|Ty| \colon 0 \leq |y| \leq x\} \text{ for all } 0 \leq x \in E.$$

If E is a normed Riesz space then $L^r(E)$ can be normed by the *regular norm* which is defined as follows: For $T \in L^r(E)$

$$\|T\|_r = \inf\{\|S\| : T = T_1 - T_2, T_i \geq 0, T_1 + T_2 \leq S\}.$$

In general $L^r(E)$ is not a Riesz space though it is a Banach algebra. But, if E is Dedekind complete, then it is a Dedekind complete Banach lattice and $\|T\|_r = \||T|\|$. The space $L^r(E)$ is then a *Banach lattice algebra*. In general, $L^r(E)$ is not closed in $L(E)$.

The positive operator T on a Riesz space E is called σ-*order continuous* if it follows from $u_n \downarrow 0$ that $Tu_n \downarrow 0$, and it is called *order continuous* if the same holds with sequences replaced by downwards directed systems.

If E is a Banach lattice, its continuous dual is denoted by E' and E' is equal to the set of all order bounded linear functionals on E. It follows that E' is a Banach lattice and we call an element $f \in E'$ order continuous (σ-order continuous) whenever f^+ and f^- are both order continuous (respectively σ-order continuous). The set of all σ-order continuous linear functionals on E is denoted by E'_o.

An operator T on the real Riesz space E extends uniquely to its complexification $E + iE$ (and is again denoted by T) by defining

$$T(x + iy) := Tx + iTy \text{ for all } x + iy \in E + iE.$$

Conversely, if T is an operator on $E + iE$ then there exist unique operators T_1, T_2 on E such that $Tx = T_1 x + iT_2 x$ for all $x \in E$. We call T a real operator if $T = T_1$ and positive if T is real and T_1 is positive. Similarly, we call T respectively regular, order bounded, σ-order continuous and order continuous whenever its real and imaginary parts T_1 and T_2 have the corresponding property. From these definitions and our remark above, it follows that $L^r(E + iE) = L^r(E) + iL^r(E)$ and therefore every element has a modulus defined as in (1) above. It is a non-trivial fact that this modulus also satisfies (2) and therefore, $|Tz| \le |T||z|$ for all $z \in E$ and for all $T \in L^r(E + iE)$.

An order bounded operator U on E is called an *orthomorphism* if it follows from $|x| \wedge |y| = 0$ that $|Ux| \wedge |y| = 0$. If E is a Banach lattice then U is an orthomorphism if and only if U belongs to the *center* $Z(E)$ of E, i.e., there exists $0 \le \lambda \in \mathbf{R}$ such that $|Ux| \le \lambda|x|$ for all $x \in E$ ([60]). If E is an f-algebra and if we define for $u \in E$ the operator U on E by $Ux = ux$ for all $x \in E$, then U is an orthomorphism on E. An operator T on the Banach lattice E is a *lattice homomorphism* if $|Tx| = T|x|$ for all $x \in E$ and it is called *disjointness preserving* whenever $|x| \wedge |y| = 0$, implies $|Tx| \wedge |Ty| = 0$.

We call an operator on the Dedekind complete Banach lattice a (abstract) kernel operator if it belongs to the band generated by the σ-order continuous finite rank operators in $L^r(E)$.

The reader can keep the spaces $C(K)$, (of continuous functions on compact Hausdorff spaces), and $L^p(X, \mu)$, (of equivalence classes of p-integrable μ-almost everywhere equal functions), in mind as examples of Banach lattices. In an $L^p(X, \mu)$ space

$(1 \leq p < \infty)$, every closed ideal is a band and it is characterized as the set of functions vanishing almost everywhere on some measurable subset of X.

Well known examples of operators on spaces $L^p(X, \mu)$ are the kernel operators. The linear map $T : L^p(Y, \nu) \to L^r(X, \mu)$ is called a *kernel operator* if there exists a $\mu \times \nu$-measurable function $t(x, y)$ on $X \times Y$ such that for all $f \in L^p(Y, \nu)$,

$$Tf(x) = \int_Y t(x, y)f(y)\, d\nu(y), \mu\text{-a.e.}$$

The kernel operator T is positive if and only of its kernel $t(.,.)$ is positive almost everywhere, and it is clear that every kernel operator can be decomposed as a difference of two positive operators. Moreover, every kernel operator is order continuous.

For further properties of Riesz spaces, Banach lattices and positive operators, we mention the texts by W.A.J. Luxemburg and A.C. Zaanen [42], H.H. Schaefer [51], A.C. Zaanen [60], C.D. Aliprantis and O. Burkinshaw [5] and P. Meyer-Nieberg [44] where the reader may find a wealth of information on these topics.

When discussing the spectral theory, we will always have in mind a complex Banach lattice which will be the complexification of a real Banach lattice. The *spectrum* $\sigma(T)$ of a continuous linear operator T on the Banach lattice E is the non-void compact set of complex numbers λ which has the property that $\lambda - T$ does not have a continuous inverse on E. If $\lambda \in \sigma(T)$ is such that $\lambda - T$ is not injective, then λ is called an *eigenvalue* of T and the set $\sigma_P(T)$ of all eigenvalues is called the *point spectrum* of T. The *approximate point spectrum* $\sigma_A(T)$ contains all λ such that $\lambda - T$ is not an isomorphism onto a closed subspace of E, and the *residual spectrum* $\sigma_R(T)$ contains all λ such that $\lambda - T$ is an isomorphism onto a closed subspace of E but distinct to E. By definition,

$$\sigma(T) = \sigma_A(T) \cup \sigma_R(T).$$

Using the theorem of Hahn-Banach one quickly sees that $\sigma_R(T) \subset \sigma_P(T')$ and we will also use the fact that $\sigma_A(T)$ contains the boundary of $\sigma(T)$. The *spectral radius* of an operator is given by the formula:

$$r(T) = \sup\{|\lambda| : \lambda \in \sigma(T)\} = \lim_{n \to \infty} \|T^n\|^{1/n}.$$

We denote the complement of $\sigma(T)$ by $\rho(T)$ and call it the resolvent set of T. On this set the operator valued function $R(\lambda, T) := (\lambda - T)^{-1}$ is an analytic function of the complex variable λ, called the *resolvent operator* of T at λ. For all $\lambda \in \rho(T)$ with $|\lambda| > r(T)$, the resolvent operator has a *Neumann expansion*

$$R(\lambda, T) = \sum_{k=0}^{\infty} \frac{T^k}{\lambda^{k+1}}.$$

It also has a Laurent series expansion about each isolated point λ_0 of $\sigma(T)$ and if this series has only a finite number of non-zero co-efficients of non-negative powers of $\lambda - \lambda_0$ then λ_0 is called a *pole* of the resolvent. In this case,

$$R(\lambda, T) = (\lambda - \lambda_0)^{-p} B_p + (\lambda - \lambda_0)^{-p+1} B_{p-1} + \ldots + (\lambda - \lambda_0)^{-1} B_1 + \sum A_n (\lambda - \lambda_0)^n$$

and the operators B_i, $i = 1, 2, \ldots p$, satisfy $B_{i+1} = (\lambda_0 - T)^i P$ with P the *residuum operator*. If T is a *Riesz operator* i.e., an asymptotically quasi-compact operator, then every point of its spectrum is a pole and an eigenvalue. Furthermore, the residuum operator P is a finite rank projection. If $B_p \neq 0$ then λ_0 is a pole of order p. For every $\lambda \in \sigma(T)$ the order of the pole in λ is also the smallest positive integer $\nu(\lambda)$ with the property that

$$N[(\lambda - T)^m] = N[(\lambda - T)^{m+1}] \text{ and } (\lambda - T)^{m+1} P = (\lambda - T)^m P = 0$$

for all $m \geq \nu(\lambda)$ (here $N(S)$ denotes the null space of the operator S). The dimension of the space $P(E) = N[(\lambda - T)^{\nu(\lambda)}]$ is called the *algebraic multiplicity* of λ for T. For further information on spectral theory of bounded operators on Banach spaces we refer the reader to [19].

3. Positivity of the spectral radius: Theorems of Andô-Krieger and De Pagter.
When studying Zaanen's proof of Jentzsch's theorem in $L^2(X, \mu)$, it soon becomes evident that the difficult part of the proof of a more general theorem would be the proof of the positivity of the spectral radius, and that new methods would be needed. These were introduced by T. Andô in 1957, when he defined the notion of an irreducible kernel operator which, when specialized to $n \times n$-matrices, reduces to the definition given above. The kernel operator T has an irreducible kernel $t(x, y)$ whenever

$$\int_{X-S} \int_S t(x, y) \, d\mu(x) \, d\mu(y) > 0$$

for all measurable subsets S in X satisfying $\mu(S) > 0$ and $\mu(X - S) > 0$.

Using the characterization of closed ideals mentioned above, this condition was generalized to the abstract case by H.H. Schaefer [51] who introduced the notion of an ideal irreducible operator and by the author [25] who considered band irreducible operators. A positive operator T on the Banach lattice E is called *ideal irreducible* (respectively *band irreducible*) if E contains no non-trivial T-invariant closed ideals (respectively bands). Since every band is also a closed ideal, every ideal irreducible operator is also band irreducible. On Banach lattices with order continuous norm, the two notions coincide, and it is an easy exercise to prove that these definitions reduce to that given above in the special case of kernel operators on function spaces with order continuous norm (respectively Dedekind complete function spaces).

In 1957 T.Andô [6] proved the following extension of Jentzsch's theorem:

THEOREM 3.1. *A positive ideal irreducible compact kernel operator on a Banach lattice with order continuous norm has a strictly positive spectral radius.*

The proof of this theorem was the object of much study. The compactness condition was removed 12 years later by H.J. Krieger [38]. In 1978 we proved [25] that order continuity of the norm can be weakened to Dedekind completeness and that ideal irreducibility can be replaced by band irreducibility. Though the compactness condition was removed from the theorem, it was in fact still very much present in the proofs. This can be seen in a very simple proof of the theorem given by the author in 1980 [26]. Using some standard arguments one can show that it may be assumed that the operator is majorized by a rank one operator. However, at least the third power of such an operator is compact by the Dodds-Fremlin theorem as expanded by Aliprantis and Burkinshaw (see [18] and [4]). Hence, the operator for which the theorem was really proved still had strong compactness properties.

A very simple idea, due to H.H. Schaefer, underlies the proof. If the operator T has the property that Tu is a quasi-interior point for every positive element u, then, in the case of a space $C(X)$, the element Tu is strictly positive. There exists then a positive number κ such that $Tu \geq \kappa u$, which implies that $r(T) \geq \kappa$. The whole exercise is then directed to proving that in general one can reduce the proof to this case.

An open question (stated by H.H. Schaefer [51]) was whether kernel operators could be replaced by compact operators in the theorem. It was recognized that if one could, the Andô-Krieger result would follow. This major breakthrough was made in 1986 by B. de Pagter [20]. Of interest is that he approached the problem from a new direction disposing of all the technicalities necessary to use the idea explained in the preceding paragraph. He based his proof on Hilden's proof of Lomonosov's invariant subspace theorem, and proved an invariant ideal theorem for positive operators. We present de Pagter's proof.

THEOREM 3.2. (B. de Pagter, 1986.) *Every compact quasi-nilpotent positive operator on a Banach lattice has a non-trivial invariant closed ideal, i.e., every ideal irreducible compact positive operator on a Banach lattice has a strictly positive spectral radius.*

PROOF: It may be assumed that the positive compact operator T is non-zero and strictly positive. If $u > 0$ the element $\sum_1^\infty 2^{-k}\|T\|^{-k}T^k u$ generates a T-invariant closed ideal which, if proper, is the desired object. If not, define

$$\mathcal{C}^+ := \{0 \leq R \in L(E) : RT = TR\},$$

$$\mathcal{J}^+ := \{0 \leq S \in L(E) : \exists R \in \mathcal{C}^+ \text{ with } 0 \leq S \leq R\},$$

$$\mathcal{J} := \{S_1 - S_2 : S_1, S_2 \in \mathcal{J}^+\}.$$

For $f \in E$ let $\mathcal{J}[f] := \{Sf : S \in \mathcal{J}\}$. It is easily seen that $\overline{\mathcal{J}[f]}$ is a closed T-invariant subspace of E, but it takes some knowledge of Banach lattice theory to prove that it is indeed an ideal. The compactness of T and its quasi-nilpotency is now invoked, analogous to the way it was done in Hilden's proof, to show that there exists an element $f \in E$ such that $\overline{\mathcal{J}[f]}$ is not equal to E : Suppose, on the contrary that for each $f \neq 0$ in E we have $\overline{\mathcal{J}[f]} = E$, and let $0 < u \in E$. Then $Tu > 0$ and for some open ball $\mathcal{U}$ centered at u we have $0 \notin \overline{\mathcal{U}}$ and $0 \notin \overline{T\mathcal{U}}$. By assumption $u \in \overline{\mathcal{J}[f]}$ for each $f \in \overline{T\mathcal{U}}$ and so there exists, for each $f \in \overline{T\mathcal{U}}$ an operator $S_f \in \mathcal{J}$ such that $S_f f \in \mathcal{U}$. Let $\mathcal{U}_f$ be an open ball centered at f such that $S_f \mathcal{U}_f \subset \mathcal{U}$. The set $\{\mathcal{U}_f : f \in \overline{T\mathcal{U}}\}$ is an open cover for the compact set $\overline{T\mathcal{U}}$ and has a finite subcover. Hence, there exist open balls $\mathcal{U}_j$ and operators $S_j \in \mathcal{J}, j = 1, \ldots, m$, such that the $\mathcal{U}_j$ cover $\overline{T\mathcal{U}}$ and $S_j\mathcal{U}_j \subset \mathcal{U}$ for all j. We now start an inductive process: for

$n = 1$, since $Tu \in \overline{TU}$, there exists $j_1, 1 \leq j_1 \leq m$, such that $S_{j_1} Tu \in \mathcal{U}$. For $n = 2$, since $TS_{j_1} Tu \in \overline{TU}$, there exists $j_2, 1 \leq j_2 \leq m$. such that $S_{j_2} TS_{j_1} Tu \in \mathcal{U}$. Continuing in this way we construct a sequence (g_n) with

$$g_n = S_{j_n} TS_{j_{n-1}} T \ldots S_{j_1} Tu \in \mathcal{U}.$$

Now each $S_j = S_j^{(1)} - S_j^{(2)}$, with $0 \leq S_j^{(i)} \leq R_j^{(i)} \in \mathcal{C}^+$, $i = 1, 2$; $j = 1 \ldots m$. It follows that

$$|g_n| \leq (R_{j_n}^{(1)} + R_{j_n}^{(2)}) \ldots (R_{j_1}^{(1)} + R_{j_1}^{(2)}) T^n u$$

and so $\|g_n\| \leq (2C)^n \|T^n\| \|u\|$, with $C = \max\{\|R_j^{(i)}\| : 1 \leq j \leq m; i = 1, 2\}$. Since T is quasi-nilpotent, $\|g_n\| \to 0$ as $n \to \infty$. This contradicts the fact that $\mathcal{U}$ was chosen such that $0 \notin \overline{\mathcal{U}}$. This completes the proof.

The reason for reproducing the proof, besides its beauty, is that there are some easy extensions of the theorem to be made following directly from the proof. This was done by V. Caselles [15] and by Y.A. Abramovich, C.D. Aliprantis and O. Burkinshaw [1] who observed that the closed invariant ideal $\overline{\mathcal{J}[f]}$ is also invariant for all positive operators C commuting with T. Calling such an ideal a *hyperinvariant ideal for* T, we see that de Pagter really proved a Banach lattice version of the Lomonosov hyperinvariant subspace theorem for compact operators on Banach spaces:

THEOREM 3.3. *Every compact quasi-nilpotent positive operator on a Banach lattice has a non-trivial hyperinvariant closed ideal.*

Applying 3.3 to the product operator ST in the next theorem and observing that for commuting operators S and T one has $r(ST) \leq r(S)r(T)$ one gets:

THEOREM 3.4. (V. Caselles, 1987.) *If an ideal irreducible positive operator T commutes with a compact positive operator S, then $r(T) > 0$ and $r(S) > 0$. In particular, if an ideal irreducible operator T has a compact power, then $r(T) > 0$.*

Caselles actually proved that 3.4 is true if $TS \leq ST$, or $ST \leq TS$. His proof follows the idea of de Pagter's proof but he observed that one can replace everywhere the commutativity by this semi-commutativity condition. He also weakens the condition that S be compact to the condition that S should dominate a compact positive operator.

It was shown by Y.A. Abramovich, C.D. Aliprantis and O. Burkinshaw (see [2]) that 3.3 can be further strengthened if the proviso that $T^2 \neq 0$ is added to the hypotheses.

THEOREM 3.5. *If a compact quasi-nilpotent positive operator* $T : E \to E$ *satisfies* $T^2 \neq 0$, *then it has a non-trivial hyperinvariant closed ideal* J *with* $T(J) \neq \{0\}$.

We observe in the next example, found by the author, that the theorem is false without the extra assumption that $T^2 \neq 0$. Here it turns out that it may happen that the absolute null ideal $N := \{x \in E : T|x| = 0\}$ is the only non-trivial hyperinvariant closed ideal of T.

Example Let $E = L^p([0,1], m), m$ the Lebesgue measure and $1 < p < \infty$. Let $U := [0, \frac{1}{2}]$, $V := [\frac{1}{2}, 1]$, and consider the rank one operator $T = \chi_V \otimes \chi_U$, where χ_X denotes the characteristic function of the set $X \subset [0,1]$. The operator T is compact and nilpotent. Suppose there exists a closed ideal $J = \{f \in L^p[0,1] : f(t) = 0 \text{ if } t \notin W\}$ which is non-trivial and hyperinvariant for T and such that $T[J] \neq \{0\}$. Since $T[E]$ is contained in the ideal generated by χ_U and J is T-invariant, $\chi_U \in J$, i.e., $\chi_U \leq \chi_W$. But, T is zero on the ideal generated by χ_U and so $m(W - U) > 0$. Moreover, since J is non-trivial, $W \neq [0,1]$ and we define the non-zero function $y := m(U)m([0,1] - W)^{-1}\chi_{([0,1]-W)}$. The operator

$$S := \chi_U \otimes \chi_U + \chi_V \otimes y$$

commutes with T since $ST = TS = m(U)\chi_V \otimes \chi_U$, but $S\chi_{(W-U)} = (\int_V \chi_{(W-U)} \, dm)y = m(W - U)y \notin J$, albeit that $\chi_{(W-U)} \in J$.

Following the terminology in [1] a positive operator T is called *strongly expanding* (also called *strongly ideal irreducible* in [60]) if Tu is a quasi-interior point for every $u > 0$. The next theorem can be proved without using 3.5.

THEOREM 3.6. (Y.A. Abramovich, C.D. Aliprantis, O. Burkinshaw.) *Let* $Q, S : E \to E$ *be two positive operators such that* Q *is strongly expanding and such that* S *dominates a compactly dominated positive operator. Then* $r(QS) > 0$.

PROOF: Let Q and S have the properties stated. Let L be positive and K be compact such that $0 < L \leq S$ and $L \leq K$. Put $C = QL$ and note that $C > 0$ and that if $Cx > 0$ for some $x > 0$, then the closed ideal generated by Cx is E. Let $N = \{x \in E : C|x| = 0\}$, let $q : E \to E/N$ denote the quotient map, and let $\overline{C} : E/N \to E$ denote the induced map. Then what we noted above amounts to the fact that the map $q\overline{C} : E/N \to E/N$ is strongly expanding and hence irreducible. Since $q\overline{C}$ has some compact iterate by the Aliprantis Burkinshaw theorem, it follows from 3.4 that $r(q\overline{C}) > 0$. A general fact from spectral theory therefore yields the fact that $r(C) = r(\overline{C}q) = r(q\overline{C}) > 0$. Hence, $r(QS) \geq r(C) > 0$.

From these theorems various interesting corollaries follow. We only state the following theorem of B. de Pagter which generalizes a theorem of V. Caselles on Harris operators.

THEOREM 3.7. (B. de Pagter, 1986.) *If T is an ideal irreducible quasi-nilpotent positive operator and if S is a compact positive operator then $[0, S] \cap [0, T^k] = \{0\}$ for every $k \geq 1$. Equivalently, if a power of an ideal irreducible positive operator T dominates a compactly dominated positive operator, then $r(T) > 0$.*

Up to this point we concentrated on ideal irreducible operators. There exist similar theorems for the more general class of band irreducible operators, but not without additional conditions. In [1] the authors give an example of a positive, band irreducible and compact operator which is quasi-nilpotent. In both [51] and [27] some of the results were proved under the added condition of σ-order continuity which seems to be the proper condition. We present two propositions which hold in our view the key to the proofs of any such extension. The argument can be found for the first time in [26] and it was also used in [27] to prove some of the results to follow.

PROPOSITION 3.8. *Let $0 \leq T$ be majorized by a positive weakly compact order continuous (respectively, σ-order continuous) operator S and let F be the closed ideal generated by the range $T[E]$ of T. If I is a non-trivial T-invariant closed ideal of F, then the band (respectively, σ-ideal) B generated by I in E is a non-trivial T-invariant band (respectively, σ-ideal).*

PROOF: We give the argument for the case that B is a band and S is order continuous. It is clear that B is again T-invariant and not $\{0\}$. Were $B = E$, then for each $0 \le x \in E$, there exist $x_\alpha \in I$ such that $x_\alpha \uparrow x$. Consequently,

$$0 \le Tx - Tx_\alpha = T(x - x_\alpha) \le S(x - x_\alpha) \downarrow 0.$$

Since S is weakly compact, it follows from Dini's theorem that $\|S(x - x_\alpha)\| \to 0$, and so $\|Tx_\alpha - Tx\| \to 0$. Hence, $Tx \in I$, since I is closed. This means that $T[E] \subset I$ and so $I = F$ contrary to the fact that I is supposed to be non-trivial. Hence, $B \ne E$ and the proof is complete.

PROPOSITION 3.9. ([27].) *If $0 \le T$ is a σ-order continuous band irreducible operator on E, then every non-zero closed T-invariant ideal in E contains a weak order unit of E.*

PROOF: Let I be a non-zero T-invariant ideal in E and let $0 < u \in I$. If $\lambda > r(T)$, put $w = R(\lambda, T)u$. Using the Neumann expansion of $R(\lambda, T)$, we see that $w \in I$ and that $Tw \le \lambda w, w > 0$. Hence, the principal ideal E_w generated by w in E is T-invariant. Let $\{w\}^{dd}$ denote the band generated by w in E and let $0 \le v \in \{w\}^{dd}$. Then $v = \sup_n v_n$, with $v_n = v \wedge nw \in E_w$, so $Tv_n \uparrow Tv$ by the σ-order continuity of T, and moreover $Tv_n \in E_w$ since E_w is T-invariant. Hence, $Tv \in \{w\}^{dd}$. This shows that $\{w\}^{dd}$ is T-invariant and so, since T is band irreducible, we have $\{w\}^{dd} = E$. The element w is therefore a weak order unit for E contained in I.

COROLLARY 3.10. *If $0 \le T$ is a σ-order continuous band irreducible operator on E which is majorized by a positive σ-order continuous weakly compact operator, then the restriction of T to the closed ideal F generated by the range of T is irreducible.*

PROOF: Let G be a closed T-invariant ideal in F. If G is non-trivial, so is, by 3.8 the σ-ideal it generates. But, by 3.9, G contains a weak order unit of E. Contradiction.

THEOREM 3.11. *Let $0 \le T : E \to E$ be a compact quasi-nilpotent operator. If T is order continuous (respectively σ-order continuous) then E contains a non-trivial T-hyperinvariant band (respectively, σ-ideal).*

PROOF: Let $N := \{x \in E : T|x| = 0\}$, then N is a T-hyperinvariant band (resp. σ-ideal).
If $N \neq \{0\}$ we are done. Suppose therefore that $N = \{0\}$. Then T is strictly positive on E
and an easy argument (see [1], Lemma 3.11) shows that every S commuting with T is order
continuous. Let F be the closed ideal generated by $T[E]$ in E. Then F is T-hyperinvariant
and T restricted to F is compact and quasi-nilpotent. It follows that there exists a T-
hyperinvariant closed non-trivial ideal I in F. By proposition 3.7 the band generated by I
is a non-trivial T-invariant band and by the order continuity of every S commuting with
T, it is indeed T-hyperinvariant.

   Both V. Caselles [15] and Y.A. Abramovich, C.D. Aliprantis and O. Burkinshaw [1] now prove analogous results from those obtained for irreducible operators for
positive σ-order continuous band irreducible operators. The main results, from which
similar corollaries as in the case of ideal irreducible operators can be derived, are:

   THEOREM 3.12. *Let E be a Banach lattice. Then the following statements*
hold.
1. *(Abramovich, Aliprantis, Burkinshaw.) Every order continuous compact positive operator T that commutes with a band irreducible positive operator has a positive spectral
radius.*
2. *Every σ-order continuous compact positive operator T that commutes with a σ-order
continuous band irreducible positive operator has positive spectral radius.*
3. *(Grobler-Schaefer.) Every σ-order continuous band irreducible compact positive operator has a positive spectral radius.*
4. *(Grobler-Schaefer.) If $T : E \to E$ is a σ-order continuous band irreducible quasi-
nilpotent positive operator and S is a compact positive operator, then $[0, S] \cap [0, T^k] = \{0\}$
for every $k \geq 1$.*

   Properties 3 and 4 were proven by H.H. Schaefer under the extra condition
that there exists a non-zero σ-order continuous linear functional on E. Independently, the
present author proved 3 and 4 without the extra condition for order continuous operators
but the proofs worked equally well for the σ-order continuous case. A.R. Schep observed
(in a private communication) that if there exists a non-zero σ-order continuous compact
operator T on E, then, for every $f \in E'_+$ such that $T'f \neq 0$, the latter functional is

necessarily σ-order continuous. This also makes it clear that the condition imposed by Schaefer was redundant.

THEOREM 3.13. (V. Caselles.) *If the positive operator T is σ-order continuous and band irreducible, if S is an operator dominating some σ-order continuous compact operator, and if either $ST \leq TS$ of $TS \leq ST$, then $r(T) > 0$ and $r(S) > 0$.*

Let us finally show how the classical Andô-Krieger theorem follows from the results obtained in this section. As mentioned in the introduction, the kernel operators are the operators in $L^r(E)$ generated by the finite rank order continuous operators. Every kernel operator is σ-order continuous. By their very definition, every positive non-zero kernel operator T dominates a non-zero positive operator U which belongs to the ideal generated by the finite rank operators. Hence $U \in [0, S]$ for some positive finite rank (and hence compact) operator S. Now apply 3.12(4) to obtain

THEOREM 3.14. (T. Andô, H.J. Krieger). *If T is a band irreducible positive kernel operator on a Dedekind complete Banach lattice, then $r(T) > 0$.*

The results of V. Caselles on Harris operators, i.e., positive operators on Dedekind complete Banach lattices such that some power of it is not disjoint from the band of kernel operators (see [14]), follow similarly.

For an extension of the Andô-Krieger theorem to the more general setting of ordered Banach spaces we refer to a paper of V. Caselles [17].

4. The peripheral spectrum. The peripheral spectrum $\sigma_r(T)$ of an operator $T : E \to E$ is defined as

$$\sigma_r(T) := \sigma(T) \cap \{z \in \mathbf{C} : |z| = r(T)\}.$$

For a positive operator $T : E \to E$ we always have $r(T) \in \sigma_r(T)$. This important fact follows immediately from the inequality

(1) $|R(\lambda, T)x| \leq R(|\lambda|, T)|x|$ for all $\lambda, |\lambda| > r(T)$.

Moreover, if $r(T)$ is a pole of the resolvent then $r(T)$ is an eigenvalue of T to which pertains a positive eigenvector, i.e., $r(T)$ is a *distinguished eigenvalue* of T. This is the well known theorem of Krein-Rutman [39]. The proof follows from the observation that if $r = r(T)$ is a pole of the resolvent of order p, then the operator

$$B_p := \lim_{\lambda \downarrow r} (\lambda - r)^p R(\lambda, T) > 0,$$

and so if $x > 0$ is such that $u = B_p x > 0$, then u is easily seen to be a positive eigenvector of T pertaining to r.

Now if α is another pole of the resolvent of T in the peripheral spectrum, then for any $t > 1$ the inequality

$$|(t\alpha - \alpha)^k R(t\alpha, T)| \leq (tr - r)^k R(tr, T)$$

shows that r is a pole of maximal order on the spectral circle.

The most general well known class of operators for which all non-zero elements in the spectrum are poles, is the class of asymptotically quasi-compact operators also called *Riesz operators*. (see [19]). This class includes the class of operators which has some compact iterate.

In the case of an AM-space with unit, it is also known that $r(T)$ is an eigenvalue of T' with a positive eigenfunction (see [44]). To summarize, we have

THEOREM 4.1. (M.G. Krein, M.A. Rutman.) *Let T be a positive operator on a Banach lattice E with spectral radius $r = r(T)$. Then $r \in \sigma_r(T)$. If r is a pole of the resolvent it is a distinguished eigenvalue of T and a pole of maximal order on the spectral circle. This holds in particular if T is a Riesz operator. If E is an AM-space with unit, it is a distinguished eigenvalue of T'.*

A subset $S \subset \mathbf{C}$ is called *cyclic* if, for all $\lambda = |\lambda|\gamma \in S$, $|\lambda|\gamma^k \in S$ for all $k \in \mathbf{Z}$. Since a positive square matrix has cyclic peripheral spectrum, it is natural to seek for conditions for this to hold in the infinite dimensional case. M.G. Krein and M.A. Rutman [39] showed that a positive compact operator T on a Banach lattice satisfying $r(T) = \|T\|$ has a cyclic peripheral spectrum. T. Andô [6] and S. Karlin [36] extended the result to more general classes or compact operators on Banach lattices. The compactness condition was removed for the first time by G.-C. Rota [48] who proved the result of Krein and Rutman in $L^1(\mu)$. By introducing the notion of irreducibility, H.H. Schaefer [49] proved that irreducible positive operators in $C(X)$ has cyclic peripheral spectrum, and for compact irreducible positive operators in general Banach lattices it was shown by F.F. Bonsall and B.J. Tomiuk [13]. The most general result is due to H.P. Lotz [40] who introduced the notion of a (G)-solvable operator. Due to other considerations we shall call it an *Abel solvable operator*. To motivate the introduction of this notion, we recall some facts from general spectral theory. If J is a closed T-invariant linear subspace of E we will denote in the sequel the restriction of T to J by $T|_J$ and the operator induced by T in the quotient E/J by T_J or, if the context is clear, by $[T]$. A few remarks concerning the relation between the spectra of these operators will be useful.

LEMMA 4.2. *Let E be a Banach space and J a closed T-invariant linear subspace, $T \in L(E)$. Then*

(i) $\rho_\infty(T) \subset \rho(T|_J) \cap \rho(T_J) \subset \rho(T)$, *where $\rho_\infty(T)$ denotes the the unbounded component of the resolvent set $\rho(T)$;*

(ii) $\sigma(T) \subset \sigma(T|_J) \cup \sigma(T_J)$;

(iii) $\sigma_A(T|_J) \subset \sigma_A(T)$;

(iv) *if $r(T_J) = r(T)$, then $\sigma_r(T_J) \subset \sigma(T)$.*

PROOF: (i) Note that if $\lambda \in \rho(T)$ and if $R(\lambda, T)$ maps J into J, then the operators $R(\lambda, T)|_J$ and $R(\lambda, T)_J$ are well defined and $R(\lambda, T)|_J = R(\lambda, T|_J)$ and $R(\lambda, T)_J = R(\lambda, T_J)$. The set $S := \{\lambda \in \rho(T) : R(\lambda, T) \text{ maps } J \text{ into } J\}$ is non-empty (any $\lambda > r(T)$ belongs to S by the Neumann expansion of $R(\lambda, T)$), is closed in $\rho(T)$ (since the map $\lambda \to R(\lambda, T)$ is continuous) and is open in $\rho(T)$ (using the expansion $R(\lambda, T) = \sum_0^\infty (\lambda - \lambda_0) R(\lambda_0, T)$ which holds in a neigbourhood of λ_0.) It follows that $\rho_\infty(T) \subset S$ and the first inclusion is proved.

To prove the second inclusion, let $\lambda \in \rho(T|_J) \cap \rho(T_J)$. If $(\lambda - T)x = 0$ we use firstly $\lambda \in \rho(T_J)$ to conclude that $x \in J$ and secondly $\lambda \in \rho(T|_J)$ to conclude that $x = 0$. Hence, $\lambda - T$ is injective. To see that it is onto, let $y \in E$, and let $x \in E$ be such that $(\lambda - T_J)[x] = [y]$. Then $(\lambda - T)x - y \in J$ and so there exists some $z \in J$ such that

$(\lambda - T|_J)z = (\lambda - T)x - y$. This shows that $(\lambda - T)(x - z) = y$, and the proof of (i) is complete.

(ii) is obtained from the second inclusion in (i) and (iii) follows immediately from the definition.

(iv) Let $t > 1$ and let $\lambda \in \sigma_r(T_J)$. Then $t\lambda \in \rho_\infty(T) \subset \rho(T)$ and $\|R(t\lambda, T_J)\| \leq \|R(t\lambda, T)\|$. If $t \downarrow 1$, the first term in this inequality tends to infinity therefore also the second term and we get the required result.

We refer the reader to [40] or to [51] for an elegant proof of the next result.

LEMMA 4.3. *Let E be a Banach lattice, T a positive operator on E and F be a closed T-invariant ideal in E. If λ_0 is a pole of order k_1 of $R(\lambda, T_F)$ and a pole of order k_2 of $R(\lambda, T|_F)$, then $R(\lambda, T)$ has a pole of order k in λ_0 with $\sup(k_1, k_2) \leq k \leq k_1 + k_2$.*

The next lemma was also used systematically in [34] to obtain a decomposition of a positive operator in a Frobenius normal form.

LEMMA 4.4. *Let T be a positive operator on a Banach lattice E and let $r = r(T)$ be a pole of order n of $R(\lambda, T)$. Then there exists a chain $\{0\} = J_0 \subset J_1 \subset \ldots \subset J_{n+1} = E$ of closed T-invariant ideals of E such that for $j = 1, 2, \ldots, n$, $\lambda = r$ is a pole of order $j - 1$ of $R(\lambda, T|_j)$ and a pole of order one of $R(\lambda, T_j)$. ($T|_j$ denoting the restriction of T to J_j and T_j denoting the operator induced by T in J_{j+1}/J_j.)*

PROOF: If Q_{-n} is the coefficient operator of $(\lambda - r)^{-n}$ in the Laurent expansion of the resolvent of T at r, then $Q_{-n} \geq 0$, the absolute kernel $J_n := \{x \in E : Q_{-n}|x| = 0\}$ of Q_{-n} is a closed T-invariant ideal of E and the resolvent of $T|_n$ has a pole of order $\leq (n - 1)$ in $\lambda = r$. Since in general $Q_{-n}Q_{-j} = 0$ for all $j > 1$, we claim that $R(\lambda, T_n)$ has a pole of order at most one in r : If on the contrary the order of the pole were $k > 1$ then again $[Q_{-k}] = \lim_{\lambda \downarrow r} (\lambda - r)^k R(\lambda, T_n) > 0$, and so for every $x \in E_+$ there exists $y \in J_n$ such that $Q_{-k}x - y \geq 0$. Hence,

$$Q_{-n}|Q_{-k}x| = Q_{-n}|Q_{-k}x - y + y| \leq Q_{-n}(Q_{-k}x - y) + Q_{-n}|y| = 0,$$

contradicting $[Q_{-k}] > 0$, and the claim is established. The preceding lemma therefore implies that r is a pole of order $n - 1$ of $R(\lambda, T|_n)$, and a pole of order 1 of $R(\lambda, T_n)$. The proof follows now by induction.

This reduction has the following bearing on our problem:

LEMMA 4.5. *Let $T \geq 0$ be a positive operator on the Banach lattice E and let $0 = J_0 \subset J_1 \subset \ldots \subset J_{n+1} = E$ be a chain of T-invariant closed ideals in E such that for $j = 0, 1, \ldots, n$ the peripheral spectra $\sigma_r(T_j)$ are cyclic. Then $\sigma_r(T)$ is cyclic.*

PROOF: Let $\alpha \in \sigma_r(T)$ and assume without loss of generality that $r(T) = 1$. Then

$$(A) \qquad\qquad \alpha \in \sigma(T|_n) \cup \sigma(T_n)$$

by Lemma 4.2(ii). If $\alpha \in \sigma(T_n)$ it follows from $r(T_n) \leq r(T)$ and $|\alpha| = 1$, that $\alpha \in \sigma_r(T_n)$ and so by hypothesis that $\alpha^k \in \sigma_r(T_n)$ for all $k \in \mathbf{Z}$. But $\sigma_r(T_n) \subset \sigma(T)$ (lemma 4.2(iv)) and we are done.

If, on the other hand, $\alpha \in \sigma(T|_n)$, we use $r(T|_n) \leq r(T)$ to conclude that $\alpha \in \sigma_r(T|_n)$, and we repeat step (A) with $T|_{n-1}$ and T_{n-1}. After $n - 1$ steps the final step follows by assumption, since $T_0 = T|_1$.

A positive operator T on a Banach lattice E is called *Abel bounded* (in the terminology of Lotz: satisfies the growth condition (G)) whenever the family of operators $(\lambda - r(T))R(\lambda, T)$ is uniformly bounded for $\lambda > r(T)$.

Clearly, T is Abel bounded whenever $r(T)$ is a simple pole of the resolvent $R(\lambda, T)$. It is also readily seen from the Neumann expansion of the resolvent that every power bounded operator T (i.e., $\|T^n\|$ is uniformly bounded for $n \in \mathbf{N}$) is Abel bounded and a result of E. Emilion [22] states that a positive operator T on a Banach lattice E is Abel bounded if and only if it is Cesàro bounded (i.e., the Cesàro sums $n^{-1} \sum_{k=0}^{n-1} T^k$ are uniformly bounded for $n \in \mathbf{N}$).

The positive operator T on the Banach lattice E is then called *Abel solvable* if there exists a chain $\{0\} \subset E_0 \subset E_1 \subset \ldots \subset E_n = E$ of closed T-invariant ideals such that the operator T_k induced by T on the quotient E_k/E_{k-1} is Abel bounded for each k, $1 \leq k \leq n$.

By the remarks above it is clear that any positive operator T which has a pole of order n is Abel solvable. Therefore, any Riesz operator (in particular any operator with some compact power) is Abel solvable. We can now state

THEOREM 4.6. (H.P. Lotz, 1968.) *Let E be a Banach lattice. Every Abel solvable positive operator on E has a cyclic peripheral spectrum.*

This beautiful and general result was proved in [40] and it poses a problem which has not been solved since 1968: there exists no example of a positive operator without cyclic peripheral spectrum.

The proof of Lotz's theorem have two ingredients which we will now discuss, since they are interesting in their own right. The first is an embedding procedure into an $\mathcal{F}$-product, due to S.K. Berberian [10] and applied by Lotz to Banach lattices. Let us denote by $l^\infty(E)$ the set of all norm bounded sequences $(x_n) \subset E$ with norm $\|(x_n)\| := \sup_n \|x_n\|$. For any filter $\mathcal{F}$ on $\mathbf{N}$ containing the filter of all subsets of $\mathbf{N}$ with finite complements, we denote by

$$c_{\mathcal{F}}(E) := \{(x_n) \in l^\infty(E) : \lim_{\mathcal{F}} \|x_n\| = 0\}.$$

The $\mathcal{F}$-product of E, denoted by $\widehat{E}$, is then defined to be the quotient

$$\widehat{E} := l^\infty(E)/c_{\mathcal{F}}(E).$$

The embedding $j : E \to \widehat{E}$, defined by $j(x) = \widehat{x} = (x, x, \ldots) + c_{\mathcal{F}}(E)$, is an isometry into. If we define for $T \in L(E)$ the operator $\widehat{T}$ by

$$\widehat{T}\bar{x} := (Tx_n) + c_{\mathcal{F}}(E), \quad \bar{x} = (x_n) + c_{\mathcal{F}}(E),$$

the mapping $T \to \widehat{T}$ is an isometric homomorphism of the Banach algebra $L(E)$ into the Banach algebra $L(\widehat{E})$ satisfying
$\sigma(T) = \sigma(\widehat{T}), \sigma_A(T) = \sigma_A(\widehat{T}) = \sigma_P(\widehat{T})$ and $R(\lambda, T)\widehat{} = R(\lambda, \widehat{T})$, for all $\lambda \in \rho(T)$.

The proofs of all these assertions are elementary and can be found in [51] and in [44]. Pertinent to our problem is the fact that all points of the peripheral spectrum of T are eigenvalues of $\widehat{T}$. Therefore, in order to show that $\sigma_r(T)$ is cyclic, one may assume without loss of generality that $\sigma_r(T) \subset \sigma_P(T)$.

The second ingredient involves the fact that any AM-space with unit e can be endowed with a multiplicative structure which turns it into an f-algebra with unit e. This follows from the Kakutani-Krein representation theorem which states that every AM-space with unit is isometrically order isomorphic to a space $C(K)$ for some compact Hausdorff space K. It can, however, also be proved without representation theory (see [32] after proposition 4.1). Since every principal ideal in a Banach lattice is an AM-space with unit, f-algebras play a significant rôle in the theory of Banach lattices. We now have an opportunity to illustrate this.

LEMMA 4.7. *Let E be a Banach lattice and let $u \in E$. Let A be the f-algebra and principal ideal $E_{|u|}$. If $0 \le \phi \in A'$ has the property that $\phi(|u|) = 1 = |\phi(u)|$, then, for every $s \in A$,*

$$\phi(us) = \phi(u)\phi(s).$$

In particular, $\phi(u^k) = \phi(u)^k$ for all $k \in \mathbf{Z}$.

PROOF: Let $\phi(u) = \exp(i\alpha)$, $u_1 := \exp(-i\alpha)u = x + iy$, with x, y real elements in A. Then $\phi(u_1) = 1$ and $\phi(|u_1|) = \phi(|u|) = 1$. Hence, $\phi(|u_1| - u_1) = 0$, which implies that $\phi(|u_1| - x) = \phi(y) = 0$. From this we get $\phi(|u_1|) \ge \phi(|x|) \ge |\phi(x)| = \phi(|u_1|)$. Also, since $|u_1| = \sup\{|x|\cos\beta + |y|\sin\beta : 0 \le \beta \le \pi/2\}$ we have for every such β that $\sin\beta\phi(|y|) \le \phi(|u_1| - |x|\cos\beta) \le (1 - \cos\beta)\phi(|u_1|)$. Now $\tan\frac{\beta}{2} \to 0$ as $\beta \to 0$ and so $\phi(|y|) = 0$. If $s \in A, |s| \le k|u|$, we get $0 \le |\phi(sy)| \le \phi(|sy|) \le k\phi(|u||y|) = \phi(|y|) = 0$ and $|\phi(s|u_1| - sx)| \le k\phi(|u_1| - x) = 0$. Hence, $\phi(su_1) = \phi(sx) + i\phi(sy) = \phi(s|u_1|) = \phi(s)$ and this finally yields $\phi(su) = \exp(i\alpha)\phi(su_1) = \phi(u)\phi(s)$.

PROPOSITION 4.8. *Let E be an AM-space with unit e and let $T \ge 0$ be a Markov operator on E, i.e., $Te = e$. Let $\alpha, |\alpha| = 1$, be an eigenvalue of T with unimodular eigenvector u (i.e., $|u| = e$.) Then α^k is an eigenvalue of T with unimodular eigenfunction u^k, the multiplication taken in the f-algebra E.*

PROOF: Let δ be a Riesz homomorphism on E satisfying $\delta(e) = 1$. Then δ is also multiplicative on the f-algebra E. Define $\phi_{(T,\delta)}(x) := \delta(Tx)$. Then $\phi_{(T,\delta)}$ is positive and $|\phi_{(T,\delta)}(u)| = \phi_{(T,\delta)}(|u|) = 1$. By the above lemma,

$$\delta(Tu^k) = \phi_{(T,\delta)}(u^k) = (\phi_{(T,\delta)}(u))^k = (\delta(\alpha u)^k = \delta(\alpha^k u^k).$$

Since the set of norm one Riesz homomorphisms separates the points of E, we arrive at the result that $Tu^k = \alpha^k u^k$ which is the required result.

Everything needed for the proof of 4.6 is now available.

PROOF OF 4.6: As remarked before, we need only proof the theorem for an Abel bounded positive linear T. Let $\alpha \in \sigma_r(T)$ and assume that $r(T) = 1$. By the embedding procedure we may assume that α is an eigenvalue of T, so suppose that $z \neq 0$ is an eigenvector pertaining to α and put $u = |z| = |\alpha z| = |Tz| \leq T|z| = Tu$. Then for all n, $T^n u \geq u$ and we get for $\lambda > 1$,

$$(\lambda - 1)\|R(\lambda, T)u\| = \|(\lambda - 1)(\sum_{n=0}^{\infty} \lambda^{-(n+1)}T^n u)\| \geq \|(\lambda - 1)(\sum_{n=0}^{\infty} \lambda^{-(n+1)}u)\| = \|u\| > 0.$$

Set

$$p(y) = \limsup_{\lambda \downarrow 1} (\lambda - 1)\|R(\lambda, T)|y|\|.$$

Then p is a continuous lattice seminorm on E and $p(u) \geq \|u\| > 0$. Also,

$$p(Tu - u) = \limsup_{\lambda \downarrow 1} (\lambda - 1)\|R(\lambda, T)(T - \lambda)u + (\lambda - 1)u\| = 0,$$

and

$$p(Ty) = \limsup_{\lambda \downarrow 1} (\lambda - 1)\|R(\lambda, T)|Ty|\|$$
$$\leq \limsup_{\lambda \downarrow 1} (\lambda - 1)\|\lambda R(\lambda, T)|y| - |y|\|$$
$$= p(y).$$

Therefore, $J := p^{-1}(0)$ is a closed T-invariant ideal, $u \notin J$, and $Tu - u \in J$. If we therefore consider $F := E/J$ and the induced operator T_J, it follows that α is an eigenvalue of T_J. From $T_J[u] = [u] > 0$ one gets that the restriction $T_J|F_{[u]}$ of T_J to principal ideal generated by $[u]$ in F is a Markov operator S on the AM-space $F_{[u]}$ and α is an eigenvalue of S. Hence, by the preceding lemma α^k is an eigenvalue of S and therefore of T_J for $k \in \mathbf{Z}$. But then, by lemma 4.2(iv), $\alpha^k \in \sigma_r(T_J) \subset \sigma(T)$. This shows that $\sigma_r(T)$ is cyclic and the proof is complete.

Remark. Our proof is the same as the original proof of Lotz [40] and is also to be found in the book [44] of P. Meyer-Nieberg. The only difference lies in the proof of lemma 4.6 and proposition 4.7 in which we avoided the use of integration theory.

Using a result of R. Emilion [22], the following result of Lotz ([40], Satz 4.12) can be slightly improved. We recall that a bounded operator T on a Banach space is called *uniformly ergodic* whenever the Cesàro means $s_n := n^{-1}(I + T + T^2 + \ldots + T^{n-1})$ converges in the uniform operator norm.

THEOREM 4.9. (H.P. Lotz.) *Let T be a positive operator on a Banach lattice E and suppose that $r(T) \geq 1$. If T is uniformly ergodic then 1 is an eigenvalue of T and an isolated point of the spectrum. Moreover, the spectrum of T is a finite union of groups of roots of 1.*

PROOF: Since $T \geq 0$ is uniformly ergodic and therefore also Cesàro bounded it follows from $T^n \leq (n+1)s_{n+1}$ that $\|T^n\| \leq (n+1)M$ and consequently that $r(T) = 1$. Hence, $1 \in \sigma(T)$ and in fact an isolated point of the spectrum (see the first half of the proof of theorem 2.7 in [37]). Now, by the result of Emilion, T is Abel bounded. The theorem of Lotz therefore implies that $\sigma_r(T)$ consists of roots of 1. To proof that 1 is an eigenvalue, note that it is an eigenvalue of $\widehat{T}$ in an $\mathcal{F}$-product of E. If $P = \lim s_n$ denotes the projection of E onto the fixed space $F = \{z \in E : Tz = z\}$ then $\widehat{P}$, being the corresponding projection in the $\mathcal{F}$-product, is non-zero. Therefore P is non-zero and the proof is complete.

In [28] it was shown that these results can also be proved in a Banach lattice algebra. If E is a Banach lattice, $L^r(E)$, the set of all regular operators, is a Banach algebra, but need not be a lattice. The spectrum of an element in this Banach algebra is called its *order spectrum* and is denoted by $\sigma_o(T)$. One may therefore also consider the question if the peripheral order spectrum of an element is cyclic. Now if E is Dedekind complete, $L^r(E)$ is also a vector lattice, and so a Banach lattice algebra. Following an idea of A.R. Schep [55], A. W. Wickstead [57] showed that if E is Dedekind complete, one can embed the algebra $L^r(E)$ into a full subalgebra of bounded linear operators on a Banach lattice, therefore deriving its spectral properties from the standard results on bounded linear operators. Hence, in the Dedekind complete case the solution of this problem follows if it can be solved in the case considered so far.

Suppose that we have a summability method (P) for a series in the sense of G. H. Hardy [29]. Following S.J. Bernau and C.B. Huijsmans, [11], an operator T on the

Banach lattice E will be called (P)-bounded if the P-method applied to the series with n-th term $T^n - T^{n-1}$ produces sequences of partial sums which are uniformly norm bounded. If the method is the Abel or the Cesàro method, the operator is in the previous terminology Abel or Cesàro bounded. One may therefore consider other boundedness conditions on T and it is still an open problem whether a more general theorem can then be proven.

We recall that that the *essential spectrum* $\sigma_{\text{ess}}(T)$ of an operator T is the set of all complex λ such that $\lambda - T$ is not a Fredholm operator. Another direction in which these results were generalized is given in a theorem by V. Caselles [16] in which he proved similar cyclicity results for the essential spectrum of dual operators on certain Banach lattices. From this he deduces the interesting result that in this case $0 \leq S \leq T$ with T a Riesz operator implies that S is a Riesz operator. The essential spectrum was also studied by A.R. Schep and B. de Pagter in [21] via measures of non-compactness. They prove inter alia that a positive AM-compact operator T one has $r_{\text{ess}}(T) \in \sigma_{\text{ess}}(T)$. We shall not pursue this study of the essential spectrum further, but the reader may consult the papers referred to as well as papers by L. Weiss and M. Wolf [56], C.B. Huijsmans and B. de Pagter [33], F. Andreu, V. Caselles, J. Martínez and J.M. Mazón [7] and J. Martínez [43].

We saw in the proof of 4.8 that if $0 \leq T$ has a cyclic peripheral spectrum and if 1 is an isolated singularity point of the spectrum, then the peripheral spectrum consists of a finite number of roots of 1. If 1 is moreover a pole of the resolvent, one may conjecture that it may have other effects on the peripheral spectrum. The theorem of F. Niiro and I. Sawashima [45] gives a satisfactory answer to this question. See also [51] and [41]. We shall prove the theorem also for the class of band irreducible and σ-order continuous operators on σ-Dedekind complete Banach lattices.

THEOREM 4.10. (F. Niiro and I. Sawashima.) *Let E be a Banach lattice and $0 \leq T$ be an operator on E for which one of the following conditions is satisfied.*

(i) *T is irreducible.*

(ii) *T is σ-order continuous, band irreducible and majorized by a weakly compact operator.*

If $r(T) = 1$, and if the resolvent $R(\lambda, T)$ has a pole in $\lambda = 1$ then the entire peripheral spectrum consists of first order poles of $R(\lambda, T)$. Moreover, the peripheral spectrum is a subgroup of the circle group consisting entirely of roots of unity.

There is a number of ingredients of the proof which we first have to obtain. Firstly a proposition which goes back to H. Wielandt [58]. In the form we present it here it was proved by H. Lotz and various other authors for special cases (see [27], [44], [60]).

PROPOSITION 4.11. (H. Wielandt.) *Let E be a Banach lattice and let $T \geq 0$. Let $x' \in E'_+$ be such that $T'x' = x'$ and $N(x') = \{0\}$. Suppose that one of the following conditions is satisfied.*

(i) *T is irreducible.*

(ii) *E is σ-Dedekind complete and T is band irreducible and σ-order continuous.*

If λ, $|\lambda| = r(T) = 1$, is an eigenvalue of T, then there exists a bijective orthomorphism U on E satisfying $\lambda T = U^{-1}TU$.

PROOF: Let u be a normalized eigenvector of T pertaining to λ. Then $|u| \leq T|u|$ and since x' is strictly positive and $T'x' = x'$, it follows that $T|u| = |u|$. The principle ideal $E_{|u|}$ generated by $|u|$ in E is, as remarked before, an f-algebra and we define on this principle ideal $Ux := ux$, with multiplication in the f-algebra. Our assumptions then simply implies that we can extend U to E and it is easily seen that U has the required properties (see [27], or [44]).

From 4.11 the next result is obtained, a result which not only contains all the ingredients for the Jentzsch-Perron and Frobenius theorems, but also the necessary facts for proving 4.10 (see also [51], [44] and [40]).

THEOREM 4.12. *Let T be a positive operator on the Banach lattice E with $r(T) = 1$ and non-void point spectrum. Suppose there exists $\phi \in E'_+$ such that $T'\phi = \phi$ and $N(\phi) = \{0\}$. If either E is σ-Dedekind complete, T band irreducible and σ-order continuous, or T is irreducible, then*

(i) *1 is an eigenvalue of T and the eigenspace pertaining to 1 is a one-dimensional sub-lattice of E spanned by a positive normalized weak order unit u; if T is irreducible, u is a quasi-interior point of E;*

(ii) *every peripheral eigenvalue α of T with normalized eigenvector x is of algebraic multiplicity one, $|x| = u$ and $\sigma(T) = \alpha\sigma(T)$;*

(iii) *the peripheral point spectrum is a subgroup of the circle group;*

(iv) *1 is the unique eigenvalue of T with a positive eigenvector.*

PROOF: (i) Let $\alpha, |\alpha| = 1$ be an eigenvalue of T with normalized eigenvector x. Then $|x| = |\alpha x| = |Tx| \le T|x|$, and consequently,

$$0 \le \phi(T|x| - |x|) = \phi(T|x|) - \phi(|x|) = T'\phi(|x|) - \phi(|x|) = 0.$$

The strict positivity of ϕ then yields $T|x| = |x|$. It follows that 1 is an eigenvalue of T and $|x|$ is a weak order unit of the eigenspace of 1. An application of this argument to any other eigenvector pertaining to 1 shows that the eigenspace of T belonging to 1 is a Riesz subspace of E in which every positive element is a weak order unit. The real part of this eigenspace is therefore linearly ordered and consequently one dimensional. If T is irreducible $|x|$ is a quasi-interior point and (i) is proved.

(ii) Again, let α, $|\alpha| = 1$, be an eigenvalue of T. Let U be the bijective orthomorphism found in the preceding theorem, satisfying $\alpha T = U^{-1}TU$. Since $\sigma(U^{-1}TU) = \sigma(T)$, $\sigma(\alpha T) = \sigma(T)$, and hence $\sigma(T) = \alpha\sigma(T)$. By what has been proved in (i) we only have to show that α has algebraic multiplicity 1. If $(I - T)^{n+1}x = 0$ then $(I - T)(I - T)^n x = 0$, i.e $(I - T)^n x = \lambda u$ and since $T'\phi = \phi$, $\phi(I - T)^n x = 0$. But this means, since $\phi(u) > 0$, that $(I - T)^n x = 0$. This holds for $n = 1, 2, \ldots$, and so the index of 1 for T is 1. This, together with the fact that the eigenspace of 1 is one dimensional, shows that 1 has algebraic multiplicity 1. From $(\alpha - T)^k = \alpha^k U(I - T)^k)U^{-1}$ it follows that the algebraic multiplicity of α is 1 as well.

(iii) If α and β are peripheral eigenvalues, and if U and V are orthomorphisms satisfying $\alpha T = U^{-1}TU, \beta T = V^{-1}TV$, and if w is any non-zero eigenvector of T pertaining to 1, $UV^{-1}w$ is an eigenvector pertaining to $\alpha\beta^{-1}$.

(iv) Let $Tv = \alpha v$ with $v \ge 0$. Applying ϕ to this equation yields $\alpha\phi(v) = T'\phi(v) = \phi(v) > 0$, and the proof is complete.

We deduce from this a weakened version of the Niiro-Sawashima result, due to H.P. Lotz [40], for the case of irreducible operators.

THEOREM 4.13. *Let T be a positive operator on the Banach lattice E with $r(T) = 1$. If either*

(i) *T is irreducible or*

(ii) E is σ-Dedekind complete and separated by E_o', T is σ-order continuous and band irreducible,

then, if 1 is a pole of $R(\lambda, T)$, all peripheral eigenvalues of T are poles of order one of $R(\lambda, T)$ and (i)–(iv) of theorem 4.12 hold. In particular, the peripheral point spectrum consists of a group of roots of unity.

PROOF: We shall only show how the result follows from the hypotheses in (ii). The other part of the proof is easier and can be found in [40].

Suppose that 1 is a pole of order k and set $Q = \lim_{\lambda \downarrow 1}(\lambda - 1)^k R(\lambda, T)$. Then $Q > 0$ and there exists $x > 0$ such that $u = Qx > 0$. It follows that $Tu = u$ and so 1 is an eigenvalue of T; the peripheral point spectrum of T is therefore non-void. Let $0 < x' \in E_o'$ be chosen so that $x'(u) > 0$. If $\phi := Q'x' = \lim_{\lambda \downarrow 1}(\lambda - 1)^k R(\lambda, T')x'$, one quickly verifies that $T'\phi = \phi$ and that $\phi(x) = x'(u) > 0$. We show that ϕ is strictly positive. It is not difficult to see that under the assumptions stated in (ii) Q is a σ-order continuous operator (see [27], lemma 4). Hence, $\phi(u_n) = Q'x'(u_n) = x'(Qu_n) \downarrow 0$, if $u_n \downarrow 0$, implying that $\phi \in E_0'$. If the absolute null-space $N(\phi) \neq \{0\}$, then $N(\phi)$ is a closed T-invariant ideal which contains a weak order unit by proposition 3.8. The σ-order continuity of ϕ implies then that it is zero, a contradiction. The conditions of theorem 4.12 are therefore fulfilled. Since 1 is an isolated singularity, it is clear that the group mentioned in (iii) must be a group of roots of unity.

Now let $F := \{z : Q|z| = 0\}$, then F is again a closed T-invariant ideal, which contains a weak order unit if it is non-zero. By the σ-order continuity of Q this would yield $Q = 0$. Hence, $F = \{0\}$. This implies that $Q^2 \neq 0$, and this is only possible if Q is the residue operator, i.e., if $k = 1$.

Suppose now that α is an eigenvalue of T with $|\alpha| = 1$. From 4.12 we get $\sigma(T) = \alpha\sigma(T)$ which implies that together with 1, α is also an isolated singularity. Moreover, if U is such that $\alpha T = U^{-1}TU$, it easily follows that $\bar{\alpha}U^{-1}R(\lambda, T)U = R(\alpha\lambda, T)$. Thus, $R(\alpha\lambda, T)$ has a pole of order 1 in $\lambda = 1$, and a simple substitution yields the result that $R(\lambda, T)$ has a pole of order 1 in $\lambda = \alpha$. This completes the proof.

PROOF OF THE THEOREM OF NIIRO AND SAWASHIMA: Suppose first that T is irreducible. Let $\mathcal{U}$ be an ultra filter on $\mathbf{N}$ and denote by $\hat{E}$ the $\mathcal{U}$-product of E. Let 1 be a pole of

$R(\lambda, T)$, then 1 is an eigenvalue with one-dimensional eigenspace spanned by a positive quasi-interior vector u. Let $P = \phi \otimes u$ be the residue in $\lambda = 1$ of $R(\lambda, T)$. Define the continuous linear functional $\widehat{\phi}$ on $\widehat{E}$ by $\widehat{\phi}(\widehat{x}) = \lim_{\mathcal{U}} \phi(x_n)$ whenever $\widehat{x} = (x_n) + c_{\mathcal{U}}(E)$. It is not difficult to check that $\widehat{P} = \widehat{\phi} \otimes \widehat{u}$. Let $R := \{\widehat{x} : \widehat{P}|\widehat{x}| = 0\} = N(\widehat{\phi})$ then R is a $\widehat{T}$-invariant closed ideal and we set $\widetilde{E} := \widehat{E}/R$. We then denote by $\widetilde{T}$ the induced operator $\widehat{T}_R$ on $\widetilde{E}$ and we set $\widetilde{x} = [\widehat{x}]$. Now $\widetilde{T}\widetilde{u} = \widetilde{u} \geq 0$ and so the closed ideal F generated by $\widetilde{u}$ is again $\widetilde{T}$-invariant. Denote finally by S the restriction of $\widetilde{T}$ to F. We have now constructed the following sequence of spaces with corresponding operators

$$
\begin{array}{ccccc}
E & \longrightarrow & \widehat{E} & \longrightarrow & \widetilde{E} & \longrightarrow & F \\
T & \longrightarrow & \widehat{T} & \longrightarrow & \widetilde{T} & \longrightarrow & S.
\end{array}
$$

Let us now consider what happens to the pole 1 of $R(\lambda, T)$:

Since $\sigma(T) = \sigma(\widehat{T})$, 1 remains an isolated singularity of $R(\lambda, \widehat{T})$, and since this resolvent equals $R(\lambda, T)\widehat{}$, we see that 1 is a pole of order 1 with residue operator $\widehat{P} = \widehat{\phi} \otimes \widehat{u}$. (Note that since $\widehat{P}^2 = \widehat{P}, \widehat{\phi}(\widehat{u}) = 1$.)

Since 1 is an isolated singularity of $R(\lambda, \widehat{T})$, and since $r(T) = 1$, there exists a reduced neighbourhood of 1 which belongs to $\rho_\infty(\widehat{T}) \subset \rho(\widehat{T}|_R) \cap \rho(\widetilde{T})$. Therefore 1 is an isolated singularity of $R(\lambda, \widetilde{T})$ and $R(\lambda, \widetilde{T}) = \widetilde{R(\lambda, \widehat{T})}$ for all λ in this neighbourhood. The residue operator of $R(\lambda, \widetilde{T})$ is therefore the operator $\widetilde{P}$ which, if we define $\widetilde{\phi}$ on $\widetilde{E}$ (as we may) by $\widetilde{\phi}(\widetilde{x}) = \widehat{\phi}(\widehat{x})$, is easily seen to be $\widetilde{\phi} \otimes \widetilde{u} \neq 0$. Therefore, 1 is a pole of order one of $R(\lambda, \widetilde{T})$.

Finally, by the same argument as above, 1 is an isolated singularity of $S = \widetilde{T}|_F$ and the residue of $R(\lambda, S)$ in 1 is $\widetilde{P}|_F \neq 0$ (apply it to $\widetilde{u}$.) Consequently, 1 is a pole of order one of $R(\lambda, S)$ in 1.

Now, although we may have lost irreducibility going from T to $\widehat{T}$, we have regained it. We claim that S is irreducible. In order to prove it, let J be a closed S-invariant ideal in F. Since $\widetilde{P} = \lim_{\lambda \downarrow 1}(\lambda - 1)R(\lambda, \widetilde{T})$, and since by the Neumann expansion J is also $R(\lambda, \widetilde{T})$-invariant for $\lambda > 1$, J is also $\widetilde{P}|_F$-invariant. Now if $0 \neq x \in G$, $\widetilde{P}|x| = (\widetilde{\phi} \otimes \widetilde{u})|x| = \widetilde{\phi}(|x|)\widetilde{u}$ which implies by the strict positivity of $\widetilde{\phi}$ that $\widetilde{u} \in G$. But F is the closed ideal generated by $\widetilde{u}$ and so $G = F$.

Let $\alpha \in \sigma(T), |\alpha| = 1$. Then α is an eigenvalue of $\widehat{T}$ and so $\widehat{T}\widehat{x} = \alpha\widehat{x}$ for some $0 \neq \widehat{x} \in \widehat{E}$. Then $\widehat{T}|\widehat{x}| \geq |\widehat{x}| > 0$ and also $\widehat{P}|\widehat{x}| \geq |\widehat{x}| > 0$. Hence, $\widehat{x} \notin R$ which implies that

α is an eigenvalue of $\widetilde{T}$ with $\widetilde{T}\widetilde{x} = \alpha\widetilde{x}$. Using the fact that $\widetilde{P} = \lim_{\lambda \to 1}(\lambda - 1)R(\lambda, \widetilde{T})$, we get $\widetilde{P}\widetilde{T} = \widetilde{P}$ and deduce that $(\widetilde{T})'\widetilde{\phi} = \widetilde{\phi}$. From $\widetilde{T}|\widetilde{x}| \geq |\widetilde{x}|$ and the strict positivity of $\widetilde{\phi}$ we arrive at $\widetilde{T}|\widetilde{x}| = |\widetilde{x}|$, which shows that $|\widetilde{x}|$ is in the range of $\widetilde{P} = \widetilde{\phi} \otimes \widetilde{u}$, i.e., $\widetilde{x} \in F$. Thus, α is an eigenvalue of S. We may now apply theorem 4.13 to S to conclude that α is a pole of order one of $R(\lambda, S)$.

We now reverse our steps to get back to E. $\widetilde{T}$ induced an operator $\widetilde{T}_F$ on $\widetilde{E}/F$ which can have a pole in $\lambda = 1$ of order at most 1. However, since $\widetilde{P}(E) \subset F$, the induced residue operator $\widetilde{P}_F = 0$ and we have that $r(\widetilde{T}|_F) < 1$. Consequently, α is in the resolvent set of $\widetilde{T}$. Using lemma 4.3 we conclude that α is a pole of order exactly 1 of $R(\lambda, \widetilde{T})$. Next, by construction the residue $\widehat{P}$ is zero on R and therefore $r(\widehat{T}|_R) < 1$; consequently, α is a pole of order one of $R(\lambda, \widehat{T})$ again by lemma 4.3. But this implies immediately that α is a pole of order one of $R(\lambda, T)$.

We have thus shown that there is a one-to-one correspondence between the elements of $\sigma_r(T)$ and the point spectrum of S. The latter is a subgroup of the circle group consisting entirely of roots of unity by 4.13. This completes the proof for the case of an irreducible T.

Suppose now that T is σ-order continuous, band irreducible and majorized by a σ-order continuous weakly compact operator. Let F be the closed ideal generated by $T[E]$. Then F is T-invariant and $T|_F$ is irreducible (compare proposition 3.7, see also [27] lemma 3). Moreover, T_F, the operator induced in the quotient E/F is zero. From the fact that $\sigma(T) \subset \sigma(T|_F) \cup \sigma(T_F)$, it therefore follows that 1 is an isolated singularity of $T|_F$ and in fact a pole. By the same argument, every $\alpha \in \sigma(T), |\alpha| = 1$ is in $\sigma(T|_F)$. Applying the result of the first part of the theorem, every such α is a pole of $R(\lambda, T|_F)$ and so by lemma 4.3 also of $R(\lambda, T)$. This completes the proof.

It is not known if this theorem remains true for band irreducible operators if the weak compactness condition is dropped.

Without the irreducibility condition above, another condition has to be added namely that the residuum P at 1 should be of finite rank and the conclusion is then also weaker, namely that $\sigma_r(T)$ consists entirely of poles of the resolvent.

If we call a point of the spectrum a *Riesz point* if it is a pole of the resolvent with finite rank residuum, a result of V. Caselles [16] shows that if S is dominated by T, if $r(T) = r(S)$ and if $r(T)$ is a Riesz point of $\sigma(T)$ then it is also a Riesz point of $\sigma(S)$.

There are various other cyclicity results for special types of operators. If T is a Riesz-homomorphism we have the following result, proved in its full generality by E. Scheffold [53] and for special cases by H.P. Lotz [40]. We refer to [44] or [51] for a proof.

THEOREM 4.14. *The spectrum of a lattice homomorphism T of a Banach lattice is cyclic. Moreover, every compact cyclic subset of $\mathbf{C}$ is the spectrum of some lattice homomorphism defined on a suitable Banach lattice.*

Another class of operators which have been investigated fairly extensively in connection with this problem is the class of disjointness preserving operators. In [9] W. Arendt and D. Hart showed that any aperiodic quasi-invertible disjointness preserving operator on a Dedekind complete Banach lattice has a rotationally invariant spectrum (for the relevant definitions see the paper). In [33] C.B. Huijsmans and B. de Pagter showed that in a Dedekind complete Banach lattice with lower semicontinuous norm any order continuous aperiodic disjointness preserving operator has a rotationally invariant peripheral spectrum; in fact, the peripheral spectrum of such an operator is the whole spectral circle. We refer the reader to this paper and the references given there for additional information. The best result is due to Y. Abramovich, E. Arenson and A. Kitover [3]. Let us denote by

$$\ell(x) = \inf_{(x_\alpha)} \sup_\alpha \|x_\alpha\|,$$

with the infimum taken over all nets $(x_\alpha) \subset E$ such that $x_\alpha \uparrow |x|$. In general ℓ is a Riesz semi-norm on E and it is a norm if for example E has the weak Fatou property for directed sets (see [60], theorem 107.5) or if the order continuous functionals separate the points of E. (The latter condition is met by every Banach function space; see [60], theorem 112.1.) The authors then prove

THEOREM 4.15. (Y. Abramovich, E. Arenson, A. Kitover.) *If ℓ is a norm on the Banach lattice E then the spectrum of any disjointness preserving operator on E with the property that all powers of the operator are mutually disjoint is rotation invariant.*

5. Theorems of Jentzsch-Perron and Frobenius. The analysis carried out in the last two sections, can now be put together in the presence of compactness conditions on the operator to yield the desired generalizations of the theorems of Jentzsch-Perron and Frobenius which was our goal. We first note a corollary to the proof of 4.13 which makes it clear when 4.12 may be applied.

PROPOSITION 5.1. *Let T be a positive operator on the Banach lattice E with $r(T) = 1$ and suppose that 1 is a pole of $R(\lambda, T)$. If either*

(i) *T is irreducible or*

(ii) *E is σ-Dedekind complete and separated by E_o', T is σ-order continuous and band irreducible,*

then the peripheral point spectrum of T is non-void and there exists a strictly positive linear functional ϕ on E satisfying $T'\phi = \phi$. If condition (ii) holds, then ϕ is σ-order continuous. If T^k is compact for some $k \in \mathbf{N}$, then the condition that E be separated by its order continuous linear functionals may be dropped from the hypothesis (ii).

PROOF: We need only look into the last claim. We use the same notation as in the proof of 4.13. Let $x' \in E_+'$ be such that $x'(u) > 0$. and define ϕ as in 4.13. Then $T'\phi = \phi$ and we need only show that ϕ is σ-order continuous. Let $u_n \downarrow 0$. The σ-order continuity if T implies that $T^k u_n \downarrow 0$ and then, since T^k is compact, $T^k u_n \to 0$ in norm. Hence, $\phi(u_n) = (T')^k \phi(u_n) = \phi(T^k u_n) \to 0$.

THEOREM 5.2. (Jentzsch-Perron.) *Let E be a Banach lattice and suppose that for $T \geq 0$, T^n is compact. If either*

(i) *T is irreducible or*

(ii) *T is σ-order continuous and band irreducible,*

then $r(T) > 0$ and $r(T)$ is an eigenvalue of T of algebraic multiplicity one. The eigenspace is spanned by a unique normalized weak order unit u. If T is irreducible, u is a quasi-interior point.

We conclude with the generalized Frobenius theorem. Note that we do not have the hypothesis that E be σ-Dedekind complete as in [27] and [44]. This is due to the compactness condition on the operator T.

THEOREM 5.3. (Frobenius) *Let E be a Banach lattice and let the positive operator T be either*

(i) *ideal irreducible or*

(ii) *σ-order continuous and band irreducible.*

If T^k is compact for some $k \in \mathbf{N}$, then $r(T) > 0$ and if $\lambda_1, \ldots, \lambda_m$ are the different eigenvalues of T satisfying $|\lambda_j| = r = r(T)$ for $j = 1, \ldots, m$, every λ_j is a root of the equation $\lambda^m - r^m = 0$. All these eigenvalues are of algebraic multiplicity one and the spectrum of T is invariant under a rotation of the complex plane by the angle $2\pi/m$, multiplicities included. Moreover, $r(T)$ is the only distinguished eigenvalue of T.

PROOF: It is clear from 5.1 and 4.12 that the theorem holds for T irreducible. Let T be σ-order continuous and band irreducible, and let F be the closed ideal generated in E by $T^k[E]$. F is then a closed T invariant ideal and we claim that $T|_F$ is irreducible: Let I be a closed T-invariant ideal in F. By 3.9 I contains a weak order unit. Therefore, if $0 < x \in E$, there exists a sequence $(x_n) \subset I$ such that $x_n \uparrow x$. But T^k is also σ-order continuous since T is, and so $T^k(x - x_n) \downarrow 0$. By the compactness of T^k, it follows that $T^k x_n \to T^k x$ in norm and since I is closed and T-invariant, $T^k x \in I$. Hence, $T^k[E] \subset I$, and by the definition of F, we get $I = F$. This establishes the claim. Denote now by T_F the operator induced by T in the quotient E/F. Clearly, $T_F^k = 0$ and therefore, $\sigma(T_F) = \{0\}$. Moreover, the compactness of T^k implies that $\rho(T) = \rho_\infty(T)$ and by 4.2, we get $\sigma(T) = \sigma(T|_F) \cup \sigma(T_F) = \sigma(T|_F) \cup \{0\}$. It is therefore clear that the spectrum of T appears exactly as the spectrum of the irreducible operator $T|_F$. We need only look at the question of the order of the poles and the multiplicities. Every $\alpha \neq 0$ in $\sigma(T|_F)$ is a pole of the resolvent $R(\lambda, T|_F)$ and also a pole of $R(\lambda, T)$ of the same order by 4.3. In order to see that the algebraic multiplicities of points in $\sigma(T)$ are the same as in $\sigma(T|_F)$, we need to look at the dimensions of the null-spaces $M_n := \{x \in E : (\alpha - T)^n x = 0, \alpha \neq 0\}$. We claim that $M_n \subset T^m[E]$ for all $m, n \in \mathbf{N}$: Fix m and note that $M_1 \subset T^m[E]$. If for some k, $M_{k-1} \subset T^m[E]$ and if $x \in M_k$, then $(\alpha - T)x \in M_{k-1} \subset T^m[E]$ and we infer that $x \in T[E] + T^m[E] \subset T[E]$ But then again, from $(\alpha - T)x \in M_{k-1} \subset T^m[E]$ we get $x \in T^2[E] + T^m[E] \subset T^2[E]$. After m steps we finally arrive at $x \in T^m[E] + T^m[E] \subset T^m[E]$. This establishes the claim. The final conclusion is therefore that the null spaces considered are exactly the same for T and for $T|_F$. Therefore, the algebraic multiplicity of any $\alpha \neq 0$ is the same in both spectra. Since the theorem holds for the irreducible $T|_F$, it therefore also holds for T.

6. Primitivity. A positive operator T is called *primitive* if $r(T)$ is the only eigenvalue on the spectral circle. In some applications this is an important property. Our first result extends a result of H. H. Schaefer [51].

THEOREM 6.1. *Let E be a Banach lattice and $T > 0$ be an operator on E. Suppose furthermore that there exists a positive linear functional $\phi \in E'$ such that $T'\phi = \phi$. Then T is primitive in each of the following cases.*

(i) *(Schaefer.) For each $x > 0$ there exists a $k \in \mathbf{N}$ such that $T^k x$ is a quasi-interior point in E.*

(ii) *T is σ-order continuous and majorized by a σ- order continuous weakly compact operator and for each $x > 0$ there exists a $k \in \mathbf{N}$ such that $T^k x$ is a weak order unit in E.*

(iii) *T and ϕ are σ-order continuous and for each $x > 0$ there exists a $k \in \mathbf{N}$ such that $T^k x$ is a weak order unit in E.*

PROOF: The proof of (i) can be found in [51, theorem V.5.6], and the proof of (iii) follows by the same arguments. To see that (ii) follows from (i), let F again denote the closed ideal generated by $T[E]$ in E and let, for given $x > 0$, I_j denote the ideal generated in E by $T^j x$. If $T^k x$ is a weak order unit for E, then the band generated by I_k is equal to E. Therefore, if $0 \leq y \in E$, there exists a sequence $0 \leq y_n \in I_k$ such that $y_n \uparrow y$ (note that I_k is a principal ideal) and so by the same argument used in 3.8, $Ty_n \to Ty$ in norm. But, $Ty_n \in I_{k+1}$, and so $Ty \in \overline{I_{k+1}}$; this means that $T[E] \subset \overline{I_{k+1}}$ which implies by definition of F that $\overline{I_{k+1}} = F$. Thus, $T^{k+1}x$ is a quasi-interior point of F. By (i) $T|_F$ is primitive and since $\sigma(T_F) = \{0\}$, T is primitive (see the proof of theorem 4.10).

Let $E = \mathbf{C}^n$ and let $0 < T$ be irreducible. If there exist more than one eigenvalue of T on the spectral circle, an application of the fact that the spectrum of T is rotation invariant yields the result that the sum of all the eigenvalues of T (counted according to multiplicities) is zero. This sum is the trace of the matrix representing T and since the entries of this matrix are non-negative numbers, the matrix has only zeros on its principal diagonal. Therefore, if an irreducible matrix T has a non-zero entry on its principal diagonal, then $r(T)$ is the only eigenvalue on the spectral circle. This observation can be generalized to operators belonging to a so called trace class as follows. We recall that a linear functional τ on an operator ideal is called a *trace* if it satisfies the additional assumptions that $\tau(P) = 1$ for all one dimensional projections P and that $\tau(TL) = \tau(LT)$ for all continuous operators L and all elements T in the operator ideal. (See [47] for further information on operator ideals.)

THEOREM 6.2. *Let E be a Banach lattice and let $0 < T$ be either*

(i) *irreducible or*

(ii) *band irreducible and σ-order continuous and E σ-Dedekind complete.*

Suppose that T belongs to an operator ideal which admits a continuous trace τ. Then $r(T) > 0$; moreover, if $\tau(T) \neq 0$, then $r(T)$ is the only eigenvalue on the spectral circle.

PROOF: We note first that if T belongs to an operator ideal which admits a continuous trace, then T is integral in the sense of Grothendieck and therefore, some power of T is compact [47]. It follows that $r(T) > 0$. Assume $r(T) = 1$ and let α be a point on the spectral circle (necessarily an eigenvalue). By 5.1 and 4.11, there exists a bijective orthomorphism U on E satisfying $\alpha T = U^{-1}TU$. Using the properties of the trace, we get $\alpha\tau(T) = \tau(\alpha T) = \tau(U^{-1}TU) = \tau(T)$, and, since $\tau(T) \neq 0$, $\alpha = 1$. This proves the result.

The result has another interpretation in the finite dimensional case. If there is a non-zero entry on the diagonal of the positive matrix T, it means that the component of the operator in the center is non-zero, i.e., $T \wedge I \neq 0$, with I the identity operator. This gives us another criterion for primitivity. The observation is due to B. de Pagter (private communication).

THEOREM 6.3. *Let E be a Dedekind complete Banach lattice and let $0 < T$ be a Riesz operator satisfying either*

(i) *T is irreducible or*

(ii) *T is σ-order continuous and band irreducible and E is separated by E'_o.*

If $T \wedge I > 0$, then $r(T) > 0$, and $r(T)$ is the only possible eigenvalue on the spectral circle.

PROOF: Our assumption that $T \wedge I > 0$ implies that $T \geq V$ for some non-zero orthomorphism V. Therefore, $r(T) \geq r(V) = \|V\| > 0$. Let $r(T) = 1$ and let α be an eigenvalue on the spectral circle. Since T is a Riesz operator, 1 is a pole of the resolvent and it follows as in the proof of 4.13 that the conditions of 4.11 are satisfied. Applying 4.11, there exists a bijective orthomorphism U on E satisfying $\alpha T = U^{-1}TU$. Let $T = V + W$ be the unique decomposition of T with V an orthomorphism and $W \wedge I = 0$. Then

$$\alpha V + \alpha W = \alpha T = U^{-1}TU = U^{-1}(V + W)U = U^{-1}VU + U^{-1}WU = V + U^{-1}WU.$$

But, since U and U^{-1} both belong to the center, $U^{-1}WU \leq kW$ for some positive k and so it follows from $W \wedge I = 0$ that $U^{-1}WU \wedge I = 0$. We therefore have in the above line a unique decomposition of αT. It follows that $\alpha V = V$ and since $V \neq 0$, we get $\alpha = 1$. This completes the proof.

We observe in connection with this result, that it holds in particular if T is compact. However, A.R. Schep ([54], Theorem 1.6) and also W. Arendt ([8], Satz 1.25) proved that if E is a Banach lattice which does not contain any atoms and if an orthomorphism V on E is majorized by a compact operator, then $V = 0$. Hence, in a Banach lattice without atoms, our assumption on T can only be fulfilled if T is not compact.

7. Open Problems. We finally state some problems which are still not solved.

1. Generalize the theorem of H.P. Lotz (theorem 4.6). Is it true that every positive operator on a Banach lattice has cyclic peripheral spectrum? If not, are there classes of infinite dimensional Banach lattices for which it is true?

2. Let T be a positive operator on the Banach lattice E and suppose that $\sigma(T)$ contains only the point 1. Is it true that $T \geq I$?

3. Let S be a positive operator on the Banach lattice E, which is disjoint to I. Is it true that $r(I - S) \geq 1$?

The latter two questions were posed by C.B. Huijsmans and B. de Pagter as problem 11 in [31]. They both have affirmative answers in the following cases:

(1.) $\dim E < \infty$;

(2.) $T^m \in Z(E)$;

(3.) T is a lattice homomorphism.

Moreover, if 1 is a pole of the resolvent question 2 has an affirmative answer and the same is true for question 3 if T^n and I are disjoint for all $n \in \mathbf{N}$.

8. Bibliography.

1.   Abramovich, Y.A., Aliprantis, C.D., Burkinshaw, O.: On the spectral radius of positive operators, *Math. Z.*, **211**, 593–607 (1992).

2.   ————— Corrigendum: On the spectral radius of positive operators, Preprint, (1993).

3.   Abramovich, Y.A., Arenson, E., Kitover, A.: *Banach C(K)-modules and operators preserving disjointness*, Pitman Research Notes in Mathematics Series, vol.277 Longman Scientific & Technical, Harlow, (1992).

4.   Aliprantis, C. D., Burkinshaw, O.: Positive compact operators on Banach lattices, *Math. Z.*, **174**, 289–298 (1980).

5.   ————— *Positive operators*, Academic Press, London, (1985).

6.   Andô, T.: Positive operators in semi-ordered linear spaces, *J. Fac. Sci. Hokkaido Univ.,Ser.1* **13**, 214–228 (1957).

7.   Andreu, F., Caselles, V., Martínez, J., Mazón, J.M.: The essential spectrum of AM-compact operators, *Indag. Math. N.S.*, **2**(2), 149–158 (1991).

8.   Arendt, W.: *Über das Spektrum regulärer Operatoren*, thesis, Tübingen, (1979).

9.   Arendt, W., Hart, D.B.: The spectrum of quasi-invertible disjointness preserving operators, *J. Functional Anal.*, **68**, 149–167 (1986).

10.   Berberian, S.K.: Approximate proper vectors, *Proc. Amer. Math. Soc.*, **13**, 111–114 (1962).

11.   Bernau, S.J., Huijsmans, C.B.: On the positivity of the unit element in a normed lattice ordered algebra, *Stud. Math.*, **97**, 143–149 (1990).

12.   Beukers, F., Huijsmans, C.B., de Pagter, B.: Unital embedding and complexification of f-algebras, *Math. Z.*, **183**, 131– 144 (1983).

13.   Bonsall, F.F., Tomiuk, B.J.: The semi-algebra generated by a compact linear operator, *Proc. Edinburgh Math. Soc.* **14**, 177–196 (1965).

14.   Caselles, V.: On irreducible operators on Banach lattices, *Indagationes Math.*, **48**, 11–16 (1986).

15.   ————— On band irreducible operators on Banach lattices, *Quaest. Math.*, **10**, 339–350 (1987).

16.   ————— On the peripheral spectrum of positive operators, *Israel J. Math.*, **58**, 144–160 (1987).

17.   ————— An extension of Ando-Krieger's theorem to ordered Banach spaces,

Proc. Amer. Math. Soc., **103**, (1988).

18.   Dodds, P.G., Fremlin, D.H.: Compact operators in Banach lattices, *Israel J. Math.*, **34**, 287–320 (1979).

19.   Dowson, H.R.: *Spectral theory of linear operators*, Academic Press, London-New York-San Francisco, (1978).

20.   De Pagter, B.: Irreducible compact operators, *Math. Z.* , **192**, 149–153 (1986).

21.   De Pagter, B., Schep, A.R.: Measures of non-compactness of operators in Banach lattices, *J. Functional Anal.*, **78**, 31–55 (1988).

22.   Emilion, R.: Mean-bounded operators and mean ergodic theorems, *J. Functional Anal.*, **61**, 1–14 (1985).

23.   Frobenius, G.: Über Matrizen aus positiven Elementen. *Sitz.- Berichten Kgl. Preuß. Akad. Wiss. Berlin*, 471–476 (1908) 514– 518 (1909).

24.   ———— Über Matrizen aus nicht-negativen Elementen. *Sitz.-Berichten Kgl. Preuß. Akad. Wiss. Berlin*, 456–477 (1912).

25.   Grobler, J.J.: On the spectral radius of irreducible and weakly irreducible operators in Banach lattices, *Quaestiones Math.* **2** 495–506 (1978).

26.   ———— A short proof of the Andô-Krieger theorem, *Math. Z.* **174**, 61–62 (1980).

27.   ———— Band irreducible operators, *Indagationes Math.*, **48**, 405–409 (1986).

28.   ———— The zero-two law in Banach lattice algebras, *Israel J. of Math.*, **64**, 32–37 (1988).

29.   Hardy, G.H.: *Divergent series*, Oxford Univ. Press, London (1956).

30.   Hopf, E.: Über lineare Integralgleichungen mit positivem Kern, *Sitzungsberichte Akad. Berlin*, 233–245 (1928).

31.   Huijsmans, C.B., Luxemburg, W.A.J. (eds.): *Positive operators and semigroups on Banach lattices, Proceedings of a Caribbean Mathematics Foundation Conference, 1990*, Kluwer Academic Publishers, Dordrecht-Boston-London, (1992).

32.   Huijsmans, C.B., de Pagter, B.: Subalgebras and Riesz subspaces of an f-algebra, *Proc. London Math. Soc.*, **48**, 161–174 (1984).

33.   ———— Disjointness preserving and diffuse operators, *Compos. Math.*, **79**, 351–374 (1991).

34.   Jang-Lewis, R.-J., Victory, H.D.: On the ideal structure of positive eventually compact linear operators on Banach lattices, *Pac. J. Math.*, **157**, 57–85 (1993).

35.   Jentzsch, R.: Über Integralgleichungen mit positivem Kern, *J. für Math.*,

141, 235–244 (1912).

36.   Karlin, S.: Positive operators, *J. Math. Mech.*, **8**, 907–938 (1955).

37.   Krengel, U.: *Ergodic theorems*, W. de Gruyter, Berlin–New York, (1985).

38.   Krieger, H.J.: Beiträge zur Theorie positiver Operatoren, *Schriftenr. Inst. Math.*, Reihe A, Heft 6, Berlin, Akademie-Verlag (1969).

39.   Krein, M.G. and Rutman, M.A.: Linear operators leaving invariant a cone in a Banach space, *Uspehi Mat. Nauk (N.S.)* **3**, no. 1(23), 3–95 (1948). Also Amer. Math. Sci. Transl. no. 26 (1950).

40.   Lotz, H.P.: Über das Spektrum positiver Operatoren. *Math. Z.* **108**, 15–32 (1968).

41.   Lotz, H.P. and Schaefer, H.H.: Über einen Satz von F. Niiro and I. Sawashima, *Math,. Z.*, **108**, 33–36 (1968).

42.   Luxemburg, W.A.J., Zaanen, A.C.: *Riesz spaces I*, North-Holland, Amsterdam-New York-Oxford, (1971).

43.   Martínez, J.: The essential spectral radius of dominated positive operators, *Proc. Amer. Math. Soc.*, **118**(2), 419–426 (1993).

44.   Meyer-Nieberg, P.: *Banach Lattices*, Springer-Verlag, Berlin-Heidelberg-New York, (1991).

45.   Niiro, F., Sawashima, I.: On spectral properties of positive irreducible operators in an arbitrary Banach-lattice and problems of H.H. Schaefer, *Sci. Papers College General Educ. Univ. Tokyo* **16**, 145–183 (1966).

46.   Perron, O.: Zur Theorie der Matrices, *Math. Ann.* **64**, 248–263 (1907).

47.   Pietsch, A.: Operator ideals with a trace, *Math. Nach.*, **100**, 61–91 (1981).

48.   Rota, G.-C.: On the eigenvalues of positive operators, *Bull. A.M.S.*, **67**, 556–558 (1961).

49.   Schaefer, H.H.: Spektraleigenschaften positiver Operatoren, *Math. Z.*, **82**, 303–313 (1963).

50.   ———— Topologische Nilpotenz irreduzibler Operatoren, *Math. Z.*, **117**, 135–140 (1970).

51.   ———— *Banach lattices and positive operators*, Springer-Verlag, Berlin-Heidelberg-New York (1974).

52.   ———— On theorems of de Pagter and Andô-Krieger, *Math. Z.*, **192**, 155–157 (1986).

53.   Scheffold, E.: Das Spektrum von Verbandsoperatoren in Banachverbände, *Math.Z.*, **123**, 177–190 (1971).

54.   Schep, A.R.: Positive diagonal and triangular operators, *J. Operator Theory*, **3**, 165–178 (1980).

55.   ———— A remark on the uniform zero-two law for positive contractions, *Semesterbericht Funktionalanalysis*, **13**, 189–194 (1988).

56.   Weis, L. and Wolff, M.: On the essential spectrum of operators on L^1, Semesterbericht Funktionalanalysis, Tübingen, Sommersemester (1984).

57.   Wickstead, A.W.: An embedding of the algebra of order bounded operators on a Dedekind complete Banach lattice, *Math. Z.*, **208**, 161–166 (1990).

58.   Wieland, H.: Unzerlegbare nicht-negative Matrizen, *Math. Z.*, **52**, 642–648 (1950).

59.   Zaanen, A.C.: *Linear analysis*, North-Holland, Groningen, (1953).

60.   ———— *Riesz spaces II*, North-Holland, Amsterdam-New York-Oxford, (1983).

1991 Mathematics Subject Classification: 47B65, 46B42, 46A40.

J.J. Grobler,
Potchefstroom University for C.H.E.,
Potchefstroom 2520,
South Africa.

Operator Theory:
Advances and Applications, Vol. 75
© 1995 Birkhäuser Verlag Basel/Switzerland

DISJOINTNESS PRESERVING OPERATORS
ON BANACH LATTICES

C.B. Huijsmans

Dedicated to Professor A.C. Zaanen on the occasion of his 80-th birthday

In this survey on disjointness preserving (bandpreserving, central) operators we will be mainly concerned with automatic order and norm boundedness properties of these classes of operators.

1. INTRODUCTION. Let E be a Banach lattice and F a normed vector lattice. It is well–known that an order bounded linear operator $T : E \to F$ is automatically norm bounded, although, strangely enough, in most textbooks ([4, Theorem 12.3], [25, Proposition 1.3.5.], [29, II 5.3], [33, Theorem 83.12]) this theorem is only proved for positive (or regular) operators (a result that goes back to G. Birkhoff for positive linear functionals [10] and seems to be due to I.A. Bahtin, M.A. Krasnoselskii and V.Y. Stecenko [7] in the operator case). For the sake of completeness we present here the simple proof (the so–called "n–cube trick"). Let $T : E \to F$ be order bounded and assume by way of contradiction that T is not norm bounded. Then there exist $x_n \in E, \|x_n\| = 1$ such that $\|Tx_n\| \geq n^3 (n = 1, 2, \ldots)$. Since E is norm complete, the series $\sum_{n=1}^{\infty} \frac{|x_n|}{n^2}$ is norm convergent with sum, say, $x \in E^+$. Now $\frac{|x_n|}{n^2} \in [0, x] (n = 1, 2, \ldots)$ combined with the order boundedness of T implies that $\frac{|Tx_n|}{n^2} \in [0, y] (n = 1, 2, \ldots)$ for appropriate $y \in F^+$. But this yields that $\|y\| \geq \frac{\|Tx_n\|}{n^2} \geq n (n = 1, 2, \ldots)$, a contradiction. We emphasize in this connection that conversely (even if E and F are Banach lattices) norm bounded operators between E and F need not be order bounded (see e.g. [4, Example 15.1]).

An immediate corollary of this result was already known by H. Nakano [26], namely that in Banach lattices the norm topology is uniquely determined by the order in the following sense: if E and F are Banach lattices and $T : E \to F$ is a lattice isomorphism from E onto F (so $T \geq 0$ and $T^{-1} \geq 0$), then T is a norm isomorphism as well (i.e., T and T^{-1} are norm bounded).

In the paper [3] Y.A. Abramovich, A.I. Veksler and A.V. Koldunov introduce the notion of d–isomorphism. The linear operator T from the vector lattice E into the vector lattice F is called a d–isomorphism whenever

$$x_1 \perp x_2 \ \ (\text{i.e., } |x_1| \wedge |x_2| = 0) \ \ \text{in } E \Leftrightarrow Tx_1 \perp Tx_2 \ \text{in } F.$$

Note that T is automatically injective. In that paper they generalize the above Nakano result to the effect that a d–isomorphism T from a Banach lattice E onto a normed vector lattice F is automatically norm bounded. Observe that the surjective d–isomorphisms are precisely those operators T for which both T and T^{-1} are disjointness preserving.

In this respect it is worthwile to observe that in general disjointness preserving operators (d–homomorphisms) between Banach lattices need not be order or norm bounded (see [1, Example 5.1]), whereas band preserving operators on a Banach lattice always are [3, Theorem 2].

Another result on the connection between the order and the norm topology in Banach lattices is due to Y.A. Abramovich. Actually, he shows in [2] that a positive isometry T from a Banach lattice E onto a Banach lattice F has necessarily a positive inverse T^{-1}.

More recently, K. Jarosz showed in [18] that an invertible disjointness preserving operator from a $C(X)$–space onto a $C(Y)$–space (X, Y compact and Hausdorff) is necessarily norm bounded.

These results led to the following question posed by Y.A. Abramovich in the problem section of [15]: if E and F are vector lattices and $T : E \to F$ is invertible and disjointness preserving, is T^{-1} disjointness preserving as well? The same question can be asked for band preserving and central operators.

In this paper we will present a survey of results on disjointness preserving operators (including the band preserving and central operators). Particularly, we will be concerned with the above question of Y.A. Abramovich and with automatic order boundedness and norm boundedness properties of these classes of operators.

2. PRELIMINARIES.

Throughout this paper E and F are *always* Archimedean vector lattices and all mappings between E and F are supposed to be *linear*; $L(E, F)$ stands for the vector space of all linear operators between E and F; as usual $L(E, E)$ will be denoted by $L(E)$. The operator $T \in L(E, F)$ is called *disjointness preserving* (or a *d–homomorphism*) whenever T preserves disjointness (d for disjointness), i.e.,

$$x_1 \perp x_2 \text{ in } E \ \Rightarrow T x_1 \perp T x_2 \text{ in } F.$$

We will denote this by $T \in \mathcal{D}(E, F)$. Such operators are sometimes also called *separating* because in the $C(X) - C(Y)$ case functions with disjoint cozero–sets are mapped by T onto functions with disjoint cozero–sets. The vector space of all order bounded operators between E and F will be denoted by $\mathcal{L}_b(E, F)$. The set of all order bounded disjointness preserving operators between E and F will be denoted by $\mathcal{D}_b(E, F)$ (so $\mathcal{D}_b(E, F) = \mathcal{D}(E, F) \cap \mathcal{L}_b(E, F)$). Recall that an operator is termed order bounded whenever it maps order intervals into order intervals.

The operator $T \in L(E)$ is called *band preserving* (notation $T \in \mathcal{B}(E)$) if

$$x_1 \perp x_2 \text{ in } E \Rightarrow T x_1 \perp x_2 \text{ in } E$$

(equivalently, $T, (B) \subset B$ for all (principal) bands B in E) and $T \in L(E)$ is called a *central operator* ($T \in Z(E)$) whenever there exists $\lambda > 0$ such that

$$|Tx| \leq \lambda |x|$$

for all $x \in E$. Band preserving operators are also called *unextending* because in the $C(X)$–case they have the local property of not extending the supports of functions, or sometimes also *multipliers* since in the $C(X)$–case they are precisely the multiplications with a fixed $C(X)$– function and in the L_p–case multiplication with an L_∞–function. Order bounded band preserving operators are termed orthomorphisms (we write $\operatorname{Orth}(E) = \mathcal{B}(E) \cap \mathcal{L}_b(E)$). Obviously,

$$Z(E) \subset \operatorname{Orth}(E) \subset \mathcal{D}_b(E),$$

all inclusions being proper in general. Although central operators are (almost by definition) order bounded, band preserving operators or disjointness preserving operators on arbitrary vector lattices need not be. The following counterexample is due to M. Meyer [23]. Let $E = F = PL[0, 1)$, the vector lattice of all piecewise linear functions f on $[0, 1)$, i.e., there exists a partition

$$0 = x_0 < x_1 < \ldots < x_{n-1} < x_n = 1$$

(n depending of f) such that f is linear on each interval $[x_{i-1}, x_i)$ ($i = 1, \ldots, n$). The function f may have jumps therefore in the partition points. Define

$$(Df)(x) = f_r'(x),$$

the right derivative of f in every $x \in [0, 1)$. It is easily verified that $D \in \mathcal{B}(PL[0, 1))$, but that D is not order bounded.

If $T \in \mathcal{D}(E, F)$ is positive (denoted by $T \geq 0$ and meaning that $T(E^+) \subset F^+$, where E^+ and F^+ are the positive cones of E and F respectively), then T is a lattice homomorphism ($T \in \text{Hom}(E, F)$).

It is well–known (see e.g. [33, Theorem 140.9]) that (with respect to the point-wise addition and scalar multiplication, the composition as multiplication and the pointwise supremum and infimum on E^+) the set $\text{Orth}(E)$ is an (Archimedean and hence commutative) f–algebra with unit element I (the identity mapping on E) and that the center $Z(E)$ is an (in general proper) sub–f–algebra of $\text{Orth}(E)$. If E is a Banach lattice, then

$$Z(E) = \text{Orth}(E) = \mathcal{B}(E)$$

(see [3, Theorem 2], [20, Theorem 7.12] and [31, Proposition 4.1]). It follows from Meyer's example, that these equalities fail to hold if E is not norm complete.

In contrast to $\text{Orth}(E)$ and $Z(E)$, the set $\mathcal{D}(E, F)$ is far from being even a vector space. In fact, the sum of two lattice homomorphisms need not even be a lattice homomorphism. By way of example, take $E = F = C([0, 1])$;

$$Sf = f(0)\mathbf{1}, \quad Tf = f(1)\mathbf{1}$$

for all f, where $\mathbf{1}(x) = 1$ ($0 \leq x \leq 1$). Now,

$$S, T \in \text{Hom}(C([0, 1])), \quad S + T \notin \text{Hom}(C([0, 1])).$$

For a description of finite sums of lattice homomorphisms (and more generally, of sums of order bounded disjointness preserving operators) we refer to [9].

In this paper we will restrict ourselves mainly to real vector lattices and Banach lattices, although most of the results extend without difficulty to the complexification. For unexplained terminology, notations and results we refer to the standard text books [4], [21], [25], [29] and [33].

3. CHARACTERIZATION; EXAMPLES.

In the first result of this section we characterize disjointness preserving operators. Although not very difficult to prove, statements (iii) and (iv) seem to be new. Let
E, F be vector lattices.

THEOREM 1. For $T \in L(E, F)$ the following are equivalent.

(i) $T \in \mathcal{D}(E, F)$

(ii) $|Tx| = |T|x||$ for all $x \in E$

(iii) $|Tx_1| \wedge |Tx_2| \leq |T(x_1 \wedge x_2)|$ for all $x_1, x_2 \in E$

(iv) $|Tx_1| \wedge |Tx_2| \leq |T(|x_1| \wedge |x_2|)$ for all $x_1, x_2 \in E$.

PROOF. By splitting up into positive and negative parts, it is easily verified
that $T \in \mathcal{D}(E, F)$ if and only if $Tx_1 \perp Tx_2$ for all x_1, x_2 in E satisfying $x_1 \wedge x_2 = 0$. If we
put $x = x_1 - x_2$, then $x_1 = x^+$ and $x_2 = x^-$. Hence, $T \in \mathcal{D}(E, F)$ if and only if $Tx^+ \perp Tx^-$
for all $x \in E$. In its turn, the latter is equivalent to

$$|Tx^+ - Tx^-| = |Tx^+ + Tx^-|$$

for all $x \in E$, i.e., $|Tx| = |T(|x|)|$ for all $x \in E$. This takes care of (i) $\Longleftrightarrow$ (ii). The
implications (iii) $\Rightarrow$ (i) and (iv) $\Rightarrow$ (i) are trivial.

(i) $\Rightarrow$ (iii) It follows from $(x_1 - x_1 \wedge x_2) \wedge (x_2 - x_1 \wedge x_2) = 0$ that $(Tx_1 - T(x_1 \wedge x_2)) \perp (Tx_2 - T(x_1 \wedge x_2))$. Hence,

$$|Tx_1| \wedge |Tx_2| - |T(x_1 \wedge x_2)| =$$
$$(|Tx_1| - |T(x_1 \wedge x_2)|) \wedge (|Tx_2| - |T(x_1 \wedge x_2)|) \leq .$$
$$|Tx_1 - T(x_1 \wedge x_2)| \wedge |Tx_2 - T(x_1 \wedge x_2)| = 0$$

(i) $\Rightarrow$ (iv) By (ii) and (iii),

$$|Tx_1| \wedge |Tx_2| = |T|x_1|| \wedge |T|x_2|| \leq |T(|x_1| \wedge |x_2|)|.$$

EXAMPLE 2. Let X, Y be compact Hausdorff spaces. In [18], K. Jarosz
presents a complete description of operators $T \in \mathcal{D}(C(X), C(Y))$. To be more precise, Y
is the union of three disjoint sets Y_1, Y_2 and Y_3 with Y_2 open and Y_3 closed; there exists a
continuous map $\varphi : Y_1 \cup Y_2 \to X$ and a continuous, non-vanishing function ψ on Y_1 such
that for all $f \in C(X)$

$$(Tf)(y) = \psi(y) \cdot (f \circ \varphi)(y) \quad (y \in Y_1)$$
$$= 0 \qquad\qquad\qquad (y \in Y_3).$$

Furthermore, the set $\psi(Y_2)$ is finite and all functionals of the form $f \longmapsto (Tf)(y)(y \in Y_2)$ are discontinuous.

If $T \in \mathcal{D}(E,F)$ and T is order bounded we will write $T \in \mathcal{D}_b(E,F)$; such an operator T is termed a *Lamperti operator* by W. Arendt in [6]. The following important result provides us with a description of order bounded disjointness preserving operators. It is due to M. Meyer [23] who uses representations in his proof. More elementary intrinsic proofs are due to S. Bernau [8] and B. de Pagter [28]. For the complex analogue of Theorems 1 and 3, see M. Meyer [24].

THEOREM 3. Let $T \in \mathcal{D}_b(E,F)$. Then
(i) there exist $T^+, T^- \in \text{Hom}(E,F)$ such that $T = T^+ - T^-$ and

$$(Tx)^+ = T^+x, (Tx)^- = T^-x(x \in E^+)$$

(ii) $|T|$ exists (i.e., $\sup(|Ty| : |y| \le x)$ exists for all $x \in E^+$ and is equal to $|T|x$); $|T| \in \text{Hom}(E,F), |T| = T^+ + T^-$ and

$$|Tx| = ||T|x| = |T|x|| = |T||x| \ (x \in E)$$

In fact, both (i) and (ii) are necessary and sufficient for T to be an order bounded disjointness preserving operator. Notice that $T^+ = T\vee 0, T^- = (-T)\vee 0$ and $|T| = T\vee(-T)$ in the standard partial ordering in $L(E,F)$. Moreover, if $x_1, x_2 \in E^+$, then $(Tx_1)^+ \wedge (Tx_2)^- = 0$. This follows immediately from

$$0 \le (Tx_1)^+ \wedge (Tx_2)^- \le T^+(x_1 + x_2) \wedge T^-(x_1 + x_2) = 0.$$

In the following theorem we present yet some other characterizations of Lamperti operators. The first three statements are for the (real or complex) Banach lattice case due to W. Arendt [6, Theorem 2.4], whereas (v) can be found in the paper [1, Proposition 3.2] by Y.A. Abramovich.

Theorem 4. For $T \in L(E,F)$ the following are equivalent.
   (i) $T \in \mathcal{D}_b(E,F)$
   (ii) $|x_1| \le |x_2|$ in $E \Longrightarrow |Tx_1| \le |Tx_2|$ in F
   (iii) T is order bounded and $|x_1| = |x_2|$ in E implies $|Tx_1| = |Tx_2|$ in F
   (iv) $|T(|x_1| \wedge |x_2|)| = |Tx_1| \wedge |Tx_2|$ for all $x_1, x_2 \in E$
   (v) $|T(|x_1| \vee |x_2|)| = |Tx_1| \vee |Tx_2|$ for all $x_1, x_2 \in E$.

Proof. The proof is heavily based on the result of Theorem 3.

(i) $\implies$ (ii) Since $|T| \geq 0$, $|x_1| \leq |x_2|$ implies $|T||x_1| \leq |T||x_2|$. By Theorem 3 (ii), $|Tx_1| \leq |Tx_2|$.

(ii) $\implies$ (iii) Trivial.

(iii) $\implies$ (i) For all $x \in E$ we have $|x| = ||x||$ and hence $|Tx| = |T|x||$. By Theorem 1, $T \in \mathcal{D}_b(E, F)$.

(i) $\implies$ (iv) $|Tx_1| \wedge |Tx_2| = |T||x_1| \wedge |T||x_2| = |T|(|x_1| \wedge |x_2|) = |T(|x_1| \wedge |x_2|)|$, where we use that $|T| \in \mathrm{Hom}(E, F)$.

(iv) $\implies$ (i) Taking $x_1 = x_2 = x$ in (iv) we get $|T|x|| = |Tx|$, so $T \in \mathcal{D}(E, F)$. If $|x_1| \leq |x_2|$, then $|Tx_1| = |T|x_1|| = |T(|x_1| \wedge |x_2|)| = |Tx_1| \wedge |Tx_2|$, so $|Tx_1| \leq |Tx_2|$.

(i) $\implies$ (v) $\implies$ (i) Similarly.

Operators satisfying the property of statement (v) are termed quasi–lattice homomorphisms by A.V. Koldunov [19].

EXAMPLE 5.

(i) This example is taken from M. Wolff [32] for the case of lattice homomorphisms and was extended by W. Arendt [6] for Lamperti operators. Let X, Y be compact Hausdorff spaces and $T \in L(C(X), C(Y))$. Then $T \in \mathcal{D}_b(C(X), C(Y))$ if and only if there exists $\varphi : Y \to X, h \in C(Y)$ such that

$$Tf = h \cdot (f \circ \varphi) \ (f \in C(X)),$$

where $h = T1_X$. The function φ is uniquely determined and continuous on $\{y \in Y : h(y) \neq 0\}$. Therefore, the Lamperti operators are in this case the so–called *weighted composition operators*.

(ii) Let (X, Ω, μ) be a measure space and $\varphi : X \to X$ a measure preserving transformation. Pick $h \in L_\infty(X)$ and define $T : L_p(X) \to L_p(X)$ by

$$Tf = h \cdot (f \circ \varphi) \ (f \in L_p).$$

Then $T \in \mathcal{D}_b(L_p) \ (1 \leq p \leq \infty)$.

4. INVERSES OF LAMPERTI OPERATORS, ORTHOMORPHISMS AND CENTRAL OPERATORS.

We start our considerations with invertible orthomorphisms, so let E be a vector lattice and $T \in \mathrm{Orth}(E)$. Since the null space $N(T)$ of T satisfies $N(T) = R(T)^d$ (with $R(T)$ the range of T), it follows that T is $1-1$ if and only if $R(T)^{dd} = E$. It is also true that $T \in \text{Orth } (E)$ is injective if and only if T is a weak order unit in Orth (E). Indeed, if T is $1-1$ and $S \in \text{Orth } (E)$ satisfies $S \perp T$, then $TS = 0$, so $TSx = 0$ for all $x \in E$. By the injectiveness of T we have $Sx = 0$ for all $x \in E$, i.e., $S = 0$. This shows that T is a weak order unit in Orth (E). Conversely, if T is a weak order unit in Orth (E), then $n|T| \wedge I \uparrow I$. If $Tx = 0$, then $|T||x| = 0$, so $n|T||x| \wedge |x| \uparrow |x|$ shows that $x = 0$. Consequently, T is $1-1$. The first equivalence is used in the following Proposition.

PROPOSITION 6. The following are equivalent.
(i) T is bijective
(ii) T is surjective
(iii) T is injective and $R(T)$ is a band.

PROOF. (i) $\Longrightarrow$ (ii) Trivial
(ii) $\Longrightarrow$ (iii) Obviously, $R(T) = E$ is a band. Hence, $R(T)^{dd} = E$ implies that T is injective
(iii) $\Longrightarrow$ (i) Since T is $1-1$, we have that $R(T)^{dd} = E$. But $R(T)$ is a band, so $R(T) = R(T)^{dd}$. Hence, $R(T) = E$, so T is surjective as well.

Let E be a Banach lattice and denote by $\mathcal{L}(E)$ the Banach algebra of all norm bounded linear operators on E. It was shown by W. Arendt in his thesis [5, Lemma 1.14] and independently by A.R. Schep [30, Theorem 1.8] that $\text{Orth}(E) = Z(E)$ is a full ($=$ inverse closed) subalgebra of $\mathcal{L}(E)$, i.e., if $T \in \text{Orth}(E)$ is invertible (so $T^{-1} \in \mathcal{L}(E)$ by the bounded inverse theorem), then even $T^{-1} \in \text{Orth}(E)$. This result was generalized by G.J.H.M. Buskes and A.C.M van Rooy to the effect that if E is a relatively uniformly complete vector lattice and $T \in \text{Orth}(E)$ is bijective, then $T^{-1} \in \text{Orth}(E)$ as well (see [11, 3.7 (i)]). In the next Theorem we will show that the condition of "relatively uniform completeness" in the latter result is redundant ([14, Theorem 3] or [25, Theorem 3.1.10].

THEOREM 7. Let E be a vector lattice and $T \in \text{Orth}(E)$ invertible. Then $T^{-1} \in \text{Orth}(E)$ as well, so $\text{Orth}(E)$ is a full subalgebra of $L(E)$. Moreover, $|T^{-1}| = |T|^{-1}$.

PROOF. By Theorem 3, $|T|$ exists and satisfies

$$||T|x| = |T|x|| = |T||x| = |Tx|$$

for all $x \in E$. The proof is divided into several steps.
Step 1. We assert that $|T|$ is bijective. Indeed, if $|T|x = 0$, then by the above $Tx = 0$. But T is $1-1$, so $x = 0$, showing that $|T|$ is injective. It remains to show that $|T|$ is onto. To this end, choose $y \in E$ arbitrary. Then there exist $x_1, x_2 \in E$ such that

$$Tx_1 = y^+, Tx_2 = y^-,$$

as T is surjective. Hence

$$|T||x_1| = |Tx_1| = y^+, |T||x_2| = |Tx_2| = y^-,$$

so

$$|T|(|x_1| - |x_2|) = y^+ - y^- = y.$$

<u>Step 2.</u> Next, we show that $|T^{-1}x| = |T|^{-1}|x|$ for all $x \in E$. Indeed, this is direct from

$$|T||T^{-1}x| = |T(T^{-1}x)| = |x|$$

for all $x \in E$.

<u>Step 3.</u> Since $|T| \in \text{Hom}(E)$ is invertible, $|T|$ is a lattice isomorphism. Then so is $|T|^{-1}$. In particular, $|T|^{-1} \geq 0$.

<u>Step 4.</u> We show that T^{-1} is order bounded. For this purpose, fix $x \in E^+$. Then we have for all $y \in E$ with $|y| \leq x$ that

$$|T^{-1}y| = |T|^{-1}|y| \leq |T|^{-1}x$$

by the positivity of $|T|^{-1}$.

<u>Step 5.</u> In order to prove that $T^{-1} \in \text{Orth}(E)$ it remains to verify that $T^{-1} \in \mathcal{B}(E)$. To this end, pick $x, y \in E, x \perp y$. Since $|T| \in \text{Orth}(E)$ we have $|x| \wedge |T||y| = 0$. Hence,

$$|T^{-1}x| \wedge |y| = |T|^{-1}|x| \wedge |y| =$$
$$|T|^{-1}(|x| \wedge |T||y|) = |T|^{-1}(0) = 0$$

where we use that $|T|^{-1} \in \text{Hom}(E)$.

<u>Step 6.</u> By Theorem 3, $|T^{-1}|$ exists and satisfies $|T^{-1}||x| = |T^{-1}x|$ for all $x \in E$. By Step 2, $|T^{-1}||x| = |T|^{-1}|x|$ for all $x \in E$. Applying this to x^+ and x^- we get $|T^{-1}|x = |T|^{-1}x$ for all $x \in E$, i.e., $|T^{-1}| = |T|^{-1}$.

It follows immediately from Theorem 7 that if E is a Banach lattice and $T \in Z(E)$ is invertible, then $T^{-1} \in Z(E)$ as well. This result, however, no longer holds for arbitrary vector lattices.

EXAMPLE 8. Let $E = C(\mathbf{R})$, the relatively uniformly complete vector lattice of all real continuous functions on the real line. Define $x : \mathbf{R} \to \mathbf{R}$ by

$$x(t) = e^{-t^2} \quad (t \in \mathbf{R}).$$

Let $T \in L(E)$ be defined by $Ty = x \cdot y$ for all $y \in E$. It follows from $0 \leq x \leq 1_{\mathbf{R}}$ that $0 \leq T \leq I$, so $T \in Z(E)$. Obviously, T is bijective, but $T^{-1} \notin Z(E)$. Hence, $Z(E) \neq \mathrm{Orth}(E)$ in this case, showing that one cannot define a lattice norm in $C(\mathbf{R})$ that makes $C(\mathbf{R})$ into a Banach lattice. This example shows that $Z(C(\mathbf{R}))$ is not inverse closed in $\mathrm{Orth}(C(\mathbf{R})), \mathcal{L}_b(C(\mathbf{R}))$ or $L(C(\mathbf{R}))$.

The question arises immediately of whether it is possible to find a necessary and sufficient condition that a bijective central operator be invertible in the center. In [12, Lemma 1.7] M. Duhoux and M. Meyer show that if E is a relatively uniformly complete vector lattice and $T \in Z(E)$ satisfies $|T| \geq \lambda I$ for some $\lambda > 0$, then T has necessarily an inverse in $Z(E)$ (actually, it is sufficient to assume here that $Z(E)$ is relatively uniformly complete). They use in their proof the representation of $Z(E)$ as a $C(X)$– space (with X compact and Hausdorff). In the next theorem we will present a representation-free proof of this result. Moreover, we will show that the converse statement holds as well.

THEOREM 9. Let E be a relatively uniformly complete vector lattice and $T \in Z(E)$. Then the following are equivalent.
  (i) T has an inverse in $Z(E)$
  (ii) $|T| \geq \lambda I$ for some $\lambda > 0$.

PROOF. (i) $\Longrightarrow$ (ii) For this implication we do not need the relatively uniform completeness of E. Suppose that $T \in Z(E), T$ is invertible and $T^{-1} \in Z(E)$. Then there exists $\mu > 0$ such that $|T^{-1}| \leq \mu I$. Hence, $|T^{-1}x| = |T^{-1}|x \leq \mu x$ for all $x \in E^+$. Applying $|T|$ on the left and right hand side we find

$$|T||T^{-1}x| = x \leq \mu |T|x$$

for all $x \in E^+$, showing that $|T| \geq \frac{1}{\mu}I$.
(ii) $\Longrightarrow$ (i) It is shown without representations in [13, Theorem 3.4] that in a relatively uniformly complete f–algebra A with unit element e every element $a \geq e$ has an inverse $a^{-1} \in A$. This result generalizes easily to the case that $|a| \geq e$. Indeed, $a^2 = |a|^2 \geq e$ as well, so $(a^2)^{-1}$ exists in A. Hence, a^{-1} exists and satisfies $a^{-1} = a \cdot (a^2)^{-1} = (a^2)^{-1} \cdot a$.
        In our case $Z(E)$ is a relatively uniformly complete f–algebra with unit element I, so $|T| \geq \lambda I(\lambda > 0)$ implies that T is invertible in $Z(E)$.

In the implication (ii) $\Longrightarrow$ (i) above one cannot do without the relatively uniform completeness, as is shown in the next example.

EXAMPLE 10. Take $E = \{x \in C([0,1]) : x$ piecewise polynomial $\}$ (with finitely many pieces). Then E is an f–algebra with unit element which is not relatively uniformly complete. Define

$$(Tx)(t) = (t + 1)x(t) \ (0 \leq t \leq 1)$$

for all $x \in E$. Then $I \leq T \leq 2I$ (so $T \in Z(E)$), T is injective but T is not surjective (e.g., 1 does not have an original under T in E).

In precisely the same manner as for Theorem 7 the following Theorem is verified (see for the Banach lattice case [6, Proposition 2.7] and the general case [17, Theorem 1]).

THEOREM 11. Let E, F be vector lattices and $T \in \mathcal{D}_b(E, F)$ be invertible. Then $T^{-1} \in \mathcal{D}_b(F, E)$ and $|T^{-1}| = |T|^{-1}$.

5. AUTOMATIC ORDER AND NORM BOUNDEDNESS OF DISJOINTNESS PRESERVING AND BAND PRESERVING OPERATORS

As before, E and F are vector lattices. Recall that an operator $T \in L(E, F)$ is said to be *regular* ($T \in \mathcal{L}_r(E, F)$) whenever T is the difference of two positive operators. Any regular operator is evidently order bounded so $\mathcal{L}_r(E, F) \subset \mathcal{L}_b(E, F)$. If F is Dedekind complete, then $\mathcal{L}_r(E, F) = \mathcal{L}_b(E, F)$ but in general (even in the case that E and F are Banach lattices) an order bounded operator is not necessarily regular. However, by Theorem 3, an order bounded disjointness preserving operator is also regular. In the following paragraphs we will consider the connection between order boundedness and norm boundedness of disjointness preserving operators.

Previous versions of the next theorem involving two sequences are due to Y.A. Abramovich [1, Proposition 3] and B. de Pagter [28, Corollary 3]. The present somewhat more general result can be found in the paper by P.T.N. McPolin and A.W. Wickstead [22, Theorem 2.1].

THEOREM 12. Let E, F be vector lattices and $T \in \mathcal{D}(E, F)$. If, for each sequence $\{x_n\}_{n=1}^{\infty}$ in E^+ which converges to 0 relatively uniformly,

$$\bigwedge_{n=1}^{\infty} |Tx_n| = 0, \quad \text{then} \quad T \in \mathcal{D}_b(E, F).$$

From this Theorem it follows that every norm bounded disjointness preserving operator from a normed vector lattice E into a normed vector lattice F is order bounded (i.e., $\mathcal{D}(E,F)\cap\mathcal{L}(E,F) \subset \mathcal{D}_b(E,F)$), a result first shown by Y.A. Abramovich [1, Corollary 1]). Indeed, if $T \in \mathcal{D}(E,F)\cap\mathcal{L}(E,F)$ and $x_n \to 0$ (relatively uniformly in E^+), then there exists $y \in E^+, \varepsilon_n \downarrow 0$ such that $0 \le x_n \le \varepsilon_n y (n = 1, 2, \ldots)$. If $0 \le z \le |Tx_n|$ $(n = 1, 2, \ldots)$, then

$$\|z\| \le \|Tx_n\| \le \|T\| \cdot \|x_n\| \le \varepsilon_n \|T\| \cdot \|y\|$$

$(n = 1, 2, \ldots)$ implies $z = 0$. Consequently, $\bigwedge_{n=1}^{\infty} |Tx_n| = 0$. As observed in the introduction, any order bounded operator from a Banach lattice E into a normed vector lattice F is norm bounded. Hence we have the following result. If E is a Banach lattice, F a normed vector lattice and $T \in \mathcal{D}(E,F)$, then the following are equivalent:

(i) $T \in \mathcal{L}(E,F)$ (T norm bounded)
(ii) $T \in \mathcal{L}_b(E,F)$ (T order bounded)
(iii) $T \in \mathcal{L}_r(E,F)$ (T regular)

In the remainder of this section we will pay attention to the question of Y.A. Abramovich as described in the introduction. First we will consider the so–called (order) *ideal of order boundedness*, as introduced by B. de Pagter in [28] (see also [4, Section 8]). Let E, F be vector lattices and $T \in L(E,F)$. Put

$$A_T = \text{ the ideal of order boundedness of } T$$
$$= \{x \in E : T[0, |x|] \text{ is order bounded }\}.$$

Then A_T is the largest order ideal in E on which T is order bounded. Notice that if A, B are ideals in E and the restrictions T/A and T/B are order bounded, then so is $T/(A+B)$. Therefore, A_T is the union of all ideals A in E such that T/A is order bounded. Observe that if $T \in \mathcal{D}(E,F)$, then $x_1, x_2 \in A_T, |x_1| \le |x_2|$ implies $|Tx_1| \le |Tx_2|$ (use Theorem 4). The following, by no means easy, result is crucial in the sequel and is due to B. de Pagter [28, Theorem 8].

PROPOSITION 13. If E is a relatively uniformly complete vector lattice and F is a vector lattice with the property that for each disjoint sequence $\{w_n\}_{n=1}$ in F^+ there exist $\lambda_n > 0$ such that $\{\lambda_n w_n\}_{n=1}^{\infty}$ is not order bounded and if $T \in \mathcal{D}(E,F)$, then A_T is order dense (so for each $0 < x \in E$ there exists $y \in A_T$ with $0 < y \le x$).

Notice that if F is a normed vector lattice, then F satisfies the condition of Proposition 13. Indeed, if $0 < w_n \in F^+$ and $\lambda_n = \frac{n}{\|w_n\|}$, then $\{\lambda_n w_n\}_{n=1}^{\infty}$ is not norm bounded, so not order bounded. Since any Banach lattice is relatively uniformly complete Proposition 13 holds in particular for a Banach lattice E and a normed vector lattice F.

Using Proposition 13 it is not difficult any more to prove the following result. For details we refer to [16].

THEOREM 14. E and F as in Proposition 13. If $T \in \mathcal{D}(E,F)$ is $1-1$, then

$$Tx_1 \perp Tx_2 \quad \text{in } F \Longrightarrow x_1 \perp x_2 \text{ in } E.$$

COROLLARY 15. E and F as in Proposition 13. If $T \in \mathcal{D}(E,F)$ is invertible, then $T^{-1} \in \mathcal{D}(F,E)$.

With Corollary 15 as a starting point the following Theorem can be shown [16].

THEOREM 16. E and F as in Proposition 13. If $T \in \mathcal{D}(E,F)$ is invertible, then $T \in \mathcal{D}_b(E,F)$.

Rather surprisingly we need the result of Corollary 15 before being able to show Theorem 16. Indeed, Corollary 15 would be a consequence of Theorem 16 by virtue of Theorem 11.

COROLLARY 17. If E is a Banach lattice, F a normed vector lattice and $T \in \mathcal{D}(E,F)$ is bijective, then $T \in \mathcal{D}_b(E,F)$. Hence, $T \in \mathcal{L}(E,F)$ and $T^{-1} \in \mathcal{D}_b(F,E)$. If F is also a Banach lattice, then $T^{-1} \in \mathcal{L}(F,E)$ as well.

This Corollary generalizes of course the K. Jarosz result [18] for the $C(X)-C(Y)$ case as mentioned in the introduction. It also shows that in the result of Y.A. Abramovich, A.I. Veksler and A.V. Koldunov [3, Theorem 4,5 and Added in proof] the assumption that $T^{-1} \in \mathcal{D}(F,E)$ is redundant. Furthermore, it gives a partial answer to the question of Y.A. Abramovich as described in the introduction.

The following Example is a modification of the previously recorded Example of M. Meyer (see also [16]). It shows that the relatively uniform completeness of E cannot be omitted in Theorem 16.

EXAMPLE 18. Let $E = F = PL[0,1)$ and let D be the right derivative (observe that $D^2 = 0$). Then E is a normed vector lattice with respect to the supremum norm which is not relatively uniform complete. The operator $T = I + D$ satisfies $T \in \mathcal{D}(E), T$ is invertible, $T^{-1} = I - D$ (as $D^2 = 0$) and $T^{-1} \in \mathcal{D}(E)$. However, T is not order bounded. Observe that even $T \in \mathcal{B}(E)$ and $T^{-1} \in \mathcal{B}(E)$.

Another situation in which the answer to the question of Y.A. Abramovich affirmative is also is dealt with in [16]. We confine ourselves to stating the result. Recall that the vector lattice E is said to be discrete (atomic) whenever E contains a maximal disjoint system consisting of atoms (i.e., the band generated by these atoms is E). Examples include all the classical sequence spaces.

THEOREM 19. If E and F are vector lattices, E is discrete and $T \in \mathcal{D}(E, F)$ is invertible, then $T^{-1} \in \mathcal{D}(F, E)$.

Finally, we devote some words to invertible band preserving operators. It was shown by B. de Pagter [28, Proposition 6] (see also [4, Theorem 8.8]) that for $T \in \mathcal{B}(E)$ (with E a vector lattice) the ideal of order boundedness A_T is always a band. It follows from Proposition 13 that if E is in addition relatively uniformly complete with the property that for each disjoint sequence $\{w_n\}_{n=1}^{\infty}$ in E^+ there exist $\lambda_n > 0$ such that $\{\lambda_n w_n\}_{n=1}^{\infty}$ is not order bounded and $T \in \mathcal{B}(E)$, then A_T is order dense. Hence, $A_T = A_T^{dd} = E$ showing that T is order bounded (so $T \in \text{Orth}(E)$). Particularly, if E is a Banach lattice, then $\mathcal{B}(E) = \text{Orth}(E)(= Z(E))$ as observed already in Section 2.

It follows immediately from the latter result combined with Theorem 7 that an invertible band preserving operator on a Banach lattice has a band preserving inverse. Recently, this result was generalized in [17, theorem 3] in the following manner.

THEOREM 20. If E is a relatively uniformly complete vector lattice and $\Gamma \in \mathcal{B}(E)$ is invertible, then $T^{-1} \in \mathcal{B}(E)$.

Another situation in which the result of Theorem 20 holds true is in the case the vector lattice E has the principal projection property. In this case it is easily verified that the linear operator T on E satisfies $T \in \mathcal{B}(E)$ if and only if T commutes with all band projections. Operators with this property were already considered by H. Nakano [27] under the name dilatators.

THEOREM 21. Let E be a vector lattice with the principal projection property and $T \in \mathcal{B}(E)$ invertible. Then $T^{-1} \in \mathcal{B}(E)$.

PROOF. Since $TP = PT$ for all order projections P, the same is true for $T^{-1} : T^{-1}P = PT^{-1}$ for all order projections P. By the above remark, T^{-1} is also band preserving.

REMARK 22. Recently we found out that our result of Corollary 17 was proved independently (by completely different means) by A.V. Koldunov in 'Hammerstein operators preserving disjointness', Proc. Amer. Math. Soc., to appear.

REFERENCES

1.   Abramovich, Y.A.: Multiplicative representation of operators preserving

disjointness, Indag. Math. 45 (1983), 265-279.

2.   Abramovich, Y.A.: Some results on the isometries in normed lattices, Optimization (Novosibirsk) 43 (1988), 74-80.

3.   Abramovich, Y.A., A.I. Veksler and A.V. Koldunov: On operators preserving disjointness, Soviet Math. Dokl. 20 (1979), 1089-1093.

4.   Aliprantis, C.D. and O. Burkinshaw: Positive operators, Academic Press, Orlando, 1985.

5.   Arendt, W.: Ueber das Spektrum regulärer Operatoren, Dissertation, Tübingen, 1979.

6.   Arendt, W.: Spectral properties of Lamperti operators, Indiana Univ. Math. J. 32 (1983), 199-215.

7.   Bahtin, I.A., M.A. Krasnoselskii and V.Y. Stecenko: On the continuity of positive linear operators, Sibirsk.Mat.Z. 3(1962), 156-160 (Russian).

8.   Bernau, S.J.: Orthomorphisms of Archimedean vector lattices, Math.Proc. Cambridge Phil.Soc. (1981), 119-128.

9.   Bernau, S.J., C.B. Huijsmans and B. de Pagter: Sums of lattice homomorphisms, Proc. A.M.S. 115 (1992), 151-156.

10.   Birkhoff, G.: Lattice theory, A.M.S. Coll.Publ. 25, 1st edition, 1940.

11.   Buskes, G.J.H.M. and A.C.M. van Rooy: Small Riesz spaces, Math.Proc.Cambridge Phil.Soc. 105(1989), 523-536.

12.   Duhoux, M. and M. Meyer: Extension and inversion of extended orthomorphisms on Riesz spaces, J.Austr.Math.Soc. 37 (1984), 223-242.

13.   Huijsmans, C.B. and B. de Pagter: Ideal theory in f-algebras, Trans. A.M.S. 269 (1982), 225-245.

14.   Huijsmans, C.B. and B. de Pagter: The inverse of an orthomorphism preprint W88-8, Leiden university, 1988.

15.   Huijsmans, C.B. and W.A.J. Luxemburg (eds.): Positive operators and semigroups on Banach lattices, Kluwer, Dordrecht, 1992.

16.   Huijsmans, C.B. and B. de Pagter: Invertible disjointness preserving operators, Proc. Edinburgh Math. Soc. 37 (1993), 125-132.

17.   Huijsmans, C.B. and A.W. Wickstead: The inverse of band preserving
       operators, Indag. Math. N.S. 3 (1992), 179-183.

18.   Jarosz, K.: Automatic continuity of separating linear isomorphisms,
       Canad.Math.Bull. 33 (1990), 139-144.

19.   Koldunov, A.V.: Linear mappings close to lattice homomorphisms,
       Zbornik modern algebra, Leningrad, vjpusk 3 (1975), 70-73 (Russian).

20.   Luxemburg, W.A.J.: Some aspects of the theory of Riesz spaces,
       Univ.of Arkansas Lecture Notes 4, Fayetteville, 1979.

21.   Luxemburg, W.A.J. and A.C. Zaanen: Riesz Spaces I,
       North–Holland, Amsterdam, 1971.

22.   McPolin, P.T.N. and A.W. Wickstead: The order boundedness of band
       preserving operators on uniformly complete vector lattices,
       Math.Proc.Cambridge Phil.Soc. 97 (1985), 481-487.

23.   Meyer, M. : Le stabilisateur d'un espace vectoriel réticulé,
       C.R.A.S. Paris 283 (1976), 249-250.

24.   Meyer, M.: Les homomorphismes d'espaces vectoriels réticulés complexes,
       C.R.A.S. Paris 292 (1981), 793-796.

25.   Meyer–Nieberg, P.: Banach lattices, Universitext, Springer,
       Berlin, 1991.

26.   Nakano, H.: Modulared semi–ordered linear spaces,
       Maruzen, Tokyo, 1950.

27.   Nakano, H.: Teilweise geordnete algebra, Japan J. Math.
       17 (1941), 425-511.

28.   de Pagter, B.: A note on disjointness preserving operators,
       Proc. A.M.S. 90 (1984), 543-549.

29.   Schaefer, H.H.: Banach lattices and positive operators,
       Grundlehren 215, Springer, Berlin, 1974.

30.   Schep, A.R.: Positive diagonal and triangular operators,
       J.Operator Theory 3 (1980), 165-178.

31.   Wickstead, A.W.: Representation and duality of multiplication operators
       on Archimedean Riesz spaces, Compositio Math. 35 (1977), 225-238.

32.   Wolff, M.: Ueber das Spektrum von Verbandshomomorphismen in $C(X)$,
           Math. Ann. 182 (1969), 161-169.

33.   Zaanen, A.C.; Riesz Spaces II, North–Holland, Amsterdam, 1983.

1991 Mathematics Subject Classification: 47B65, 46B42, 46A40.

C.B. Huijsmans
Department of Mathematics
University of Leiden
P.O. Box 9512
2300 RA Leiden
The Netherlands

Operator Theory:
Advances and Applications, Vol. 75
© 1995 Birkhäuser Verlag Basel/Switzerland

THE DANIELL-STONE-RIESZ REPRESENTATION THEOREM

Dedicated to Professor A.C. Zaanen on the occasion of his 80th Birthday

HEINZ KÖNIG

We present a new version of the Daniell-Stone representation theorem for certain lattice cones of $[0, \infty[$-valued functions. It contains a new version of the Riesz representation theorem on Hausdorff topological spaces. The latter result characterizes those elementary integrals, defined on certain lattice cones of upper semicontinuous $[0, \infty[$-valued functions concentrated on compact subsets, which come from Radon measures.

Introduction

The conventional Riesz representation theorem is for locally compact Hausdorff topological spaces X. It establishes a one-to-one correspondence between the isotone (=: positive) linear functionals, defined on the vector lattice of the continous functions $X \to \mathbb{R}$ with compact support, and the Radon measures on X; see for example Bauer [1992]. As soon as one considers Radon measures on more comprehensive classes of Hausdorff topological spaces X, the above correspondence breaks down and has to be filled with new substance. The respective task also arises for the Daniell-Stone representation theorem in abstract measure theory; for this theorem we refer to Zaanen [1961] and [1967] Chapters 3 and 4. Basic contributions to the two problems are due to Pollard-Topsøe [1975] and Topsøe [1976], and Anger-Portenier [1992a] [1992b].

In the present paper we shall obtain the Riesz representation theorem in a new version for all Hausdorff topological spaces X. Our theorem *characterizes* those isotone and positive-linear functionals, defined on certain lattice cones of upper semicontinuous functions

$X \to [0, \infty[$ concentrated on compact subsets, which come from the Radon measures on X. A characterization of this kind has been an issue ever since the respective notion of Radon measures became clear; see Topsøe [1983] §2. Before this we shall establish the Daniell-Stone representation theorem in the respective spirit, and in σ (= sequential) and τ (= nonsequential) versions. The Riesz representation theorem will then be a special case of the τ theorem.

Our treatment is based on the contribution of the present author [1985] [1992b] to the new theories of construction of measures from primitive set functions, defined on lattices of subsets, as initiated in Topsøe [1970a] [1970b], Kelley-Srinivasan [1971] and Kelley-Nayak-Srinivasan [1973], and Ridder [1971] [1973]. Our contribution permits to *characterize* those set functions which can be extended to measures with certain prescribed regularity and smoothness properties. The basic tool is a little collection of new envelopes for set functions, of σ or τ and inner or outer type, which are relatives of the Carathéodory outer measure but quite different. In the former literature the conventional $\star$ envelopes have been used instead; this means that the problem has been transferred to the nonsmooth situation, with the result that one obtains *sufficient* conditions for extendable set functions but *not equivalent* conditions.

In the present context we concentrate on the inner situation, which is in accordance with the definition of Radon measures. As above we form inner σ and τ envelopes associated with the functional to be represented. In terms of these envelopes we shall obtain the desired characterizations. In the earlier papers Pollard-Topsøe [1975] and Topsøe [1976], and in Anger-Portenier [1992a] the problem has been transferred to the nonsmooth situation as above, with the result that one obtains *sufficient* conditions for representable functionals but *not equivalent* conditions. We note that the initial assumptions of the above authors are sometimes weaker than ours in two respects, first in the Stone-type condition on the basic lattice cone, and second in the inner regularity requirement on the measures which are to form the desired representation. However, we do not want to load our treatment with the additional complications which thus arise, and therefore remain within what we consider to be the natural frame.

A final word concerns the nonsmooth or *abstract Riemann* situation to which we have referred above. In the context of construction of measures, this time of contents, the sufficient condition for extendable set functions in Topsøe [1970b] becomes an equivalent one, and is identical to that in König [1992b]. In contrast, in the context of Daniell-Stone representation of functionals, the sufficient conditions for representable functionals in Topsøe [1976] and Anger-Portenier [1992a] do not become equivalent conditions, and we have not been able to produce a characterization this time. Thus we have to realize that in the present context the nonsmooth or abstract Riemann situation is more complicated than the σ and τ smooth situations.

1. Preliminaries on measures and integration

We use the standard notions and notations with a few additions. Let X be a nonvoid set and $\mathfrak{S}$ be a nonvoid system of subsets of X. $\mathfrak{S}$ is called a lattice iff it is stable under finite unions and intersections. Let $\mathfrak{S}_* \subset \mathfrak{S}_\sigma \subset \mathfrak{S}_\tau$ consist of the intersections of the finite/countable/arbitrary nonvoid subsystems of $\mathfrak{S}$. For $\mathfrak{S}$ and $\mathfrak{T}$ we define the *transporter*

$$\mathfrak{S} \top \mathfrak{T} := \{A \subset X : A \cap S \in \mathfrak{T} \ \forall S \in \mathfrak{S}\};$$

see König [1991].

Let $\mathfrak{S}$ be a lattice. A set function $\varphi : \mathfrak{S} \to [0, \infty]$ is called modular iff

$$\varphi(A \cup B) + \varphi(A \cap B) = \varphi(A) + \varphi(B) \ \forall A, B \in \mathfrak{S};$$

and supermodular/submodular iff this holds true with $\geq$ / $\leq$. An isotone set function $\varphi : \mathfrak{S} \to \bar{\mathbb{R}}$ is called downward σ smooth iff

$$\varphi(S_l) \downarrow \varphi(A) \ \forall (S_l)_l \text{ in } \mathfrak{S} \text{ with } S_l \downarrow A \in \mathfrak{S};$$

and *almost* downward σ smooth iff this holds true whenever $\varphi(S_l) < \infty \ \forall l \in \mathbb{N}$. It is called downward τ smooth iff

$$\inf_{S \in \mathfrak{M}} \varphi(S) = \varphi(A) \ \forall \mathfrak{M} \subset \mathfrak{S} \text{ nonvoid downward directed with } \mathfrak{M} \downarrow A \in \mathfrak{S};$$

and *almost* downward τ smooth iff this holds true whenever $\varphi(S) < \infty\ \forall S \in \mathfrak{M}$. In the upward counterparts the exceptional value is of course $-\infty$.

Let $\mathfrak{S}$ be a lattice with $\emptyset \in \mathfrak{S}$ and $\varphi : \mathfrak{S} \to [0, \infty[$ be isotone with $\varphi(\emptyset) = 0$. We define the *inner envelopes* $\varphi_*, \varphi_\sigma, \varphi_\tau : \mathfrak{P}(X) \to [0, \infty]$ to be

$$\varphi_*(A) = \sup\{\varphi(S) : S \in \mathfrak{S} \text{ with } S \subset A\},$$

$$\varphi_\sigma(A) = \sup\{\lim_{l \to \infty} \varphi(S_l) : (S_l)_l \text{ in } \mathfrak{S} \text{ with } S_l \downarrow \text{ some set } \subset A\},$$

$$\varphi_\tau(A) = \sup\{\inf_{S \in \mathfrak{M}} \varphi(S) : \mathfrak{M} \subset \mathfrak{S} \text{ nonvoid with } \mathfrak{M} \downarrow \text{ some set } \subset A\},$$

where of course $\mathfrak{M} \downarrow$ includes that $\mathfrak{M}$ is downward directed. An obvious shorthand notation is

$$\varphi_\bullet(A) = \sup\{\inf_{S \in \mathfrak{M}} \varphi(S) : \mathfrak{M} \subset \mathfrak{S} \text{ nonvoid } \bullet \text{ with } \mathfrak{M} \downarrow \text{ some set } \subset A\},$$

where $\bullet = *\sigma\tau$.

1.1 Properties   1) $\varphi_\bullet$ *is isotone.* 2) $\varphi_\bullet$ *is inner regular* $\mathfrak{S}_\bullet$. 3) φ *super-modular* $\Rightarrow \varphi_\bullet$ *supermodular.* 4) $\varphi_* \leq \varphi_\sigma \leq \varphi_\tau$ *and* $\varphi = \varphi_*|\mathfrak{S} \leq \varphi_\sigma|\mathfrak{S} \leq \varphi_\tau|\mathfrak{S}$. 5) *For* $\bullet = \sigma\tau : \varphi = \varphi_\bullet|\mathfrak{S} \Leftrightarrow \varphi$ *is downward* $\bullet$ *smooth.*

We also define for $\bullet = *\sigma\tau$ the *satellites* $\varphi_\bullet^B : \mathfrak{P}(X) \to [0, \infty]$ with $B \in \mathfrak{S}$ to be

$$\varphi_\bullet^B(A) = \sup\left\{ \begin{array}{l} \inf_{S \in \mathfrak{M}} \varphi(S) : \mathfrak{M} \subset \mathfrak{S} \text{ nonvoid } \bullet \text{ with } \mathfrak{M} \downarrow \text{ some set } \subset A \\ \qquad\qquad\qquad\qquad\qquad \text{and } S \subset B\ \forall S \in \mathfrak{M} \end{array} \right\}.$$

The reason is that these formations appear in 1.3 condition ii') below. We formulate the connection with the $\varphi_\bullet$ themselves.

1.2 Properties   1) *For* $B \in \mathfrak{S}$ *we have* $\varphi_\bullet^B \leq \varphi(B) < \infty$. 2) $\varphi_\bullet(A) = \sup\{\varphi_\bullet^B(A) : B \in \mathfrak{S}\}\ \forall A \subset X$.

We come to the basic inner extension theorem referred to in the Introduction. The full theorem for $\bullet = *\sigma\tau$ appears in König [1992b] Theorem 5.3 with 5.5 without proof.

It is proved for $\bullet = *$ in Topsøe [1970b] Theorem 4.1 and for $\bullet = \sigma$ in König [1985] Theorem 3.4. The complete proof is in the lecture notes König [1993] which are planned to appear as a textbook.

To formulate the theorem we define an *inner $\bullet$ extension* of $\varphi : \mathfrak{S} \to [0, \infty[$ for $\bullet = *\sigma\tau$ to be a content $\alpha : \mathfrak{A} \to [0, \infty]$ on an algebra $\mathfrak{A}$ on X which extends φ, and which satisfies $\mathfrak{S}_\bullet \subset \mathfrak{A}$ with

1) α is inner regular $\mathfrak{S}_\bullet$;

2) for $\bullet = \sigma\tau : \alpha|\mathfrak{S}_\bullet$ is downward $\bullet$ smooth (note that $\alpha|\mathfrak{S}_\bullet < \infty$).

We also recall the Carathéodory construction: Assume that the set H carries an associative and commutative addition $+$ with neutral element $0 \in H$. For a set function $\phi : B(X) \to H$ with $\phi(\emptyset) = 0$ define

$$\mathfrak{C}(\phi) := \{A \subset X : \phi(M) = \phi(M \cap A) + \phi(M \cap A') \ \forall M \subset X\},$$

where A' denotes the complement of A. Then $\mathfrak{C}(\phi)$ is an algebra on X, and we have

$$\psi(M \cup A) + \psi(M \cap A) = \psi(M) + \phi(A) \ \forall M \subset X \text{ and } A \subset \mathfrak{C}(\phi).$$

1.3 Inner Extension Theorem *Let $\mathfrak{S}$ be a lattice with $\emptyset \in \mathfrak{S}$ and $\varphi : \mathfrak{S} \to [0, \infty[$ be isotone supermodular with $\varphi(\emptyset) = 0$. Then for each $\bullet = *\sigma\tau$ the following are equivalent:*

(i)   *There exists an inner $\bullet$ extension of φ.*

(ii) $\varphi_\bullet|\mathfrak{S} = \varphi$, *and* $\varphi(B) \le \varphi(A) + \varphi_\bullet(B \backslash A) \ \forall A, B \in \mathfrak{S} \text{ with } A \subset B$.

(ii)' $\varphi_\bullet(\emptyset) = 0$ *(that is for $\bullet = \sigma\tau : \varphi$ is downward $\bullet$ smooth at $\emptyset$ in the obvious sense), and* $\varphi(B) \le \varphi(A) + \varphi_\bullet^B(B \backslash A) \ \forall A, B \in \mathfrak{S} \text{ with } A \subset B$.

(iii) $\varphi_\bullet|\mathfrak{S} = \varphi$ *and* $\mathfrak{S} \subset \mathfrak{C}(\varphi_\bullet)$; *that is* $\varphi_\bullet|\mathfrak{C}(\varphi_\bullet)$ *is an extension of φ.*

Under these conditions we have

1. $\varphi_\bullet|\mathfrak{C}(\varphi_\bullet)$ *is an inner $\bullet$ extension of φ, and each inner $\bullet$ extension of φ is a restriction of $\varphi_\bullet|\mathfrak{C}(\varphi_\bullet)$.*

2. *For* $\bullet = \sigma\tau : \varphi_\bullet|\mathfrak{C}(\varphi_\bullet)$ *is a measure on the* σ *algebra* $\mathfrak{C}(\varphi_\bullet)$.

3. $\mathfrak{STS}_\bullet \subset \mathfrak{C}(\varphi_\bullet)$; *and* $\varphi_\bullet|\mathfrak{STS}_\bullet$ *is almost downward* $\bullet$ *smooth (obvious except for* $\bullet = \tau$).

A set function $\varphi : \mathfrak{S} \to [0, \infty[$ as above is called an *inner* $\bullet$ *premeasure* iff it fulfills the equivalent conditions 1.3.(i), (ii), (ii)', (iii). It is called *inner* $\bullet$ *tight* iff it fulfills the second partial condition in (ii)'. Note that the two partial conditions in (ii) and in (ii)' are of opposite type: in the succession $\bullet = *\sigma\tau$ the first one becomes sharper, while the second one becomes weaker. Also note that (ii)' becomes false when formulated with the weaker second part of (ii). In fact, there is an isotone modular $\varphi : \mathfrak{S} \to [0, \infty[$ with $\varphi(\emptyset) = 0$ such that $\varphi_\sigma(\emptyset) = 0$ but $\varphi_\sigma(A) = \infty$ $\forall$ nonvoid $A \subset X$.

We turn to the preliminaries on integration. First we explain the notion of *inner integral* which will be used. Let $\alpha : \mathfrak{A} \to [0, \infty]$ be a content on an algebra $\mathfrak{A}$ on X. For $f \in [0, \infty]^X$ the inner integral $\int f d\alpha \in [0, \infty]$ is defined to mean

$$\int f d\alpha = \sup \left\{ \sum_{l=1}^r t_l \alpha(A(l)) : \begin{array}{c} t_1, \cdots, t_r > 0 \text{ and } A(1), \cdots, A(r) \in \mathfrak{A} \text{ pairwise} \\[2mm] \text{disjoint with } f|A(l) \geq t_l (l = 1, \cdots, r) \end{array} \right\};$$

equivalent is

$$\int f d\alpha = \sup \left\{ \sum_{l=1}^r t_l \alpha(A(l)) : \begin{array}{c} t_1, \cdots, t_r > 0 \text{ and } A(1), \cdots, A(r) \in \mathfrak{A} \\[2mm] \text{with } f \geq \sum_{l=1}^r t_l \chi_{A(l)} \end{array} \right\}.$$

Further define $US(\mathfrak{A})$ to consist of the functions $f \in [0, \infty]^X$ such that $[f \geq t] \in \mathfrak{A}$ $\forall t > 0$. Note that the sum of two members of $US(\mathfrak{A})$ need not be in $US(\mathfrak{A})$, but that it does when $\mathfrak{A}$ is stable under countable intersections.

1.4 Properties 1) $\int d\alpha$ is isotone in f. 2) $\int \chi_A d\alpha = \alpha(A)$ $\forall A \in \mathfrak{A}$. 3) *We have*

$$\int (f + g) d\alpha \geq \int f d\alpha + \int g d\alpha \quad \forall f, g \in [0, \infty]^X,$$

with = *if at least one of the functions is in* $US(\mathfrak{A})$. *4)* $t\alpha([f \geq t]) \leq \int (f \wedge t) d\alpha \ \ \forall f \in US(\mathfrak{A})$

and $t > 0$. *5) We have*

$$\int f d\alpha = \int_{0\leftarrow}^{\rightarrow \infty} \alpha([f \geq t]) dt \ \ \forall f \in US(\mathfrak{A}),$$

where the integral on the right means an improper Riemann integral (the integrand is mono-tone decreasing and ≥ 0).

In this connection we recall a useful lemma from König [1992a] which will also be needed later on.

1.5 Lemma   *For* $f \in \bar{\mathbb{R}}^X$ *and real numbers* $a = t(0) < t(1) < \cdots < t(r) = b$ *we have*

$$\sum_{l=1}^{r}(t(l) - t(l-1))\chi_{[f \geq t(l)]} \leq (f-a)^+ \wedge (b-a) \leq \sum_{l=1}^{r}(t(l) - t(l-1))\chi_{[f \geq t(l-1)]};$$

also the same with $[f > \cdot]$ *instead of* $[f \geq \cdot]$.

We conclude with a result on downward τ smoothness; the identical result on downward σ smoothness is contained in the Beppo Levi theorem.

1.6 Proposition   *Let* $\mathfrak{S}$ *be a lattice with* $\emptyset \in \mathfrak{S}$ *and* $\varphi : \mathfrak{S} \to [0, \infty[$ *be an inner* $\bullet$ *premeasure for some* $\bullet = \sigma\tau$ *with* $\alpha := \varphi_\bullet |\mathfrak{C}(\varphi_\bullet)$. *Define* $V \subset [0, \infty]^X$ *to consist of the functions* $f : X \to [0, \infty]$ *such that* $[f \geq t] \in \mathfrak{S}T\mathfrak{S}_\bullet \ \forall t > 0$ *and* $\int f d\alpha < \infty$. *Then the functional* $V \to [0, \infty[: f \mapsto \int f d\alpha$ *is downward* $\bullet$ *smooth in the obvious sense.*

Proof   We can assume $\bullet = \tau$. We fix a nonvoid subset $M \subset V$ with $M \downarrow F$; thus $F \in V$ as well. To be shown is

$$\inf\{\int f d\alpha : f \in M\} \leq \int F d\alpha.$$

1) Fix $P \in V$ and $\varepsilon > 0$, and then $0 < a < b < \infty$ such that

$$\int_{0\leftarrow}^{a} \alpha([P \geq t]) dt \leq \varepsilon \text{ and } \int_{b}^{\rightarrow \infty} \alpha([P \geq t]) dt \leq \varepsilon.$$

Note that 1.4.5 can be applied in view of 1.3.3). For the $f \in M$ with $f \leq P$ then

$$\int f d\alpha = \int_{0\leftarrow}^{\rightarrow\infty} \alpha([f \geq t])dt \leq \int_a^b \alpha([f \geq t])dt + 2\varepsilon.$$

2) After the definition of the Riemann integral we can fix $a = t(0) < t(1) < \cdots < t(r) = b$ such that

$$\sum_{l=1}^r (t(l) - t(l-1))\alpha([F \geq t(l-1)]) \leq \int_a^b \alpha([F \geq t])dt + \varepsilon.$$

3) For each $t \geq 0$ we have $\{[f \geq t] : f \in M\} \downarrow [F \geq t]$. From 1.3.3 it follows that

$$\inf \{\alpha([f \geq t]) : f \in M\} = \alpha([F \geq t]) \quad \forall t > 0;$$

here we have used 1.4.4). Thus there exist $f_l \in M (l = 1, \cdots, r)$ such that

$$\alpha([f_l \geq t(l-1)]) \leq \alpha([F \geq t(l-1)]) + \frac{\varepsilon}{b-a}.$$

4) Let now $f \in M$ with $f \leq f_1, \cdots, f_r, P$. Then

$$\begin{aligned}
\int f d\alpha &\leq \int_a^b \alpha([f \geq t])dt + 2\varepsilon \quad \text{from 1).} \\
&\leq \sum_{l=1}^r (t(l) - t(l-1))\alpha([f \geq t(l-1)]) + 2\varepsilon \\
&\leq \sum_{l=1}^r (t(l) - t(l-1))\alpha([f_l \geq t(l-1)]) + 2\varepsilon \\
&\leq \sum_{l=1}^r (t(l) - t(l-1))\alpha([F \geq t(l-1)]) + 3\varepsilon \quad \text{from 3).} \\
&\leq \int_a^b \alpha([F \geq t])dt + 4\varepsilon \quad \text{from 2).}
\end{aligned}$$

The assertion is now clear.

2. Function cones and elementary integrals

Let X be a nonvoid set. We consider cones $E \subset [0, \infty[^X$ of functions $f : X \rightarrow [0, \infty[$. Further assumptions on E will be (i) the *lattice* condition

$$u, v \in E \Rightarrow u \vee v := \text{Max}\,(u, v) \in E \text{ and } u \wedge v := \text{Min}\,(u, v) \in E,$$

and (ii) the *Stone* condition

$$f \in E \text{ and } t > 0 \Rightarrow f \wedge t \in E \text{ and } (f - t)^+ \in E.$$

In view of $u = u \wedge t + (u-t)^+$ $\forall u, t \in \mathbb{R}$ condition (ii) appears to be the natural counterpart of the familiar Stone condition for linear subspaces of $\mathbb{R}^X$. Simple examples show that the conditions (i), (ii) are independent. Note that the difference $v - u$ of $u, v \in E$ with $u \leq v$ need not be in E; thus E need not be the positive cone of some linear subspace of $\mathbb{R}^X$.

2.1 Examples   Let X be a Hausdorff topological space. The class $USC^+(X)$ of the upper semicontinous functions $f : X \to [0, \infty[$ is a Stonean lattice cone; likewise the classes

$USCK^+(X)$: the $f \in USC^+(X)$ which vanish outside some compact subset;

$USC\infty^+(X)$: the $f \in USC^+(X)$ such that $[f \geq t]$ is compact $\forall t > 0$.

2.2 Remark   *Let $E \subset [0, \infty[^X$ be a Stonean cone. Then $f \wedge t - f \wedge s \in E$ $\forall f \in E$ and $0 < s < t$. This follows from $f \wedge t - f \wedge s = f \wedge t - (f \wedge t) \wedge s = (f \wedge t - s)^+ \in E$.*

With the cone $E \subset [0, \infty[^X$ we associate the two set systems

$$\mathfrak{T}(E) := \{[f \geq t] : f \in E \text{ and } t > 0\} = \{[f \geq 1] : f \in E\};$$

$$\mathfrak{t}(E) := \{A \subset X : \chi_A \in E\}.$$

Thus $\emptyset \in \mathfrak{t}(E) \subset \mathfrak{T}(E)$. For example, for the Stonean lattice cone $E := \{f \in C(\mathbb{R}, [0, \infty[) : f(x) \to 0 \text{ for } x \to \pm \infty\}$ on $\mathbb{R}$ we have $\mathfrak{t}(E) = \{\emptyset\}$, while $\mathfrak{T}(E)$ consists of all compact subsets of $\mathbb{R}$. If E is a lattice cone then $\mathfrak{T}(E)$ and $\mathfrak{t}(E)$ are lattices. If E is Stonean then $\mathfrak{T}(E) = \{[f = 1] : f \in E \text{ with } f \leq 1\}$.

We also associate with the lattice cone $E \subset [0, \infty[^X$ the two function classes

$$E_\bullet := \{ \inf_{f \in M} f : M \subset E \text{ nonvoid } \bullet\} \text{ for } \bullet = \sigma\tau;$$

one can of course restrict oneself to the $M \downarrow$. Thus

$$E_\sigma = \{ \lim_{l \to \infty} f_l : (f_l)_l \text{ in } E \text{ antitone }\};$$

$$E_\tau = \{F \in [0, \infty[^X : F = \inf_{f \in E, f \geq F} f\}.$$

The $E_\bullet$ are lattice cones as well with $E \subset E_\sigma \subset E_\tau$, and the Stone condition carries over from E to the $E_\bullet$.

2.3 Remark   Let $E \subset [0, \infty[^X$ be a lattice cone. For $\bullet = \sigma\tau$ then 1) $(\mathfrak{T}(E))_\bullet = \mathfrak{T}(E_\bullet)$ and $(\mathfrak{t}(E))_\bullet \subset \mathfrak{t}(E_\bullet)$; and 2) if E ist Stonean then $(\mathfrak{T}(E))_\bullet = \mathfrak{T}(E_\bullet) = \mathfrak{t}(E_\bullet)$.

2.4 Lemma   For real numbers $0 \le x \le 1$ and $0 \le t < 1$ define

$$x_{(t)} := \frac{1}{1-t}(x - t)^+.$$

Then 1) $0 \le x_{(t)} \le 1$. 2) Then function $(x, t) \mapsto x_{(t)}$ is continuous. 3) For fixed $0 \le t < 1$ the funtion $x \mapsto x_{(t)}$ is monotone increasing with

$$x_{(t)} = 0 \ for \ 0 \le x \le t \ and \ x_{(t)} = 1 \ for \ x = 1.$$

4) For fixed $0 \le x \le 1$ the function $t \mapsto x_{(t)}$ is monotone decreasing, and for $t \uparrow 1$ we have $x_{(t)} \downarrow 0$ if $0 \le x < 1$ and $x_{(t)} \downarrow 1$ if $x = 1$.

Proof of 2.4 Immediate verifications.

Proof of 2.3 1) First we have for $M \subset E$ nonvoid $\bigcap_{f \in M}[f \ge 1] = [\inf_{f \in M} f \ge 1]$. This proves $(\mathfrak{T}(E))_\bullet = \mathfrak{T}(E_\bullet)$. Second we have for $A \subset X$ and $\mathfrak{M} \subset \mathfrak{P}(X)$ nonvoid $A = \bigcap_{M \in \mathfrak{M}} M \Rightarrow \chi_A = \inf_{M \in \mathfrak{M}} \chi_M$. This proves $(\mathfrak{t}(E))_\bullet \subset \mathfrak{t}(E_\bullet)$.

2) i) We first prove $\mathfrak{T}(E) \subset \mathfrak{t}(E_\sigma)$ and hence $\subset \mathfrak{t}(E_\tau)$. Let $A \in \mathfrak{T}(E)$, that is $A = [f = 1]$ for some $f \in E$ with $f \le 1$. For $0 \le t < 1$ now $f_{(t)} \in E$, and for $t \uparrow 1$ we have $f_{(t)} \downarrow \chi_A$ from 2.4.4). Hence $\chi_A \in E_\sigma \subset E_\tau$. ii) From i) we obtain $\mathfrak{T}(E_\bullet) \subset \mathfrak{t}((E_\bullet)_\sigma) = \mathfrak{t}(E_\bullet)$ since $E_\bullet$ is Stonean. The assertion follows.

After this we write proposition 1.6 in the form which will be needed later.

2.5 Proposition   Let $E \subset [0, \infty[^X$ be a lattice cone. Let $\varphi : \mathfrak{T}(E) \to [0, \infty[$ be an inner $\bullet$ premeasure for some $\bullet = \sigma\tau$ with $\alpha := \varphi_\bullet \mid \mathfrak{C}(\varphi_\bullet)$, and assume that $\int f d\alpha < \infty \ \forall f \in E$. Then all $f \in E_\bullet$ are integrable α; and the functional $E_\bullet \to [0, \infty[: f \mapsto \int f d\alpha$ is downward $\bullet$ smooth in the obvious sense.

Proof   For $f \in E_\bullet$ it is clear that $\int f d\alpha < \infty$. For $t > 0$ we have from 2.3.1)
and 1.3.3)

$$[f \geq t] \in \mathfrak{T}(E_\bullet) = (\mathfrak{T}(E))_\bullet \subset \mathfrak{T}(E)\top(\mathfrak{T}(E))_\bullet \subset \mathfrak{C}(\varphi_\bullet).$$

Thus f is measurable $\mathfrak{C}(\varphi_\bullet)$ and hence integrable α. Furthermore f is in $V \subset [0, \infty]^X$ as
defined in 1.6. The assertion follows.

We turn to the functionals on function cones $E \subset [0, \infty[^X$ which are to be
represented. We call $I : E \rightarrow [0, \infty[$ an *elementary integral* iff it is isotone and positive-
linear, that is additive and fulfills $I(cf) = cI(f) \; \forall f \in E$ and $c \geq 0$. With the elementary
integral $I : E \rightarrow [0, \infty[$ we associate the conventional *inner* and *outer* envelopes I_*, I^* :
$[0, \infty]^X \rightarrow [0, \infty]$, defined to be

$$I_*(f) = \sup\{I(u) : u \in E \text{ with } u \leq f\},$$
$$I^*(f) = \inf\{I(u) : u \in E \text{ with } u \geq f\}.$$

In our context I^* is a basic formation, while I_* is much less important than the new inner
envelopes announced in the Introduction, which will be postponed until the next section.
We list the obvious properties of I^*.

2.6 Properties   1) I^* *is isotone.* 2) I^* *is sublinear in the obvious sense,*
with the usual convention $0\infty := 0$. 3) *If* E *is a lattice cone then*

$$I^*(u \vee v) + I^*(u \wedge v) \leq I^*(u) + I^*(v) \quad \forall u, v \in [0, \infty]^X.$$

4) $I^*|E = I$. 5) $I^*(\chi_A) < \infty \; \forall A \in \mathfrak{T}(E)$.

2.7 Proposition   *Let* E *be a lattice cone and* $\bullet = \sigma\tau$. 1) *If* I *is downward*
$\bullet$ *smooth then* $I^*|E_\bullet$ *is an elementary integral and downward* $\bullet$ *smooth as well.* 2) *If* I *is*
downward $\bullet$ *smooth at 0 then* $I^*|E_\bullet$ *is downward* $\bullet$ *smooth at 0 as well, as before in the*
obvious sense.

2.8 Lemma   Let E be a lattice cone. σ) If $(f_l)_l$ in E_σ with $f_l \downarrow f$ then $f \in E_\sigma$; and there exists $(u_l)_l$ in E such that $u_l \geq f_l$ $\forall l \in \mathbb{N}$ and $u_l \downarrow f$. τ) If $M \subset E_\tau$ nonvoid with $M \downarrow f$ then $f \in E_\tau$; and $N := \{u \in E : u \geq$ some member of $M\} \subset E$ is nonvoid and stable under $\wedge$ with $N \downarrow f$.

Proof of 2.8 σ) For $n \in \mathbb{N}$ let $(u_n^l)_l$ in E with $u_n^l \downarrow f_n$. Put $u_l := u_1^l \wedge \cdots \wedge u_l^l \in E$ $\forall l$. Then i) $u_l \downarrow$ and hence $u_l \downarrow g \in E_\sigma$. Now ii) $u_l \geq f_1 \wedge \cdots \wedge f_l = f_l$ $\forall l$ and hence $g \leq f$; and iii) $1 \leq n \leq l \Rightarrow u_l \leq u_n^l$; for $l \to \infty$ thus $g \leq f_n$ $\forall n$ and hence $g \leq f$. It follows that $g = f$. τ) It is clear that $N \subset E$ is nonvoid and stable under $\wedge$, hence downward directed. Now

$$\inf_{u \in N} u = \inf_{h \in M} \left(\inf_{u \in E, u \geq h} u \right) = \inf_{h \in M} h = f;$$

the assertion follows.

Proof of 2.7 2) Obvious from 2.8 applied to $f = 0$. 1) i) We first prove that $I^*|E_\bullet$ is downward $\bullet$ smooth. Fix $f \in E_\bullet$ and $M \subset E_\bullet$ nonvoid $\bullet$ with $M \downarrow f$. From 2.8 we obtain $N \subset E$ nonvoid $\bullet$ with $N \downarrow f$, and such that for each $u \in N$ there exists $h \in M$ with $u \geq h$. Now fix $v \in E$ with $v \geq f$. Then $\{u \vee v : u \in N\} \subset E$ is nonvoid $\bullet$ with $\downarrow f \vee v = v$. It follows that

$$\begin{aligned}
\inf_{u \in N} I(u) &\leq \inf_{u \in N} I(u \vee v) = I(v) \text{ and hence } \leq I^*(f), \\
\inf_{h \in M} I^*(h) &\leq \inf_{u \in N} I(u) \leq I^*(f).
\end{aligned}$$

The assertion is now clear. ii) The nontrivial assertion is that $I^*|E_\bullet$ is additive. But this is clear in view of i).

The envelopes $I_\bullet, I^* : [0, \infty]^X \to [0, \infty]$ contain set functions $\vartheta, \Theta : \mathfrak{T}(E) \to [0, \infty[$ defined to be

$$\vartheta(A) = I_\bullet(\chi_A) \text{ and } \Theta(A) = I^*(\chi_A) \quad \forall A \in \mathfrak{T}(E).$$

Their finiteness follows from 2.6.5). Thus ϑ and Θ are isotone set functions with $\vartheta \leq \Theta$ and $\vartheta(\emptyset) = \Theta(\emptyset) = 0$.

We conclude with a representation theorem in the spirit of the socalled non-additive theory. The theorem is due to Greco [1982] in an even more comprehensive version; see also Denneberg [1992] Theorem 13.2.

Let $I : E \to [0, \infty[$ be an elementary integral. An isotone set function $\varphi : \mathfrak{T}(E) \to [0, \infty[$ with $\varphi(\emptyset) = 0$ is said to *represent* I iff

$$I(f) = \int_{0\leftarrow}^{\to\infty} \varphi([f \geq t])dt \quad \forall f \in E;$$

note that the integrand on the right is monotone decreasing.

2.9 Remark   *Let $I : E \to [0, \infty[$ be an elementary integral on a Stonean function cone E. If there exists $\varphi : \mathfrak{T}(E) \to [0, \infty[$ which represents I then*

(0) $I(f \wedge t) \downarrow 0$ *for* $t \downarrow 0$ *and* $I(f \wedge t) \uparrow I(f)$ *for* $t \uparrow \infty$ $\forall f \in E$.

This follows from

$$I(f \wedge t) = \int_{0\leftarrow}^{t} \varphi([f \geq s])ds \quad \forall t > 0.$$

2.10 Theorem   *Let $I : E \to [0, \infty[$ be an elementary integral on a Stonean function cone E which fulfills condition (0). 1) An isotone set function $\varphi : \mathfrak{T}(E) \to [0, \infty[$ with $\varphi(\emptyset) = 0$ represent I iff $\vartheta \leq \varphi \leq \Theta$. 2) If an isotone $\varphi : \mathfrak{T}(E) \to [0, \infty[$ with $\varphi(\emptyset) = 0$ represents I and is downward σ smooth then $\varphi = \Theta$.*

Proof   (i) For $0 < s < t$ one verifies that

$$\chi_{[f\geq t]} \leq \frac{1}{t-s}((f \wedge t) - (f \wedge s)) \leq \chi_{[f\geq s]},$$

and the middle term is in E by 2.2; thus

$$\Theta([f \geq t]) = I^*(\chi_{[f\geq t]}) \leq \frac{1}{t-s}(I(f \wedge t) - I(f \wedge s)) \leq I_*(\chi_{[f\geq s]}) = \vartheta([f \geq s]).$$

For $0 < a < b < \infty$ and $a = t(0) < t(1) < \cdots < t(r) = b$ it follows that

$$\sum_{l=1}^{r}(t(l) - t(l-1))\Theta([f \geq t(l)]) \leq I(f \wedge b) - I(f \wedge a) \leq \sum_{l=1}^{r}(t(l) - t(l-1))\vartheta([f \geq t(l-1)]),$$

$$\int_{a}^{b}\Theta([f \geq t])dt \leq I(f \wedge b) - I(f \wedge a) \leq \int_{a}^{b}\vartheta([f \geq t])dt,$$

and hence = both times. For $a \downarrow 0$ and $b \uparrow \infty$ it follows that both ϑ and Θ represent I. Therefore all isotone $\varphi : \mathfrak{T}(E) \to [0, \infty[$ with $\vartheta \leq \varphi \leq \Theta$ do the same.

(ii) We fix an isotone $\varphi : \mathfrak{T}(E) \to [0, \infty[$ with $\varphi(\emptyset) = 0$ which represents I and show that $\vartheta \leq \varphi \leq \Theta$. Let $A \in \mathfrak{T}(E)$. For $f \in E$ with $f \leq \chi_A$ we have

$$[f \geq t] \subset A \text{ if } t > 0 \text{ and } [f \geq t] = \emptyset \text{ if } t > 1,$$

hence $I(f) \leq \varphi(A)$; it follows that $\vartheta(A) = I_*(\chi_A) \leq \varphi(A)$. Likewise for $f \in E$ with $f \geq \chi_A$ we have

$$[f \geq t] \supset A \text{ if } 0 < t \leq 1,$$

hence $I(f) \geq \varphi(A)$; it follows that $\Theta(A) = I^*(\chi_A) \geq \varphi(A)$.

(iii) We fix an isotone $\varphi : \mathfrak{T}(E) \to [0, \infty[$ with $\varphi(\emptyset) = 0$ which represents I and is downward σ smooth. For fixed $f \in E$ we put

$$P : P(t) = \varphi([f \geq t]) \text{ and } Q : Q(t) = \Theta([f \geq t]) \text{ for } t > 0.$$

Thus P and Q are monotone decreasing with $P \leq Q$ from (ii), and with $\int_a^b P(t)dt = \int_a^b Q(t)dt$ $\forall 0 < a < b < \infty$ from (i) (ii). It follows that $P(t-) = Q(t-) \geq Q(t)$ $\forall t > 0$; since P is left continuous we obtain $P(t) \geq Q(t)$ $\forall t > 0$. Thus $P = Q$ for each $f \in E$, and hence $\varphi = \Theta$.

3. The Daniell-Stone representation theorem

Let $E \subset [0, \infty[^X$ be a lattice cone on the nonvoid set X and $I : E \to [0, \infty[$ be an elementary integral. We define the *inner envelopes* $I_\sigma, I_\tau : [0, \infty]^X \to [0, \infty]$ to be

$$I_*(f) = \sup \{ \inf{}_{u \in M} I(u) : M \subset E \text{ nonvoid } \bullet \text{ with } M \downarrow \text{ some function } \leq f \}.$$

We also define the *satellites* $I_\sigma^v, I_\tau^v : [0, \infty]^X \to [0, \infty]$ with $v \in E$ to be

$$I_\bullet^v(f) = \sup \left\{ \begin{array}{l} \inf_{u \in M} I(u) : M \subset E \text{ nonvoid } \bullet \text{ with } M \downarrow \text{ some function } \leq f \\ \text{and } u \leq v \; \forall u \in M \end{array} \right\}.$$

As before the reason is that these formations appear in 3.7 condition (ii)' below.

3.1 Properties

1) $I_\bullet(f)$ and $I_\bullet^v(f)$ are isotone in f.

2) $I_* \leq I_\sigma \leq I_\tau$ and $I_\sigma^v \leq I_\tau^v$. Furthermore $I_*(f) \leq I_\bullet^v(f) \; \forall f \leq v$.

3) $I_\bullet^v(f)$ is isotone in v and $I_\bullet^v \leq I(v) < \infty$.

4) $I_\bullet(f) = \sup\{I_\bullet^v(f) : v \in E\} \; \forall f \in [0, \infty]^X$.

3.2 Remark
For $\bullet = \sigma\tau$ we have 1) $I_\bullet|E = I \iff I$ is downward $\bullet$ smooth; and 2) $I_\bullet(0) = 0 \iff I$ is downward $\bullet$ smooth at 0.

Proof 2) is obvious. 1) From the definition $I_\bullet|E \geq I$; and $I_\bullet|E \leq I$ means that for $f \in E$ and $M \subset E$ nonvoid $\bullet$ with $M \downarrow$ some function $\leq f$ one has $\inf_{u \in M} I(u) \leq I(f)$. It is clear that this means that I is downward $\bullet$ smooth.

3.3 Proposition
Assume that I is downward $\bullet$ smooth for some $\bullet = \sigma\tau$.
1) $I_\bullet(f) = I_\bullet^v(f) \; \forall f \leq v$. 2) $I_\bullet(f) = I^*(f) \; \forall f \in E_\bullet$. 3) $I_\bullet = I_*$ when $E_\bullet = E$.

Proof 1) To be shown is $I_\bullet(f) \leq I_\bullet^v(f)$. Fix $M \subset E$ nonvoid $\bullet$ with $M \downarrow$ some function $\leq f$. Then $\{u \vee v : u \in M\} \subset E$ is nonvoid $\bullet$ with $\downarrow v$, and hence $\inf_{u \in M} I(u \vee v) = I(v)$.

Now let $P, Q \in M$ and then $u \in M$ with $u \leq P, Q$. We obtain

$$\inf_{h \in M} I(h) + I(v) \leq I(u) + I(v) = I(u + v) = I(u \vee v) + I(u \wedge v) \leq I(P \vee v) + I(Q \wedge v),$$

and hence from the above

$$\inf_{h \in M} I(h) \leq \inf_{Q \in M} I(Q \wedge v) \leq I_\bullet^v(f).$$

The assertion follows. 2) $I_\bullet(f) \geq I^*(f)$ is clear from the definition; thus to be shown is $I_\bullet(f) \leq I^*(f)$, that is $I_\bullet(f) \leq I(v)$ $\forall v \in E$ with $f \leq v$. But from 1) and 3.1.3) we have in fact $I_\bullet(f) = I_\bullet^v(f) \leq I(v)$. 3) is clear.

After these preparations we approach the Daniell-Stone representation theorem. We first absolve the more formal part which does not involve the new envelopes. We assume $E \subset [0, \infty[^X$ to be a Stonean lattice cone.

Let $I : E \to [0, \infty[$ be an elementary integral. A measure $\alpha : \mathfrak{A} \to [0, \infty]$ on a σ algebra $\mathfrak{A}$ on X is called a *representing measure* for I iff $\mathfrak{T}(E) \subset \mathfrak{A}$ (which means that all members of E are measurable $\mathfrak{A}$) and $I(f) = \int f d\alpha$ $\forall f \in E$. The latter formula means that $\alpha | \mathfrak{T}(E)$ is $< \infty$ and represents I in the sense of the previous section; this follows from 1.4.4) and 5). Thus we obtain from 2.9 and 2.10 the equivalence below.

3.4 Equivalence   *If $I : E \to [0, \infty[$ has some representing measure then it fulfills the former condition*

(0) $I(f \wedge t) \downarrow 0$ *for* $t \downarrow 0$ *and* $I(f \wedge t) \uparrow I(f)$ *for* $t \uparrow \infty$ $\forall f \in E$.

Under the condition (0) the representing measures for I coincide with the measures $\alpha : \mathfrak{A} \to [0, \infty]$ which extend the set function $\Theta : \mathfrak{T}(E) \to [0, \infty[$.

For $\bullet = \sigma\tau$ we define an *inner $\bullet$ representing measure* for I to be a representing measure $\alpha : \mathfrak{A} \to [0, \infty]$ for I which satisfies $(\mathfrak{T}(E))_\bullet \subset \mathfrak{A}$ with

1) α is inner regular $(\mathfrak{T}(E))_\bullet$;

2) $\alpha | (\mathfrak{T}(E))_\bullet$ is downward $\bullet$ smooth.

Note that for $\bullet = \sigma$ the formulation becomes much simpler: An inner σ representing measure for I is a representing measure $\alpha : \mathfrak{A} \to [0, \infty]$ for I which is inner regular $(\mathfrak{T}(E))_\sigma$. Thus in the notations of the first section we obtain the equivalence below.

3.5 Equivalence *Under the condition (0) the inner • representing mea-sures for I coincide with the measures* $\alpha : \mathfrak{A} \to [0,\infty]$ *which are inner • extensions of* $\Theta : \mathfrak{T}(E) \to [0,\infty[$.

We combine the above with part of 1.3 and a few additions.

3.6 Theorem *Let* $I : E \to [0,\infty[$ *be an elementary integral on the Stonean lattice cone* E. *Then for each* • $= \sigma\tau$ *the following are equivalent.*

(i) There exists an inner • representing measure for I.

(iii) The set function $\Theta : \mathfrak{T}(E) \to [0,\infty[$ *is an inner • premeasure; and I fulfills (0).*

Under these conditions we have

1) $\Theta_\bullet|\mathfrak{C}(\Theta_\bullet)$ *is an inner • representing measure for I, and each inner • representing measure for I is a restriction of* $\Theta_\bullet|\mathfrak{C}(\Theta_\bullet)$.

2) I is downward • smooth.

3) Each inner • representing measure $\alpha : \mathfrak{A} \to [0,\infty]$ *for I satisfies* $I_\bullet(f) = \int f d\alpha \ \forall f \in [0,\infty]^X$, *and hence from 3.3.2) in particular* $I^*(f) = \int f d\alpha \ \forall f \in E_\bullet$.

Proof The equivalence (i) $\Leftrightarrow$ (iii) and 1) are immediate consequences of the results cited above. 2) then follows from 2.5. It remains to prove 3). We fix a function $f : X \to [0,\infty]$. (i) We have

$$\int f d\alpha = \sup \left\{ \sum_{l=1}^r t_l \alpha(A(l)) : \begin{array}{l} t_1, \cdots, t_r > 0 \text{ and } A(1), \cdots, A(r) \in \mathfrak{A} \text{ pairwise} \\[1ex] \text{disjoint with } f|A(l) \geq t_l (l = 1, \cdots, r) \end{array} \right\}$$

$$= \sup \left\{ \sum_{l=1}^r t_l \alpha(A(l)) : \begin{array}{l} t_1, \cdots, t_r > 0 \text{ and } A(1), \cdots, A(r) \in (\mathfrak{T}(E))_\bullet \text{ pairwise} \\[1ex] \text{disjoint with } f|A(l) \geq t_l (l = 1, \cdots, r) \end{array} \right\},$$

since α is inner regular $(\mathfrak{T}(E))_\bullet$. Thus $\int f d\alpha$ is independent of the particular $\alpha : \mathfrak{A} \to [0, \infty]$, so that we can assume that $\alpha = \Theta_\bullet | \mathfrak{C}(\Theta_\bullet)$. (ii) Let $M \subset E$ be nonvoid $\bullet$ with $M \downarrow F \leq f$. Then $F \in E_\bullet$. From 2.5 we conclude that

$$\mathrm{Inf}_{u \in M} I(u) = \mathrm{Inf}_{u \in M} \int u d\alpha = \int F d\alpha \leq \int f d\alpha.$$

Therefore $I_\bullet(f) \leq \int f d\alpha$. (iii) In the situation of the second bracket $\{\cdots\}$ in (i) we have $A(l) \in (\mathfrak{T}(E))_\bullet = \mathfrak{T}(E_\bullet) = \mathfrak{t}(E_\bullet)$ from 2.3.2), that is $\chi_{A(l)} \in E_\bullet (l = 1, \cdots, r)$. Hence

$$F := \sum_{l=1}^{r} t_l \chi_{A(l)} \in E_\bullet \text{ with } F \leq f \text{ and } \int F d\alpha = \sum_{l=1}^{r} t_l \alpha(A(l)).$$

It follows that

$$\int f d\alpha = \sup \{\int F d\alpha : F \in E_\bullet \text{ with } F \leq f\}.$$

For $F \in E_\bullet$ with $F \leq f$ we have from 2.5

$$\int F d\alpha = \inf \{\int u d\alpha = I(u) : u \in E \text{ with } u \geq F\} \leq I_\bullet(f).$$

Therefore $\int f d\alpha \leq I_\bullet(f)$. The proof is complete.

We come to the main characterization theorem. Combined with 3.6 it corresponds to the inner extension theorem 1.3 relative to the construction of measures.

3.7 Theorem   *Let $I : E \to [0, \infty[$ be an elementary integral on the Stonean lattice cone E. Then for each $\bullet = \sigma\tau$ the following are equivalent.*

   (i)   *There exists an inner $\bullet$ representing measure for I.*

   (ii)   *I is downward $\bullet$ smooth, and*

$$I(v) \leq I(u) + I_\bullet(v - u) \quad \forall u, v \in E \text{ with } u \leq v.$$

   (ii)'   *I is downward $\bullet$ smooth at 0, and*

$$I(v) \leq I(u) + I_\bullet^v(v - u) \quad \forall u, v \in E \text{ with } u \leq v.$$

An elementary integral $I : E \to [0, \infty[$ on a lattice cone E is called *inner $\bullet$ tight* for $\bullet = \sigma\tau$ iff it fulfills the second partial condition in (ii)'. It will be called *inner $*$ tight* iff

$$I(v) \leq I(u) + I_*(v - u) \quad \forall u, v \in E \text{ with } u \leq v.$$

Thus in the succession $\bullet = *\sigma\tau$ the notion of inner $\bullet$ tightness becomes weaker.

Proof (i) $\Rightarrow$ (ii) I is downward $\bullet$ smooth by 3.6.2). For $u, v \in E$ with $u \leq v$ and an inner $\bullet$ representing measure $\alpha : \mathfrak{A} \to [0, \infty]$ for I we obtain from 3.6.3) and 1.4.3)

$$I(u) + I_\bullet(v - u) = \int u d\alpha + \int (v - u) d\alpha = \int v d\alpha = I(v).$$

(ii) $\Rightarrow$ (ii)' is obvious in view of 3.3.1). It remains to prove the implication (ii)' $\Rightarrow$ (i).

1) Let $B \in \mathfrak{T}(E)$ and $v \in E$ with $v \leq 1$ such that $B = [v = 1]$. In the notation of 2.4 we have $v_{(t)} \in E$ $\forall 0 \leq t < 1$ and $v_{(t)} \downarrow \chi_B$ for $t \uparrow 1$. We claim that $I(v_{(t)}) \downarrow I^*(\chi_B) = \Theta(B)$ for $t \uparrow 1$. 1.1). We fix $u \in E$ with $u \geq \chi_B$. To be shown is

$$\lim_{t \uparrow 1} I(v_{(t)}) \leq I(u).$$

We can assume that $u \leq v \leq 1$. Also fix $\varepsilon > 0$. 1.2) From the inner $\bullet$ tightness of I applied to $u \leq v$ we obtain an $M \subset E$ nonvoid $\bullet$ with $M \downarrow F \leq v - u$ and $h \leq v$ $\forall h \in M$ such that

$$I(h) \geq I(v) - I(u) - \varepsilon \quad \forall h \in M.$$

1.3) The set $\{h \wedge v_{(t)} : h \in M \text{ and } 0 \leq t < 1\} \subset E$ is nonvoid $\bullet$ and downward directed $\downarrow F \wedge \chi_B \leq (v - u) \wedge \chi_B$, which is $= 0$ since $u = v = 1$ on B. Since I is downward $\bullet$ smooth at 0 we obtain

$$\inf \left\{ I(h \wedge v_{(t)}) : h \in M \text{ and } 0 \leq t < 1 \right\} = 0.$$

1.4) For $h \in M$ and $0 \leq t < 1$ we have

$$I(v) - I(u) - \varepsilon + I(v_{(t)}) \leq I(h) + I(v_{(t)})$$
$$= I(h \vee v_{(t)}) + I(h \wedge v_{(t)}) \leq I(v) + I(h \wedge v_{(t)}),$$

where we have used $h \vee v_{(t)} \leq v$, and hence

$$I(v_{(t)}) \leq \varepsilon + I(u) + I(h \wedge v_{(t)}).$$

From 1.3) it follows that $\lim\limits_{t \uparrow 1} I(v_{(t)}) \leq \varepsilon + I(u)$ $\forall \varepsilon > 0$ and hence the assertion.

2) We next show that $\Theta : \mathfrak{T}(E) \to [0, \infty[$ is modular. Let $A, B \in \mathfrak{T}(E)$ and $u, v \in E$ with $u, v \leq 1$ such that $A = [u = 1]$ and $B = [v = 1]$. Then $A \cup B = [u \vee v = 1]$ and $A \cap B = [u \wedge v = 1]$. For $0 \leq t < 1$ now

$$(u \vee v)_{\langle t \rangle} = u_{\langle t \rangle} \vee v_{\langle t \rangle} \text{ and } (u \wedge v)_{\langle t \rangle} = u_{\langle t \rangle} \wedge v_{\langle t \rangle};$$

this is clear from the definition since $x \mapsto x_{\langle t \rangle}$ is monotone increasing. Thus

$$I((u \vee v)_{\langle t \rangle}) + I((u \wedge v)_{\langle t \rangle}) = I(u_{\langle t \rangle}) + I(v_{\langle t \rangle}) \quad \forall 0 \leq t < 1.$$

For $t \uparrow 1$ from 1) the assertion follows.

3) From the definition $\Theta(A) = I^*(\chi_A) \quad \forall A \in \mathfrak{T}(E)$ and from 2.3.2) and 2.7.2) it follows that Θ is downward $\bullet$ smooth at $\emptyset$.

4) It remains to prove that Θ is inner $\bullet$ tight. Let $A, B \in \mathfrak{T}(E)$ with $A \subset B$ and $\varepsilon > 0$. 4.1) We choose $u, v \in E$ with $u, v \leq 1$ such that $A = [u = 1]$ and $B = [v = 1]$. We can achieve that

$$I(u) \leq I^*(\chi_A) + \varepsilon = \Theta(A) + \varepsilon \text{ and } I(v) \leq I^*(\chi_B) + \varepsilon = \Theta(B) + \varepsilon,$$

and then that $u \leq v$. 4.2) From the inner $\bullet$ tightness of I applied to $u \leq v$ we obtain an $M \subset E$ nonvoid $\bullet$ with $M \downarrow F \leq v - u$ and $h \leq v \quad \forall h \in M$ such that

$$I(h) \geq I(v) - I(u) - \varepsilon \geq \Theta(B) - \Theta(A) - 2\varepsilon \quad \forall h \in M.$$

4.3) For $h \in M$ and $0 < t < 1$ we have

$$h + v_{\langle t \rangle} \leq (\frac{h}{\varepsilon} \wedge v)_{\langle t \rangle} + \frac{\varepsilon}{t} v + v;$$

because this holds true

for $v \leq t$ in view of $v_{\langle t \rangle} = 0$ and $h \leq v$,

for $v \geq t$ and $h \leq \varepsilon$ in view of $v_{\langle t \rangle} \leq v$ and $h \leq \varepsilon \leq \frac{\varepsilon}{t} v$,

for $v \geq t$ and $h \geq \varepsilon$ in view of $\frac{h}{\varepsilon} \wedge v = v$ and $h \leq v$.

It follows that

$$I(h) + I(v_{\langle t \rangle}) \leq I((\frac{h}{\varepsilon} \wedge v)_{\langle t \rangle}) + \frac{\varepsilon}{t} I(v) + I(v).$$

For $t \uparrow 1$ we obtain from 1)

$$I(h) + \Theta(B) \leq \Theta([\tfrac{h}{\varepsilon} \wedge v = 1]) + \varepsilon I(v) + I(v)$$
$$= \Theta([h \geq \varepsilon] \cap B) + \varepsilon I(v) + I(v).$$

Combined with 4.1) and 4.2) it follows that

$$\Theta(B) - \Theta(A) \leq \Theta([h \geq \varepsilon] \cap B) + \varepsilon(\Theta(B) + \varepsilon + 3).$$

This holds true for all $h \in M$. 4.4) Now the set system $\{[h \geq \varepsilon] \cap B : h \in M\} \subset \mathfrak{T}(E)$ is nonvoid $\bullet$ and downward directed $\downarrow$ $[F \geq \varepsilon] \cap B \subset [v - u \geq \varepsilon] \cap B$, which is $\subset B \backslash A$ since $u = v = 1$ on A; and all members of the set system are $\subset B$. Therefore from the definition

$$\Theta(B) - \Theta(A) \leq \Theta_\bullet^B(B \backslash A) + \varepsilon(\Theta(B) + \varepsilon + 3).$$

This holds true for all $\varepsilon > 0$. The assertion follows. Thus the proof of 3.7 is complete.

3.8 Remark In theorem 3.7 condition (i) does not enforce that I is inner
$*$ tight, at least for $\bullet = \sigma$; see the example below. Thus conditions (ii) and (ii)', when
formulated in terms of inner $*$ tightness of I, are *sufficient but not equivalent* conditions for
(i). However after 3.3.3) and 1), the modified (ii) and (ii)' remain equivalent conditions in
the special case that $E_\bullet = E$. As mentioned in the Introduction, the former results are all
in the frame of inner $*$ tightness.

3.9 Example On $X =]0, 1[$ define $E \subset [0, \infty[^X$ to consist of the bounded
lower semicontinuous functions $f : X \to [0, \infty[$. One verifies that E is a Stonean lattice
cone. (i) We have $\mathfrak{T}(E) = \mathfrak{V}_\sigma$, with $\mathfrak{V}$ the system of the open subsets of X; this is a
simple verification. Thus $\mathfrak{T}(E)$ contains the system $\mathfrak{K}$ of the compact subsets of X. (ii)
Let $I : I(f) = \int f d\lambda \ \forall f \in E$ with $\lambda :$ Bor $(X) \to [0, \infty[$ the Borel-Lebesgue measure on
X. Then I is an elementary integral on E; and λ is an inner σ representing measure for I
since it is inner regular $\mathfrak{K}$. (iii) I is not inner $*$ tight. In fact, let $v = 1$ and $u = \chi_A$, where
$A \subset X$ is a dense open subset with $\lambda(A) < 1$. Then $I_*(v - u) = 0$, since for each $f \in E$ with

$f \leq v - u$ we have $[f > 0]$ open $\subset [v - u > 0] = A'$ and hence $[f > 0] = \emptyset$ or $f = 0$. On the other hand $I(v) - I(u) = 1 - \lambda(A) > 0$. The assertion follows.

4. The Riesz representation theorem

In the present section we assume X to be a nonvoid Hausdorff topological space. We consider a Stonean lattice cone $E \subset USC^+(X)$; for the notations see 2.1. We even assume that $E \subset USC\infty^+(X)$, which means that $\mathfrak{T}(E) \subset \text{Komp}(X)$, the system of the compact subsets of X. We also need a richness condition for E.

4.1 Remark   *For a Stonean lattice cone $E \subset USC\infty^+(X)$ the following are equivalent. (i) $(\mathfrak{T}(E))_\tau = \text{Komp}(X)$. (ii) $\chi_K \in E_\tau \; \forall K \in \text{Komp}(X)$. (iii) $USCK^+(X) \subset E_\tau$. In fact, the equivalence* (i) $\Leftrightarrow$ (ii) *follows from 2.3.2.* (iii) $\Rightarrow$ (ii) *is obvious, and* (ii) $\Rightarrow$ (iii) *is a simple manipulation.*

We define E to be *rich* iff it satisfies the equivalent conditions 4.1.(i), (ii), (iii). It is a standard fact that

$$CK^+(X) := \{f \in C(X, [0, \infty[) : f \text{ vanishes outside of some } K \in \text{Komp}(X)\}$$

is rich iff X is locally compact.

We next recall the definition of Radon measures. A Borel measure $\alpha : \text{Bor}(X) \to [0, \infty]$ is called a *Radon measure* iff $\alpha|\text{Komp}(X) < \infty$ and α is inner regular $\text{Komp}(X)$. Basic properties of Radon measures are contained in our inner extension theorem 1.3 applied to $\mathfrak{S} := \text{Komp}(X)$, in particular the extension theorem of Kisyński; see Berg-Christensen-Ressel [1984]. We restrict ourselves to the proposition below which contains what we need at present.

4.2 Proposition   *Let $\alpha : \text{Bor}(X) \to [0, \infty]$ be a content such that $\alpha|\text{Komp}(X) < \infty$ and α is inner regular $\text{Komp}(X)$. Then α is a measure and thus a Radon measure, and $\alpha|\text{Komp}(X)$ is downward τ smooth.*

Proof   We use the implications in 1.3 for several times. Let $\varphi := \alpha|$ Komp $(X) < \infty$. (i) φ is isotone modular with $\varphi(\emptyset) = 0$. By definition α is an inner $*$ extension of φ. (ii) φ is an inner $*$ premeasure; hence an inner τ premeasure as well, because on the particular lattice Komp (X) downward τ smoothness at $\emptyset$ is automatic. (iii) $\varphi = \alpha|$ Komp (X) is downward τ smooth. By definition α is an inner τ extension of φ. (iv) α is a restriction of $\varphi_\tau|\mathfrak{C}(\varphi_\tau)$, that is Bor $(X) \subset \mathfrak{C}(\varphi_\tau)$ and $\alpha = \varphi_\tau|$ Bor (X). Hence α is a measure. (v) We also obtain that the restriction $\alpha|\text{Cl}(X)$ of α to the class $\text{Cl}(X)$ of the closed subsets of X is almost downward τ smooth.

After this the Riesz representation theorem is an immediate consequence of the Daniell-Stone theorem in the τ version. We need one more remark.

4.3 Remark   *For an elementary integral $I : E \to [0, \infty[$ on a Stonean lattice cone $E \subset USC\infty^+(X)$ the following are equivalent.*

(i) I is downward τ smooth at 0.

(ii) $I(f \wedge t) \downarrow 0$ for $t \downarrow 0$ $\forall f \in E$.

The latter condition is obvious for a rich cone $E \subset USCK^+(X)$, but need not be true for a rich cone $E \subset USC\infty^+(X)$.

Proof   To be shown is (ii) $\Rightarrow$ (i). Let $M \subset E$ be nonvoid with $M \downarrow 0$. Fix $f \in M$ and $\varepsilon > 0$. Then $K := [f \geq \varepsilon] \in$ Komp (X) and hence from the Dini Theorem $\inf_{u \in M} \sup (u|K) = 0$. Now for $u \in M$ with $u \leq f$ we have

$$u \leq \tfrac{1}{\varepsilon} \sup (u|K)f + f \wedge \varepsilon,$$
$$I(u) \leq \tfrac{1}{\varepsilon} \sup (u|K)I(f) + I(f \wedge \varepsilon).$$

It follows that $\inf_{u \in M} I(u) \leq I(f \wedge \varepsilon)$. The assertion is now clear.

4.4 Theorem   *For an elementary integral $I : E \to [0, \infty[$ on a rich Stonean lattice cone $E \subset USC\infty^+(X)$ the following are equivalent.*

(i)   *There exists a Radon representing measure $\alpha : \text{Bor}\,(X) \to [0, \infty]$ for I.*

(ii)' $I(f \wedge t) \downarrow 0 \text{ for } t \downarrow 0 \;\; \forall f \in E$, and

$$I(v) \leq I(u) + I_\tau^v(v - u) \;\; \forall u, v \in E \text{ with } u \leq v.$$

Under these conditions we have

1)   *There is a unique Radon representing measure $\alpha : \text{Bor}\,(X) \to [0, \infty]$ for I, that is $\alpha = \Theta_\tau | \text{Bor}\,(x)$. It is of course an inner τ representing measure for I.*

2)   *I is downward τ smooth.*

3)   *$I_\tau(f) = \int f d\alpha \;\; \forall f \in [0, \infty]^X$; and $I^\bullet(f) = \int f d\alpha \;\; \forall f \in E_\tau$.*

Proof   First assume (ii)'. After 3.6 & 3.7 combined with 4.3 the set function $\Theta : \mathfrak{T}(E) \to [0, \infty[$ is an inner τ premeasure, and $\Theta_\tau | \mathfrak{C}(\Theta_\tau)$ is an inner τ representing measure for I. The transporter $\mathfrak{T}(E)\top(\mathfrak{T}(E))_\tau = \mathfrak{T}(E)\top \text{Komp}\,(X) \subset \mathfrak{C}(\Theta_\tau)$ contains the closed subsets of X; hence $\text{Bor}\,(X) \subset \mathfrak{C}(\Theta_\tau)$. Thus $\alpha := \Theta_\tau |\, \text{Bor}\,(X)$ is a representing measure for I which satisfies $\alpha | \text{Komp}\,(X) < \infty$ and is inner regular $\text{Komp}\,(X)$, therefore a Radon representing measure for I.

Now suppose (i), and let $\alpha : \text{Bor}\,(X) \to [0, \infty]$ be a Radon representing measure for I. By 4.2 $\alpha |\, \text{Komp}\,(X)$ is downward τ smooth; hence α is an inner τ representing measure for I. It follows that $\Theta : \mathfrak{T}(E) \to [0, \infty[$ is an inner τ premeasure, and α is a restriction of $\Theta_\tau | \mathfrak{C}(\Theta_\tau)$. That means $\text{Bor}\,(X) \subset \mathfrak{C}(\Theta_\tau)$ and $\alpha = \Theta_\tau |\, \text{Bor}\,(X)$. The proof is complete.

4.5 Remark In theorem 4.4 condition (ii)', when formulated in terms of inner $*$ tightness of I, that is with I_*, becomes sharper and thus a *sufficient* condition for (i); after 3.3.3) and 1), it remains an equivalent condition in the special case that $E_\tau = E$. This time it is an open question whether the modified (ii)' remains an equivalent condition in all cases.

At last we specialize our result to $E = USCK^+(X)$, which is a rich Stonean lattice cone with $E_\tau = E$.

4.6 Corollary   *There is a one-to-one correspondence between the elementary integrals $I : USCK^+(X) \to [0, \infty[$ which are inner $*$ tight and fulfill $I(f \wedge t) \to 0$ for $t \downarrow 0$, and the Radon measures $\alpha : \mathrm{Bor}\,(X) \to [0, \infty]$; it is given by $I(f) = \int f d\alpha \;\; \forall USCK^+(X)$.*

This result corresponds to Anger-Portenier [1992a] Theorem 12.5. However, this work is based on the different notion of Radon measures which includes local finiteness.

5. Remarks on the nonsmooth situation

Let as before $E \subset [0, \infty[^X$ be a Stonean lattice cone and $I : E \to [0, \infty[$ be an elementary integral. Our main theme so far has been the representation of I under the σ and τ smoothness assumptions. The present section is devoted to a few unsystematic remarks on the *nonsmooth* situation, with the aim to demonstrate that this situation is quite different and more complicated.

We define a *representing content* for I to be a content $\alpha : \mathfrak{A} \to [0, \infty]$ on an algebra $\mathfrak{A}$ on X such that $\mathfrak{T}(E) \subset \mathfrak{A}$ (which means that $E \subset US(\mathfrak{A})$) and $I(f) = \int f d\alpha \; \forall f \in E$. The latter formula means that $\alpha|\mathfrak{T}(E)$ is $< \infty$ and represents I in the sense of section 2; this follows from 1.4.4) and 5). Thus we obtain from 2.9 and 2.10.1) the equivalence below. It is different from the equivalence 3.4, because in the nonsmooth situation there is no result like 2.10.2). This will be responsible for part of the difficulties to come.

5.1 Equivalence   *If I has some representing content then it fulfills condition (0). Under the condition (0) the representing contents for I are the contents $\alpha : \mathfrak{A} \to [0, \infty]$ with $\mathfrak{T}(E) \subset \mathfrak{A}$ and $\vartheta \leq \alpha|\mathfrak{T}(E) \leq \Theta$.*

We define an *inner representing content* for I to be a representing content $\alpha : \mathfrak{A} \to [0, \infty]$ for I which is inner regular $\mathfrak{T}(E)$. Thus from the first section we obtain the equivalence below.

5.2 Equivalence   *Under the condition* (0) *the inner representing contents*
for I *are the restrictions*

$$\alpha = \varphi_*|\mathfrak{A} \text{ with } \mathfrak{T}(E) \subset \mathfrak{A} \subset \mathfrak{C}(\varphi_*)$$

from the inner $*$ *premeasures* $\varphi : \mathfrak{T}(E) \to [0,\infty[$ *with* $\vartheta \leq \varphi \leq \Theta$.

Now we ask whether the assertions which correspond to the main theorems
3.6 & 3.7 are true. 1) Are the following equivalent?

(i) There exists an inner representing content for I.

(ii) I fulfills (0) and is inner $*$ tight, that is

$$I(v) \leq I(u) + I_*(v - u) \quad \forall u, v \in E \text{ with } u \leq v.$$

(iii) I fulfills (0), and $\Theta : \mathfrak{T}(E) \to [0,\infty[$ is an inner $*$ premeasure.

2) Assume that (iii) is fulfilled and hence $\Theta_*|\mathfrak{C}(\Theta_*)$ is an inner representing content for I.
Are then all inner representing contents for I restrictions of $\Theta_*|\mathfrak{C}(\Theta_*)$? 3) Is it true that
$I_*(f) = \int f d\alpha \ \forall f \in [0,\infty]^X$ for all inner representing contents $\alpha : \mathfrak{A} \to [0,\infty]$ for I, or at
least for some inner representing content, provided that such ones exist?

The first answers below look quite unfavorable.

5.3 Proposition   1) (iii) $0 \Rightarrow$ (i). 2) (iii) $\not\Rightarrow$ (ii), *and hence* (i) $\not\Rightarrow$ (ii) *as well.*
3) *For the union of* (i) *and* (ii) *we have* (i) & (ii) $\not\Rightarrow$ (iii).

Proof   1) is obvious. 2) This follows from example 3.9. In fact, $\lambda :$ Bor
$(X) \to [0,\infty[$ is an inner σ representing measure for I; thus $\Theta = \lambda|\mathfrak{T}(E) = \lambda|\mathfrak{V}_\sigma$ is an inner
σ premeasure by 3.5, and hence an inner $*$ premeasure since $(\mathfrak{T}(E))_\sigma = \mathfrak{T}(E)$. Of course I
fulfills (0). But I is not inner $*$ tight as shown in 3.9. 3) will follow from the next example.

5.4 Example   On $X = \mathbb{R}$ define $E \subset [0, \infty[^X$ to consist of the bounded continuous functions $f : X \to [0, \infty[$ which are monotone increasing on $[0, \infty[$ and monotone decreasing on $] - \infty, 0]$ with $\lim\limits_{x \to \infty} f(x) = \lim\limits_{x \to -\infty} f(x)$. E is Stonean lattice cone.

1. $\mathfrak{T}(E)$ consists of $\emptyset$ and $\mathbb{R}$, and of the subsets $] - \infty, a]$ and $[b, \infty[$ and $] - \infty, a] \cup [b, \infty[$ with real $a < 0 < b$.

Let $I : I(f) = \lim\limits_{x \to \pm\infty} f(x) \;\; \forall f \in E$. Thus I is an elementary integral on E which fulfills (0). One verifies that

2. $I^*(f) = \sup f \;\; \forall f \in [0, \infty]^X$;

3. $I_*(f) = \liminf\limits_{|x| \to \infty} f(x) \;\; \forall f \in [0, \infty]^X$.

From 3. we see that I is inner $*$ tight and thus fulfills (ii). From 2. we see that $\Theta : \mathfrak{T}(E) \to [0, \infty[$ is $\Theta(\emptyset) = 0$ and $\Theta(A) = 1$ for $A \neq \emptyset$. Thus 1. shows that Θ is not supermodular, so that (iii) is violated. At last define $\varphi, \psi : \mathfrak{T}(E) \to [0, \infty[$ to be

$$\varphi(A) = \left\{ \begin{array}{ll} 1 & \text{if } A \text{ meets }]0, \infty[\\ 0 & \text{if not} \end{array} \right\}, \quad \psi(A) = \left\{ \begin{array}{ll} 1 & \text{if } A \text{ meets }] - \infty, 0[\\ 0 & \text{if not} \end{array} \right\}.$$

One verifies that φ and ψ are inner $*$ premeasures with $\vartheta \leq \varphi, \psi \leq \Theta$. Therefore I fulfills (i) after 5.2.

On the other hand the implication (ii) $\Rightarrow$ (i) can be seen to be true. This assertion is different in nature from all the former ones: Its proof involves a Hahn-Banach type conclusion, and the above example 5.4 shows that there is no uniqueness. Results of this type are also obtained in Anger-Portenier [1992a]. We do not present the proof this time, but shall postpone the result to another context which will also contain König [1992c] where Hahn-Banach type conclusions are vital as well. Another reason that we do not insist upon the proof of the implication (ii) $\Rightarrow$ (i) is that inner $*$ tightness of I does not appear ot be the adequate and true condition in the nonsmooth situation, as 5.3.2) makes evident.

In Pollard-Topsøe [1975] and Topsøe [1976] a certain separation condition has been introduced which is able to substitute downward smoothness in the nonsmooth situation.

5.5 Remark   *For an elementary integral* $I : E \to [0, \infty[$ *on a lattice cone*
E *the following are equivalent.*

(i) *For* $A, B \in \mathfrak{T}(E)$ *with* $A \cap B = \emptyset$ *and* $\varepsilon > 0$ *there exist* $u, v \in E$ *with* $u \geq \chi_A$ *and*
$v \geq \chi_B$ *such that* $I(u \wedge v) \leq \varepsilon$.

(ii) *For* $A, B \in \mathfrak{T}(E)$ *with* $A \cap B = \emptyset$ *we have* $\Theta(A \cup B) = \Theta(A) + \Theta(B)$.

Then I is said to fulfill *separation condition* (00).

Proof   (i) $\Rightarrow$ (ii) Θ is submodular by 2.6.3). On the other hand fix $\varepsilon > 0$
and choose first $w \in E$ with $w \geq \chi_{A \cup B}$ and $I(w) \leq \Theta(A \cup B) + \varepsilon$ and then $u, v \in E$ after (i)
with $u, v \leq w$. It follows that

$$\begin{aligned}
\Theta(A) + \Theta(B) &\leq I(u) + I(v) = I(u \vee v) + I(u \wedge v) \\
&\leq I(w) + I(u \wedge v) \leq \Theta(A \cup B) + 2\varepsilon.
\end{aligned}$$

(ii) $\Rightarrow$ (i) For $\varepsilon > 0$ let $u, v \in E$ with $u \geq \chi_A$ and $v \geq \chi_B$ such that $I(u) \leq \Theta(A) + \varepsilon$ and
$I(v) \leq \Theta(B) + \varepsilon$. Then

$$\begin{aligned}
\Theta(A \cup B) + I(u \wedge v) &\leq I(u \vee v) + I(u \wedge v) = I(u) + I(v) \\
&\leq \Theta(A) + \Theta(B) + 2\varepsilon = \Theta(A \cup B) + 2\varepsilon,
\end{aligned}$$

and hence $I(u \wedge v) \leq 2\varepsilon$.

Thus condition (00) is fulfilled whenever the function Θ is modular, in particular under condition (iii) above, and under condition 3.6 (iii) in the smooth situations as
well. Now condition (00) implies a basic uniqueness assertion.

5.6 Remark   *Let I fulfill* (0) *and* (00). *If* $\varphi : \mathfrak{T}(E) \to [0, \infty[$ *is an inner* $*$
premeasure with $\vartheta \leq \varphi \leq \Theta$ *then* $\varphi = \Theta$.

Proof   1) For $f \in E$ we have after 2.10

$$I(f) = \int_{0\leftarrow}^{\to \infty} \vartheta([f \geq t])dt = \int_{0\leftarrow}^{\to \infty} \Theta([f \geq t])dt,$$

and hence $\vartheta([f \geq t]) = \Theta([f \geq t])$ $\forall t > 0$ except on a countable subset of $]0, \infty[$. 2) We have

$$\Theta(A) - \varphi(A) \leq \Theta(B) - \varphi(B) \quad \forall A, B \in \mathfrak{T}(E) \text{ with } A \subset B.$$

In fact, we have

$$\varphi(B) - \varphi(A) \leq \sup\{\varphi(K) : K \in \mathfrak{T}(E) \text{ with } K \subset B \backslash A\},$$

and on the right each time $\varphi(K) \leq \Theta(K) = \Theta(A \cup K) - \Theta(A) \leq \Theta(B) - \Theta(A)$. 3) Now let $A \in \mathfrak{T}(E)$, and $f \in E$ with $f \leq 1$ and $A = [f = 1]$. After 1) let $B := [f \geq t]$ for some $0 < t < 1$ with $\vartheta(B) = \Theta(B)$. Then $A \subset B \in \mathfrak{T}(E)$, and from 2) we obtain $0 \leq \Theta(A) - \varphi(A) \leq \Theta(B) - \varphi(B) \leq \Theta(B) - \vartheta(B) = 0$ and hence $\varphi(A) = \Theta(A)$.

The last remark furnishes more answers to our initial questions.

5.7 Propostion 1) *The union* (i) & (00) *is equivalent to* (iii). 2) *The answer to question* 2) *is positive.* 3) *The answer to question* 3) *is negative, even in classical examples.*

Proof 1) 2) are immediate consequences of 5.2 and 5.6. 3) will follow from the next example.

5.8 Example Let X be a compact Hausdorff topological space and $E = C(X, [0, \infty[)$. Thus $\mathfrak{T}(E) = \text{Komp}(X)$. We fix a point $a \in X$ such that $\{a\}$ is not open, and define $I : I(f) = f(a)$ $\forall f \in E$. I is an elementary integral which fulfills (0). One verifies that

$$\Theta(A) = \begin{cases} 1 & \text{for } A \in \mathfrak{T}(E) \text{ with } a \in A \\ 0 & \text{for } A \in \mathfrak{T}(E) \text{ with } a \notin A \end{cases}, \Theta_*(A) = \begin{cases} 1 & \text{for } A \subset X \text{ with } a \in A \\ 0 & \text{for } A \subset X \text{ with } a \notin A \end{cases}.$$

Thus $\Theta : \mathfrak{T}(E) \to [0, \infty[$ is an inner $*$ premeasure with $\mathfrak{C}(\Theta_*) = \mathfrak{V}(X)$. We see from 5.7.2) that the inner representing contents for I are the restrictions $\alpha = \Theta_*|\mathfrak{A}$ with $\mathfrak{T}(E) \subset \mathfrak{A}$. We have each time $\int \chi_{\{a\}} d\alpha = \alpha(\{a\}) = \Theta(\{a\}) = 1$. But on the other hand $I_*(\chi_{\{a\}}) = 0$

since $\{a\}$ is not open. Thus the formula in question 3) in violated for all inner representing contents for I.

The implication (ii) $\Rightarrow$ (i) discussed above combined with 5.7.1) leads to the implication (ii) & (00) $\Rightarrow$ (iii). This latter result is contained in Topsøe [1976] Theorem 1.

Our final conclusion is that the conventional inner envelope I_* is in none of the situations $\bullet = *\sigma\tau$ an adequate formation in connection with the representation of I. But while in the smooth situations $\bullet = \sigma\tau$ the new inner envelopes $I_\bullet$ have been found to be adequate, there is no candidate visible in the nonsmooth situation $\bullet = *$ so far.

References

ANGER, B., PORTENIER, C.: Radon Integrals. PMM 103, Birkhäuser 1992.

-: Radon Integrals and Riesz Representation. In: Proc. Conf. Measure Theory, Oberwolfach 1990. Rend. Circ. Mat. Palermo, Suppl. II (28) 1992, pp. 269-300.

BAUER, H.: Mass und Integrationstheorie, 2. Aufl., de Gruyter 1992.

BERG, C., CHRISTENSEN, J.P.R., RESSEL, P.: Harmonic Analysis on Semigroups. GTM 100, Springer 1984.

DENNEBERG, D.: Lectures on Non-additive Measure and Integral. Preprint Nr. 42, Univ. Bremen 1992.

GRECO, G.H.: Sulla Rappresentazione di Funzionali mediante Integrali. Rend. Sem. Mat. Univ. Padova 66 (1982), 21-42.

KELLEY, J.L., NAYAK, M.K., SRINIVASAN, T.P.: Pre-measures on Lattices of Sets - II. In: Proc. Symp. Vector and Operator Valued Measures and Applications, Utah 1972. Academic Press 1973, pp. 155-164.

KELLEY, J.L., SRINIVASAN, T.P.: Pre-measures on Lattices of Sets. Math. Ann. 190 (1971), 233-241.

KÖNIG, H.: On the Basic Extension Theorem in Measure Theory. Math. Z. 190 (1985), 83-94.

-: The Transporter Theorem - a new Version of the Dynkin Class Theorem. Arch. Math. 57 (1991), 588-596.

-: Daniell-Stone Integration without the Lattice Condition and its Application to Uniform Algebras. Ann. Univ. Saraviensis, Ser. Math. 4 (1992), 1-91.

-: New Constructions related to inner and outer regularity of Set Functions. In: Proc. Conf. Topology, Measures, and Fractals, Warnemünde 1991. Math. Res. 66, Akademie Verlag 1992, pp. 137-146.

-: On inner/outer regular Extension of Contents. In: Proc. Conf. Measure Theory, Oberwolfach 1990. Rend. Circ. Mat. Palermo, Suppl. II (28) 1992, pp. 59-85.

-: Mass- und Integraltheorie, Lecture Notes, Univ. Saarbrücken 1993.

POLLARD, D., TOPSØE, F.: A unified Approach to Riesz Type Representation Theorems.

Studia Math. 54 (1975), 173-190.

RIDDER, J.: Dualität in den Methoden zur Erweiterung von beschränkt- wie von total-additiven Massen II, III. Indag. Math. 33 (1971), 399-410 and 35 (1973), 393-396.

TOPSØE, F.: Compactness in Spaces of Measures. Studia Math. 36 (1970), 195-212.

-: Topology and Measure. LNM 133, Springer 1970.

-: Further Results on Integral Representations. Studia Math. 55 (1976), 239-245.

-: Radon Measures, some basic Constructions. In: Proc. Conf. Measure Theory and Applications, Sherbrooke 1982. LNM 1033, Springer 1983, pp. 303-311.

ZAANEN, A.C.: A Note on the Daniell-Stone Integral. In: Proc. Coll. Analyse Fonctionelle, Louvain 1960. Centre Belge Rech. Math. 1961, pp. 63-69.

-: Integration. North-Holland 1967.

Heinz König

Fachbereich Mathematik

Universität des Saarlandes

D-66041 Saarbrücken

Germany

Operator Theory:
Advances and Applications, Vol. 75
© 1995 Birkhäuser Verlag Basel/Switzerland

DIAGONALS OF THE POWERS OF AN OPERATOR ON A BANACH LATTICE

W.A.J. Luxemburg, B. de Pagter and A. R. Schep

Dedicated to Professor Dr. A.C. Zaanen on the occasion of his eightieth birthday

INTRODUCTION

This paper is devoted to a detailed study of the properties of the band projection $\mathcal{D}$ of the complete lattice ordered algebra $\mathcal{L}_r(E)$ of the regular (or order bounded) operators of a Dedekind complete Banach lattice E onto the center $Z(E)$ of E. We recall that the center $Z(E)$ is the commutative subalgebra of $\mathcal{L}_r(E)$ of all T satisfying $|T| \leq \lambda I$, where I is the identity operator. In the finite dimensional case, with respect to the standard numerical basis, $Z(E)$ is the algebra of all diagonal matrices. For this reason the band projection $\mathcal{D}$ is called the diagonal map of E.

The band decomposition $\mathcal{L}_r(E) = Z(E) \oplus Z(E)^d$ of $\mathcal{L}_r(E)$ shows that every $T \in \mathcal{L}_r(E)$ can be written uniquely as the disjoint sum of the diagonal part $\mathcal{D}(T)$ and its disjoint complement $\mathcal{D}_\perp(T)$ which is disjoint from I.

In section 1 some of the basic definitions and terminology are given and a number of the basic properties of the diagonal map to be used later are recalled. A number of examples are presented here to illustrate the nature of the diagonal part of a number of special operators.

Section 2 deals with the properties of the sequence $\{\mathcal{D}(T^n); n = 0, 1, 2, \ldots\}$ of the diagonals of the powers of an operator. The main result of the paper is contained in Proposition 2.3 which states that if S and T are commuting positive operators normalized to be of spectral radius less than or equal to one, then there exists a sequence $\{F_n : n = 1, 2, \ldots\}$ of positive diagonal operators satisfying for all $n = 1, 2, \ldots$

$$\mathcal{D}(ST^{n-1}) - \sum_{k=0}^{n-1} F_{n-k}\mathcal{D}(T^k) \text{ and } \sum_{k=1}^{n} F_k \leq I.$$

The authors would like to express their gratitude to Phillipe Clément, who brought Kingman's book ([13]) and the theory of renewal sequences to their attention.

The interesting consequences of this result, many of which are new even in the finite dimensional case, are discussed in the remainder of the paper. For instance in section 2 it is shown to imply that if $T \geq 0$ and $r(T) \leq 1$, then the sequence $\{S_n = I + \mathcal{D}(T) + \cdots + \mathcal{D}(T^{n-1}) : n = 1, 2, \ldots\}$ is subadditive. This implies that the following type of mean ergodic theorem holds for positive operators, namely that the averages $\frac{S_n}{n}$ are order convergent to a diagonal operator $Q \in Z(E)$, where Q is the disjoint sum of the band projection of E onto the band of fixed points of $\mathcal{D}(T)$ and its disjoint complement in $Z(E)$.

Section 3 is devoted to a discussion of an important inequality satisfied by the resolvent $R(\lambda, T)$ of an order bounded linear operator T. This inequality was conjectured by T. Andô in 1977 and is called here Andô's inequality. It states that if $T \in \mathcal{L}_r(E)$, then for all $|\lambda| > r(|T|)$ we have $|TR(\lambda, T)| \leq |\lambda R(\lambda, T)|$. This inequality, among other things, has the interesting consequence that, if $r(|T|) \leq 1$, the sequence of diagonal operators $\{D_n : n = 0, \pm 1, \pm 2, \ldots\}$, defined by $D_n = \mathcal{D}(T^n)$ for $n \geq 0$ and $D_n = \overline{\mathcal{D}(T^{-n})}$ for $n < 0$, is a generalized operator valued positive definite sequence. Since $Z(E)$ can be represented as a $C(K)$-space, where K is a compact extremally disconnected Hausdorff space, the above result shows that for all $\omega \in K$, the sequence $\{D_n(\omega) : n = 0, \pm 1, \pm 2, \ldots\}$ is positive definite.

In section 4, however, by taking $S = T$ in Proposition 2.3 mentioned earlier, the surprising result is obtained that for $T \geq 0$ with $r(T) \leq 1$ and for all $\omega \in K$, the sequence $\{\mathcal{D}(T^n)(\omega) : n = 0, 1, 2, \ldots\}$ is a renewal sequence in the sense of Feller, i.e., a sequence of recurrent probabilities. This fact implies in particular the positive definite property of these sequences mentioned above in case $T \geq 0$. Other consequences of this fundamental result are presented in section 4 as well. In particular it is shown that, by using the fact that renewal sequences are periodic, every positive operator T defines a period function $d(\omega), \omega \in K$. From the main result Proposition 2.3 the following extension of Andô's inequality is deduced in Proposition 4.5, $|\mathcal{D}(SR(\lambda, T))| \leq |\mathcal{D}(\lambda R(\lambda, T))|$ for all $|\lambda| > 1 = r(T) = r(S)$, where S and T are commuting positive operators.

Section 5 deals with a detailed analysis of the properties of the diagonals of the powers of linear lattice homomorphisms and the more general disjointness preserving operators. Numerous new results are presented and known results improved or given new and simpler proofs. Generalized unilateral shifts containing the ℓ_2-weighted unilateral shifts are introduced for general Banach lattices. In particular it is shown that such operators share with their ℓ_2 counterparts the property that their spectra are closed disks.

1. PRELIMINARIES

We will assume that the reader is familiar with the basic terminology and theory of Banach lattices, as can be found e.g. in the books ([2], [15], [16] and [21]). All Banach lattices E in this paper will be assumed to be *complex Banach lattices*, i.e., $E = \operatorname{Re} E \oplus i \operatorname{Re} E$, where $\operatorname{Re} E$ is a real Banach lattice. Recall that the absolute value in $\operatorname{Re} E$ is extended to E by means of the formula $|z| = \sup\{(\cos\theta)x + (\sin\theta)y : 0 \le \theta \le 2\pi\}$, where $z = x + iy$ with $x, y \in \operatorname{Re} E$ (see e.g [21] , Sect. 91), and the norm in E satisfies $\|z\| = \||z|\|$ for all $z \in E$. *We assume throughout this paper that E is Dedekind complete* (i.e., that $\operatorname{Re} E$ is a Dedekind complete vector lattice). We denote by $\mathcal{L}_r(E)$ the space of regular operators on E, i.e., those operators which can be written as linear combinations of positive operators. Note that $\mathcal{L}_r(E)$ coincides with the space of all order bounded operators on E, and is also denoted by $\mathcal{L}_b(E)$. The space $\mathcal{L}_r(E)$ is a subspace of the space $\mathcal{L}(E)$ of all bounded linear operators on E, but in general $\mathcal{L}_r(E)$ is not a closed subspace. Under the assumptions, the space $\mathcal{L}_r(E)$ is a complex vector lattice (where the positive cone consists of all positive operators on E), and for $T \in \mathcal{L}_r(E)$ the modulus $|T|$ is given by $|T|(x) = \sup\{|Tz| : |z| \le x\}$ for all $0 \le x \in E$. For any $T \in \mathcal{L}_r(E)$, the *regular norm* is defined by $\|T\|_r := \||T|\|$, and then $(\mathcal{L}_r(E), \|\cdot\|_r)$ is a complex Banach lattice algebra. Note that $\|T\| \le \|T\|_r$ holds for all $T \in \mathcal{L}_r(E)$, but in general the two norms $\|\cdot\|$ and $\|\cdot\|_r$ are not equivalent. The *center* $Z(E)$ of E can now be defined by $Z(E) = \{T \in \mathcal{L}_r(E) : |T| \le \lambda I \text{ for some } \lambda \ge 0\}$. It is well known that $Z(E)$ is a commutative full subalgebra of the space $\mathcal{L}(E)$. If $T \in Z(E)$, then $\|T\| = \|T\|_r = \inf\{\lambda \ge 0 : |T| \le \lambda I\}$.

EXAMPLES 1.1. (i) Let $E = \mathbb{C}^n$, considered as a Banach lattice with respect to the standard positive cone and any lattice norm. A linear operator T can be represented by an $n \times n$-matrix (t_{ij}) with respect to the standard basis. Then the center $Z(E)$ corresponds to the diagonal matrices.

(ii) Let E be a Banach function space on the (σ-finite) measure space (Ω, Σ, μ). Then $Z(E)$ can be identified with $L^\infty(\mu)$ acting by multiplication on E (see [21], section 142).

The idempotent elements in $Z(E)$ are precisely the band projections in E; every operator in $Z(E)$ can be approximated, in operator norm, by linear combinations of band projections. Since $Z(E)$ is a Dedekind complete AM–space, $Z(E)$ is algebraically and isometrically isomorphic to some $C(K)$ space, where K is an extremally disconnected compact Hausdorff space (actually K is the Stone representation space of the Boolean algebra of all band projections in E).

By definition, $Z(E)$ is equal to the principal ideal generated by I in $\mathcal{L}_r(E)$. Actually, $Z(E)$ is a band in $\mathcal{L}_r(E)$ (see [21], section 142), and so we have the band decomposition

$$\mathcal{L}_r(E) = Z(E) \oplus Z(E)^d.$$

The corresponding band projection in $\mathcal{L}_r(E)$ onto $Z(E)$ will be denoted by $\mathcal{D}$. The projection onto the disjoint complement $Z(E)^d$ will be denoted by $\mathcal{D}_\perp$, i.e., $\mathcal{D}_\perp = I - \mathcal{D}$. In this paper we will call $\mathcal{D}(T)$ the *diagonal* of the operator T; $\mathcal{D}$ is called the *diagonal map*. We collect some of the known properties of the diagonal map in the following proposition.

PROPOSITION 1.2. *Let E be a Dedekind complete Banach lattice. Then the following hold.*

(i) $\mathcal{D}(\pi T) = \mathcal{D}(T\pi) = \pi \mathcal{D}(T)$ for all $T \in \mathcal{L}_r(E)$ and all $\pi \in Z(E)$;

(ii) $\mathcal{D}(ST) \geq \mathcal{D}(S)\mathcal{D}(T)$ for all $0 \leq T, S \in \mathcal{L}_r(E)$;

(iii) For $0 \leq T \in \mathcal{L}_r(E)$ we have

$$\mathcal{D}(T) = \inf\left\{ \sum_{i=1}^n P_i T P_i : \sum_{i=1}^n P_i = I, P_i^2 = P_i, P_i \geq 0, n \in \mathbb{N} \right\};$$

(iv) $0 \leq \mathcal{D}(T) \leq r(T)I$ for all $0 \leq T \in \mathcal{L}_r(E)$ *(where $r(T)$ denotes the spectral radius of T)*;

(v) For all $T \in \mathcal{L}_r(E)$ we have $\|\mathcal{D}(T)\| \leq \|T\|$.

PROOF. (i) Given $\pi \in Z(E)$ we define the operators L_π and R_π on $\mathcal{L}_r(E)$ by $L_\pi(T) = \pi T$ and $R_\pi(T) = T\pi$ for all $T \in \mathcal{L}_r(E)$. It is now easy to see that $L_\pi, R_\pi \in Z(\mathcal{L}_r(E))$. Since $\mathcal{D} \in Z(\mathcal{L}_r(E))$, the operators L_π and R_π both commute with $\mathcal{D}$, from which (i) follows.

(ii) Since $ST = S(\mathcal{D}(T) + \mathcal{D}_\perp(T)) = S\mathcal{D}(T) + S\mathcal{D}_\perp(T)$, it follows via (i) that $\mathcal{D}(ST) = \mathcal{D}(S)\mathcal{D}(T) + \mathcal{D}(S\mathcal{D}_\perp(T))$, and hence $\mathcal{D}(ST) \geq \mathcal{D}(S)\mathcal{D}(T)$.

(iii) This is proved in [17].

(iv) Using (ii) we see that we see that $0 \leq \mathcal{D}(T)^n \leq \mathcal{D}(T^n) \leq \|T^n\| \cdot I$ for all $n \geq 1$. By taking n-th roots and letting $n \to \infty$ we see that (iv) holds.

(v) This is a theorem of J. Voigt ([19]). $\square$

Property (i) in the above proposition is sometimes called the *averaging property* of the diagonal map $\mathcal{D}$. In connection with (ii) above we note that in general $\mathcal{D}$ is not multiplicative. In the next section we will obtain, in an important special case, a nontrivial upperbound for $\mathcal{D}(ST)$.

EXAMPLES 1.3. (i) Let $E = \mathbb{C}^n$ as in Example 1.1 (i). For any linear operator $T : \mathbb{C}^n \to \mathbb{C}^n$, with matrix (t_{ij}) with respect to the standard basis, $\mathcal{D}(T)$ corresponds to the diagonal matrix $\operatorname{diag}(t_{11}, \ldots, t_{nn})$.

Similarly, if $E = \ell_p$ $(1 \leq p < \infty)$ and $T : \ell_p \to \ell_p$ is a regular operator which is given by the infinite matrix (t_{ij}), then $\mathcal{D}(T)$ corresponds to the infinite diagonal matrix $\text{diag}(t_{11}, t_{22}, \dots)$. Note that these diagonal matrices can be identified with ℓ_∞-sequences, acting via multiplication on ℓ_p (as observed already in Example 1.1 (ii), $Z(\ell_p)$ can be identified in this way with ℓ_∞).

(ii) Let E be an extremally disconnected Hausdorff space, and consider the Banach lattice $E = C(K)$. In this case $Z(E)$ can be identified with the space $C(K)$ acting by multiplication on itself (see [21], section 141).

Given a continuous mapping $\phi : K \to K$, and a function $0 \leq h \in C(K)$ we define the positive linear operator $T : C(K) \to C(K)$ by $Tf(\omega) = h(\omega)f(\phi(\omega))$ for all $\omega \in K$ and all $f \in C(K)$. In order to describe the diagonal of T, we introduce the following notation. Let $\text{Fix}(\phi) = \{\omega \in K : \phi(\omega) = \omega\}$, and $F_\phi = \text{int Fix}(\phi)$. Then F_ϕ is an open and closed subset of K, so $\chi_{F_\phi} \in C(K)$. With slight abuse of notation, we denote by χ_{F_ϕ} also the operator of multiplication by χ_{F_ϕ}. Now it is not difficult to see that $\mathcal{D}(T) = \chi_{F_\phi} T$, i.e., $\mathcal{D}(T)f = \chi_{F_\phi} hf$ for all $f \in C(K)$.

(iii) Let (X, Σ, μ) be a σ-finite measure space and consider the Banach lattice $E = L_p(X, \mu)$ with $1 \leq p \leq \infty$. Suppose that $\phi : X \to X$ is a measurable map (i.e., $\phi^{-1}(A) \in \Sigma$ for all $A \in \Sigma$), which is null-preserving (i.e., if $A \in \Sigma$ and $\mu(A) = 0$, then $\mu(\phi^{-1}(A)) = 0$). Let $h \geq 0$ be a measurable function. For a measurable function f on X we define the measurable function Tf by $Tf(\omega) = h(\omega)f(\phi(\omega))$ a.e. on X. Under suitable conditions on ϕ and h, this defines a positive linear operator $T : L_p(X, \mu) \to L_p(X, \mu)$ (actually, a necessary and sufficient for this is, in case $1 \leq p < \infty$, that there exists a constant $C \geq 0$ such that $\int_{\phi^{-1}(A)} h^p \, d\mu \leq C\mu(A)$ for all $A \in \Sigma$ with $\mu(A) < \infty$; if $p = \infty$, then the only condition is that $h \in L_\infty(X, \mu)$). We introduce some terminology. A set $A \in \Sigma$ is called ϕ-stable if $A \subseteq \phi^{-1}(A)$ (neglecting, here and in what follows, μ-nullsets). A set $F \in \Sigma$ is called ϕ-fixed if all its measurable subsets are ϕ-stable. Denote the collection of all ϕ-fixed sets by $\mathcal{F}_\phi$. Then $\mathcal{F}_\phi$ is a σ-ideal in Σ, and hence, by a standard exhaustion argument, there exists a unique $F_\phi \in \Sigma$ such that $\mathcal{F}_\phi = \{F \in \Sigma : F \subseteq F_\phi\}$. We note that in case X is a separable metric space and Σ contains all Borel sets, then it is not difficult to show that $F_\phi = \{\omega \in X : \phi(\omega) = \omega\}$. Now we can describe the diagonal of the above operator T. Indeed, using that $F_\phi \cap \phi^{-1}(A) = F_\phi \cap A$ for all $A \in \Sigma$, and that $Z(E) = L_\infty(X, \mu)$ (see Example 1.1 (ii)), it follows easily that $\mathcal{D}(T) = \chi_{F_\phi} T$, i.e., $\mathcal{D}(T)f = \chi_{F_\phi} hf$ for all $f \in L_p(X, \mu)$.

2. PROPERTIES OF THE DIAGONAL MAP

In Proposition 1.2(ii) we have seen that $\mathcal{D}(ST) \geq \mathcal{D}(S)\mathcal{D}(T)$ for all $0 \leq S, T \in \mathcal{L}_r(E)$. In the next theorem we obtain, under suitable conditions, an upper bound for

$\mathcal{D}(ST)$.

THEOREM 2.1. *Let E be a Dedekind complete Banach lattice and let $0 \leq S, T \in \mathcal{L}_r(E)$ such that $ST = TS$ and $r(S), r(T) \leq 1$. Then $\mathcal{D}(ST) \leq \mathcal{D}(S)\mathcal{D}(T) + I - \sup\{\mathcal{D}(S), \mathcal{D}(T)\}$.*

PROOF. Let $U = \mathcal{D}(ST) - \mathcal{D}(S)\mathcal{D}(T)$. Then we have

$$(1) \qquad U = \mathcal{D}(ST) - \mathcal{D}(\mathcal{D}(S)T) = \mathcal{D}((S - \mathcal{D}(S))T) \leq (S - \mathcal{D}(S))T.$$

Now take $0 \leq u \in E$ and $\lambda > 1$, and put $w = R(\lambda, S)R(\lambda, T)u$. Since $SR(\lambda, S)u = \{\lambda R(\lambda, S) - I\}u \leq \lambda R(\lambda, S)u$, we have $Sw = R(\lambda, T)SR(\lambda, S)u \leq \lambda R(\lambda, T)R(\lambda, S)u = \lambda w$. Similarly $Tw \leq \lambda w$. Hence from (1) it follows that

$$Uw \leq (S - \mathcal{D}(S))Tw \leq \lambda(Sw - \mathcal{D}(S)w) \leq \lambda(\lambda w - \mathcal{D}(S)w),$$

i.e., $[U - \lambda^2 I + \lambda \mathcal{D}(S)]^+ w = 0$. Since $0 \leq u \leq \lambda^2 w$, this implies that $[U - \lambda^2 I + \lambda \mathcal{D}(S)]^+ u = 0$, i.e., $Uu \leq \lambda(\lambda - \mathcal{D}(S))u$. This holds for all $\lambda > 1$, so that $0 \leq Uu \leq (I - \mathcal{D}(S))u$ for all $0 \leq u \in E$. This shows that $0 \leq U \leq I - \mathcal{D}(S)$, which is equivalent to the conclusion of the theorem, using the symmetry between S and T. □

Applying the above result to the powers of an operator, we immediately get the following consequence.

COROLLARY 2.2. *Let E be a Dedekind complete Banach lattice and let $0 \leq T \in \mathcal{L}_r(E)$ with $r(T) \leq 1$. Then $\mathcal{D}(T^n)\mathcal{D}(T^m) \leq \mathcal{D}(T^{n+m}) \leq \mathcal{D}(T^n)\mathcal{D}(T^m) + I - \mathcal{D}(T^m)$ for all $n, m = 1, 2, \ldots$.*

Next we will obtain some more inequalities for the diagonals. For this we need the following proposition. which will also be of importance in later sections.

PROPOSITION 2.3. *Let E be a Dedekind complete Banach lattice and let $0 \leq S, T \in \mathcal{L}_r(E)$ such that $ST = TS$ and $r(S), r(T) \leq 1$. Then there exists a (unique) sequence $\{F_n\}_{n=1}^{\infty}$ in $Z(E)^+$ such that*

$$(2) \qquad \mathcal{D}(ST^{n-1}) = \sum_{k=0}^{n-1} F_{n-k}\mathcal{D}(T^k)$$

and $\sum_{k=1}^{n} F_k \leq I$ for all $n \geq 1$.

PROOF. Define the sequence $\{G_n\}_{n=1}^{\infty}$ in $\mathcal{L}_r(E)^+$ inductively by

$$G_1 = S, \qquad G_n = \mathcal{D}_\perp(G_{n-1})T \qquad (n \geq 2),$$

and let $F_n = \mathcal{D}(G_n)$. Note that $G_n T = G_{n+1} + F_n T$ for all $n \geq 1$. We first show that

(3)
$$ST^{n-1} = G_n + \sum_{k=1}^{n-1} F_{n-k} T^k$$

for all $n \geq 2$. Since $ST = \mathcal{D}_\perp(S)T + \mathcal{D}(S)T = G_2 + F_1 T$, (3) holds for $n = 2$. Assuming that (3) holds for some $n \geq 2$ we have

$$ST^n = ST^{n-1}T = \left\{ G_n + \sum_{k=1}^{n-1} F_{n-k} T^k \right\} T$$

$$= G_n T + \sum_{k=1}^{n-1} F_{n-k} T^{k+1}$$

$$= G_{n+1} + F_n T + \sum_{k=1}^{n-1} F_{n-k} T^{k+1} = G_{n+1} + \sum_{k=1}^{n} F_{n+1-k} T^k$$

which is (3) for $n + 1$. From (3) it follows immediately that

$$\mathcal{D}(ST^{n-1}) = \sum_{k=0}^{n-1} F_{n-k} \mathcal{D}(T^k)$$

for all $n \geq 2$. It remains to prove that $\sum_{k=1}^{n} F_k \leq I$ for all $n \geq 1$. To this end take $0 \leq u \in E, \lambda > 1$ and put $w = R(\lambda, S)R(\lambda, T)u$. Since $Tw \leq \lambda w$, we have $0 \leq G_n w = \mathcal{D}_\perp(G_{n-1})Tw \leq \lambda \mathcal{D}_\perp(G_{n-1})w = \lambda(G_{n-1}w - F_{n-1}w)$ for all $n \geq 2$. We claim that

(4)
$$0 \leq G_n w \leq \lambda^n w - \sum_{k=1}^{n-1} \lambda^k F_{n-k} w$$

for all $n \geq 2$. Indeed, it follows from $Sw \leq \lambda w$ that $0 \leq G_2 w \leq \lambda(G_1 w - F_1 w) = \lambda(Sw - F_1 w) \leq \lambda^2 w - \lambda F_1 w$, which is (4) for $n = 2$. Now assume that (4) holds for some $n \geq 2$. Then

$$0 \leq G_{n+1}w \leq \lambda(G_n w - F_n w) \leq \lambda \left(\lambda^n w - \sum_{k=1}^{n-1} \lambda^k F_{n-k}w \right) - \lambda F_n w$$

$$= \lambda^{n+1} w - \sum_{k=1}^{n} \lambda^k F_{n+1-k} w,$$

and this proves the claim. It now follows from (4) that

$$0 \leq F_n w \leq G_n w \leq \lambda^n w - \sum_{k=1}^{n-1} \lambda^k F_{n-k} w,$$

and hence

$$\sum_{k=1}^{n} \lambda^{n-k} F_k w \leq \lambda^n w$$

for all $n \geq 2$, i.e.,

$$\left[\sum_{k=1}^{n} \lambda^{n-k} F_k - \lambda^n I\right]^+ w = 0$$

for all $n \geq 2$. Since $0 \leq u \leq \lambda^2 w$, this implies that

$$\left[\sum_{k=1}^{n} \lambda^{n-k} F_k - \lambda^n I\right]^+ u = 0,$$

and hence $\sum_{k=1}^{n} \lambda^{n-k} F_k u \leq \lambda^n u$ for all $n \geq 2$. This holds for any $\lambda > 1$, so we may conclude that $\sum_{k=1}^{n} F_k u \leq u$ for all $0 \leq u \in E$ and all $n \geq 2$, and this completes the proof of the proposition. $\square$

THEOREM 2.4. *Let E be a Dedekind complete Banach lattice and let $0 \leq S, T \in \mathcal{L}_r(E)$ such that $ST = TS$ and $r(S), r(T) \leq 1$. Then*

$$\mathcal{D}[S(I + T + \cdots + T^n)] \leq \mathcal{D}[I + T + \cdots + T^n]$$

for all $n \geq 0$.

PROOF. Let $\{F_k\}_{k=1}^{\infty}$ be the sequence in $Z(E)^+$ from the above proposition. For $n \geq 0$ we then have

$$\mathcal{D}\left[S\left(\sum_{k=0}^{n} T^k\right)\right] = \sum_{k=0}^{n} \mathcal{D}(ST^k)$$

$$= \sum_{k=0}^{n}\sum_{j=0}^{k} F_{k+1-j}\mathcal{D}(T^j) = \sum_{j=0}^{n}\sum_{k=j}^{n} F_{k+1-j}\mathcal{D}(T^j)$$

$$= \sum_{j=0}^{n}\left(\sum_{k=1}^{n+1-j} F_k\right)\mathcal{D}(T^j) \leq \sum_{j=0}^{n} \mathcal{D}(T^j). \quad \square$$

By taking $S = T^m$ in the above theorem we immediately get the following result.

COROLLARY 2.5. *Let E be a Dedekind complete Banach lattice and let $0 \leq T \in \mathcal{L}_r(E)$ with $r(T) \leq 1$. Then*

$$\mathcal{D}\left[\sum_{k=m}^{m+n} T^k\right] \leq \mathcal{D}\left[\sum_{k=0}^{n} T^k\right]$$

for all $m, n \geq 1$.

COROLLARY 2.6. *Let E be a Dedekind complete Banach lattice and let*
$0 \leq T \in \mathcal{L}_r(E)$ *with* $r(T) \leq 1$. *For* $n = 0, 1, \ldots$ *put* $S_n = I + \mathcal{D}(T) + \cdots + \mathcal{D}(T^{n-1})$. *Then*
$S_{n+m} \leq S_n + S_m$ *for all* $n, m \geq 0$.

PROOF. Using Corollary 2.5 we have

$$S_{n+m} = I + \cdots + \mathcal{D}(T^{m-1}) + \mathcal{D}(T^m + \cdots + T^{m+n-1})$$
$$\leq S_m + \mathcal{D}(I + \cdots + T^{n-1}) = S_m + S_n. \quad \square$$

The subadditivity of the sequence $\{S_n\}_{n=1}^{\infty}$ in the last corollary has some interesting consequences for the asymptotic behavior of the diagonals $\mathcal{D}(T^n)$ as $n \to \infty$. As is well known, if $\{a_n\}_{n=1}^{\infty}$ is a subadditive sequence in $\mathbb{R}$, then $\lim_{n \to \infty} \frac{a_n}{n} = \inf_n \frac{a_n}{n}$ exists ($-\infty$ is possible). A similar result holds in any Dedekind complete (real) vector lattice. We recall some relevant facts. Let L be a (real) vector lattice. The sequence $\{x_n\}_{n=1}^{\infty}$ in L is called order convergent to y in L if there exists a sequence $\{p_n\}_{n=1}^{\infty}$ in L such that $|x_n - y| \leq p_n$ $(n = 1, 2 \ldots)$ and $p_n \downarrow 0$ (see e.g. [14],section 16). This is denoted by $o - \lim_{n \to \infty} x_n = y$ or $x_n \xrightarrow{o} y$ $(n \to \infty)$. If we assume in addition that L is Dedekind complete, and $\{x_n\}_{n=1}^{\infty}$ is an order bounded sequence in L, then it is easy to show that $x_n \xrightarrow{o} y$ $(n \to \infty)$ if and only if $\limsup_{n \to \infty} x_n = \liminf_{n \to \infty} x_n = y$. Now it is a standard argument to prove the following result.

LEMMA 2.7. *Let L be a Dedekind complete vector lattice and suppose that* $\{u_n\}_{n=1}^{\infty}$ *is a sequence in L^+ satisfying* $u_{n+m} \leq u_n + u_m$ *for all $n, m \geq 1$. Then* $\{u_n/n\}_{n=1}^{\infty}$ *is order convergent in L, and*

$$o - \lim_{n \to \infty} \frac{u_n}{n} = \inf_n \frac{u_n}{n}.$$

A combination of Corollary 2.6 and the above lemma (with $L = Z(E)$) now yields the following theorem.

THEOREM 2.8. *Let E be a Dedekind complete Banach lattice and let $0 \leq T \in \mathcal{L}_r(E)$ with $r(T) \leq 1$. Then there exists $Q \in Z(E)$ such that*

$$o - \lim_{n \to \infty} \frac{I + \mathcal{D}(T) + \cdots + \mathcal{D}(T^{n-1})}{n} = Q$$

in $Z(E)$.

REMARK 2.9. (i) The limit operator Q in the above theorem is, in general, not a projection as the following simple example shows. Let $\mathbb{T}$ denote the unit circle (identified with $[0, 2\pi]$) and let $E = L_p(\mathbb{T}, d\theta/2\pi)$. Define $T : E \to E$ by means of $Tf(\theta) = f(\theta + \pi)$.

Then $\mathcal{D}(T^{2n+1}) = 0$ and $\mathcal{D}(T^{2n}) = I$ for all $n \geq 0$. Hence $Q = \frac{1}{2}I$ in this case. This example also shows that the order limit of $\{\mathcal{D}(T^n)\}_{n=1}^{\infty}$ does not exist in general.

(ii) In case that E is a Banach function space on some measure space (X, μ), $Z(E)$ can be identified with $L_{\infty}(X, \mu)$. We note that order convergence of an order bounded sequence in $L_{\infty}(X, \mu)$ is precisely pointwise convergence a.e. on X.

If E is a Banach lattice with order continuous norm, then it is clear that an order convergent sequence in $Z(E)$ is strongly convergent.

COROLLARY 2.10. *Let E be a Dedekind complete Banach lattice with order continuous norm and let $0 \leq T \in \mathcal{L}_r(E)$ with $r(T) \leq 1$. Then there exists $Q \in Z(E)$ such that*

$$\frac{I + \mathcal{D}(T) + \cdots + \mathcal{D}(T^{n-1})}{n} \to Q$$

strongly as $n \to \infty$.

It is a natural question to ask, whether one can identify the limit Q in Theorem 2.8 in some way. One such characterization will be given in section 4 of this paper. Here we will now present an answer in case the operator T is mean ergodic. We note that the following result is not completely obvious, since the diagonal map $\mathcal{D} : \mathcal{L}_r(E) \to Z(E)$ is in general not strongly continuous.

PROPOSITION 2.11. *Let E be a Dedekind complete Banach lattice and let $0 \leq T \in \mathcal{L}_r(E)$ with $r(T) \leq 1$. Assume that the Cesàro means $A_n = \frac{I + T + \cdots + T^n}{n+1}$ converge strongly to P as $n \to \infty$, i.e., assume that T is mean ergodic. Then $\mathcal{D}(P) = Q$, where Q is the Cesàro order limit of the sequence $\mathcal{D}(T^n)$ in $Z(E)$.*

PROOF. Since $0 \leq P \in \mathcal{L}_r(E,)$ we can write $P = R + S$, where $R = \mathcal{D}(P)$ and $S = \mathcal{D}_{\perp}(P)$. Then $PT = TP = P$ implies that $T^n P = P$, and hence $T^n R + T^n S = R + S$ for all $n \geq 0$. It follows that $\mathcal{D}_{\perp}(T^n R) \leq S$ for all $n \geq 0$, so $\mathcal{D}_{\perp}(A_n R) \leq S$ for all $n \geq 1$. This implies that

(5) $A_n R \leq S + \mathcal{D}(A_n)R$

for all $n \geq 1$. For each $x \in E$ the sequence $\{A_n x\}$ converges in norm to Px, so a subsequence of $A_n x$ converges in order to Px (see e.g. [21], Theorem 100.6). Since $\mathcal{D}(A_n)x$ converges in order to Qx for all $x \in E$, it follows from (5) that $PR \leq S + QR$, i.e. $R^2 + SR \leq S + QR$. By applying the projection $\mathcal{D}$ to both sides of this last inequality we get that $R^2 \leq QR$. On the other hand $\mathcal{D}(A_n) \leq A_n$ implies that $Q \leq P$, so $Q \leq R$. From this we get that $(R - Q)^2 \leq R(R - Q) \leq 0$ in $Z(E)$, and hence $Q = R$, i.e., $Q = \mathcal{D}(P)$. $\square$

3. ANDÔ'S INEQUALITY

As before, E is a Dedekind complete complex Banach lattice. Let $0 \leq T \in \mathcal{L}_r(E)$ and $|\lambda| > r(T)$. From the Neumann series of the resolvent operator $R(\lambda, T)$ one sees that $R(\lambda, T) \in \mathcal{L}_r(E)$ and $|R(\lambda, T)| \leq R(|\lambda|, T)$. In particular $R(\lambda, T) \geq 0$ for $\lambda > r(T)$. From the equation $TR(\lambda, T) = \lambda R(\lambda, T) - I$ for λ in the resolvent set $\rho(T)$ of T, it follows that $0 \leq TR(\lambda, T) \leq \lambda R(\lambda, T)$ for all $\lambda > r(T)$. This motivates the following theorem, which was conjectured by T. Andô in 1977 at an Oberwolfach meeting. Andô proved that this result was true for atomic Banach lattices (and published the proof of the finite dimensional case in [3]) and proved that the conclusion of this theorem was true in case $|\lambda| > 3r(T)$ for general Banach lattices.

THEOREM 3.1. *Let E be a Dedekind complete Banach lattice and let $T \in \mathcal{L}_r(E)$. Then $|TR(\lambda, T)| \leq |\lambda R(\lambda, T)|$ for all $|\lambda| > r(|T|)$.*

PROOF. Since for $\lambda \in \rho(T)$ we have $TR(\lambda, T) = \lambda R(\lambda, T) - I$, it is clear that $|\mathcal{D}_\perp[TR(\lambda, T)]| = |\mathcal{D}_\perp[\lambda R(\lambda, T)]|$, and so the theorem is equivalent to the inequality $|\mathcal{D}[TR(\lambda, T)]| \leq |\mathcal{D}[\lambda R(\lambda, T)]|$ in $Z(E)$. Let

$$Q = (|\mathcal{D}[TR(\lambda, T)]| - |\mathcal{D}[\lambda R(\lambda, T)]|)^+.$$

Then $(|\mathcal{D}[TR(\lambda, T)]| - |\mathcal{D}[\lambda R(\lambda, T)]|)^- Q = 0$, and so $|\mathcal{D}[\lambda R(\lambda, T)]|Q \leq |\mathcal{D}[TR(\lambda, T)]|Q$. Since $|\mathcal{D}_\perp[\lambda R(\lambda, T)]|Q = |\mathcal{D}_\perp[TR(\lambda, T)]|Q$, we get

$$
\begin{aligned}
|\lambda R(\lambda, T)|Q &= |\mathcal{D}[\lambda R(\lambda, T)] + \mathcal{D}_\perp[\lambda R(\lambda, T)]|Q \\
&= |\mathcal{D}[\lambda R(\lambda, T)]|Q + |\mathcal{D}_\perp[\lambda R(\lambda, T)]|Q \\
&\leq |\mathcal{D}[TR(\lambda, T)]|Q + |\mathcal{D}_\perp[TR(\lambda, T)]|Q = |TR(\lambda, T)|Q.
\end{aligned}
$$

Hence $|\lambda R(\lambda, T)|Q \leq |TR(\lambda, T)|Q \leq |T||R(\lambda, T)|Q$, and so $(|\lambda|I - |T|)|R(\lambda, T)|Q \leq 0$. Since $|\lambda| > r(|T|)$ we have $R(|\lambda|, |T|) \geq 0$, which now implies that

$$|R(\lambda, T)|Q = R(|\lambda|, |T|)(|\lambda|I - |T|)|R(\lambda, T)|Q \leq 0.$$

Therefore $|R(\lambda, T)|Q = 0$, i.e., $|R(\lambda, T)Q| = 0$, so $R(\lambda, T)Q = 0$, and this implies $Q = 0$. From the definition of Q it follows that $|\mathcal{D}[TR(\lambda, T)]| \leq |\mathcal{D}[\lambda R(\lambda, T)]|$, which is, as observed above, equivalent to the desired inequality. $\square$

Before we discuss the consequences of the above theorem (which we call 'Andô's inequality'), we first make some remarks.

REMARKS 3.2. (i) The condition $|\lambda| > r(|T|)$ in the above theorem is in a certain sense sharp, as can be seen in the following simple example. Let $T : \mathbb{C}^2 \to \mathbb{C}^2$ be the operator given by the matrix

$$\begin{pmatrix} 1 & -1 \\ 1 & -1 \end{pmatrix}.$$

Then $r(T) = 0$ and $r(|T|) = 2$. The set of all $\lambda \in \mathbb{C}$ for which the inequality $|TR(\lambda, T)| \le |\lambda R(\lambda, T)|$ holds, is given by $\{\lambda \in \mathbb{C} : |\lambda + 1| \ge 1 \text{ and } |\lambda - 1| \ge 1\}$, and the smallest disk, with center at 0, in which this set is contained has indeed radius 2.

(ii) The result of Theorem 3.1 can be extended easily to the setting of Banach lattice algebras. We only indicate briefly this extension. Let A be a (complex) Banach lattice algebra with unit element $e \ge 0$ (A is not assumed to be Dedekind complete). Let A_e denote the order ideal generated by e. Then A_e is a subalgebra of A which is actually an f-algebra. Without much difficulties one can show that A_e is a projection band in A. Using these observations, an argument similar to the proof of the above theorem now shows that for $x \in A$ we have $|xR(\lambda, x)| \le |\lambda R(\lambda, x)|$ for all $|\lambda| > r(|x|)$.

(iii) If E is a Dedekind complete Banach lattice and $T \in \mathcal{L}_r(E)$ is such that $r(|T|) = 0$, then (see Proposition 1.2 (iv)) $\mathcal{D}(|T|^n) = 0$ for all $n \ge 1$, and hence $\mathcal{D}(T^n) = 0$ for all $n \ge 1$. In this situation most of what follows is trivial, and therefore we will assume that $r(|T|) > 0$, and actually we shall use the normalization $r(|T|) = 1$.

(iv) Suppose that E is a Dedekind complete Banach lattice, $T \in \mathcal{L}_r(E)$ and $r(|T|) = 1$. As observed in the proof of Theorem 3.1, Andô's inequality is equivalent to the inequality $|\mathcal{D}[TR(\lambda, T)]| \le |\mathcal{D}[\lambda R(\lambda, T)]|$ for $|\lambda| > 1$. We shall indicate now that this latter inequality can be improved slightly. For this purpose we define $U(z) = \sum_{n=0}^{\infty} z^n \mathcal{D}(T^n)$ for $|z| < 1$. Then $U(0) = I$ and

$$U(z) = \mathcal{D}\left[\frac{1}{z}R(\frac{1}{z}, T)\right], \qquad 0 < |z| < 1.$$

Using Theorem 3.1 we find that

$$|U(z) - I| = \left|\mathcal{D}\left[TR(\frac{1}{z}, T)\right]\right| \le \left|\mathcal{D}\left[\frac{1}{z}R(\frac{1}{z}, T)\right]\right| = |U(z)|$$

for all $0 < |z| < 1$, and hence $|U(z) - I| \le |U(z)|$ for all $|z| < 1$. Identifying $Z(E)$ with a space $C(K)$, where K is an extremally disconnected compact Hausdorff space, we denote the value of the function $U(z)$ at the point $\omega \in K$ by $U(z, \omega)$. For $\omega \in K$ and $|z| < 1$ we define

$$V(z, \omega) = \frac{U(z, \omega) - I}{U(z, \omega)}.$$

For fixed $\omega \in K$, the function $V(\cdot, \omega)$ is analytic on $|z| < 1$ and satisfies $|V(z, \omega)| \leq 1$ for all $|z| < 1$ and $V(0, \omega) = 0$. From Schwartz's lemma it follows that $|V(z, \omega)| \leq |z|$ for all $|z| < 1$. Hence $|U(z) - I| \leq |zU(z)|$ in $Z(E)$ for all $|z| < 1$. Putting $z = 1/\lambda$, this implies that $|\mathcal{D}[TR(\lambda, T)]| \leq |\mathcal{D}[R(\lambda, T)]|$ for $|\lambda| > 1$.

Now we shall discuss some special properties of the sequence $\{\mathcal{D}(T^n)\}_{n=0}^{\infty}$ of diagonals of the powers of an operator T, which are consequences of Theorem 3.1.

PROPOSITION 3.3. *Let E be a Dedekind complete Banach lattice and let $T \in \mathcal{L}_r(E)$ with $r(|T|) = 1$. Then*

(6)
$$\mathrm{Re}\left\{ \mathcal{D}[\lambda R(\lambda, T)] - \frac{1}{2}I \right\} \geq 0$$

for all $|\lambda| > 1$.

PROOF. By Theorem 3.1 we have $|\mathcal{D}[TR(\lambda, T)]| \leq |\mathcal{D}[\lambda R(\lambda, T)]|$ for all $|\lambda| > 1$. Since $TR(\lambda, T) = \lambda R(\lambda, T) - I$, this implies that

(7)
$$|\mathcal{D}[\lambda R(\lambda, T)] - I| \leq |\mathcal{D}[\lambda R(\lambda, T)]|$$

for all $|\lambda| > 1$. Identifying $Z(E)$ with a $C(K)$–space, and using that for $z \in \mathbb{C}$ the inequality $|z - 1| \leq |z|$ is equivalent to $\mathrm{Re}\, z \geq 1/2$, it follows from (7) that $\mathrm{Re}\, \mathcal{D}[\lambda R(\lambda, T)] \geq \frac{1}{2}I$. $\square$

REMARK 3.4. Before proceeding, we first recall some classical results which go back to the work of Caratheodory, Toeplitz and Herglotz (see e.g. [6] and the references in there). Suppose that the analytic function $f(z)$ is given by $f(z) = \sum_{n=0}^{\infty} a_n z^n$ for $|z| < 1$, satisfying $f(0) = 1$ and $\mathrm{Re}\, f(z) \geq 1/2$ for all $|z| < 1$. For $0 \leq r < 1$ let $F_r(\theta) = 2\,\mathrm{Re}\{f(re^{i\theta}) - 1/2\}$, $0 \leq \theta \leq 2\pi$. Then $F_r(\theta) \geq 0$ and

$$F_r(\theta) = \sum_{n=-\infty}^{\infty} \hat{a}_n r^{|n|} e^{in\theta},$$

where $\hat{a}_n = a_n$ if $n \geq 0$ and $\hat{a}_n = \bar{a}_{-n}$ if $n < 0$. For $0 \leq r < 1$ we define $0 \leq \phi_r \in C(\mathbb{T})^*$ by

$$\phi_r(g) = \frac{1}{2\pi} \int_0^{2\pi} g(\theta) F_r(\theta)\, d\theta, \qquad g \in C(\mathbb{T}).$$

Note that $\|\phi_r\| = 1/(2\pi) \int_0^{2\pi} F_r(\theta)\, d\theta = 1$ for all $0 \leq r < 1$. If we denote $e_k(\theta) = e^{-ik\theta}$, then $\phi_r(e_k) = \hat{a}_k r^{|k|}$ for all $0 \leq r < 1$ and all $k \in \mathbb{Z}$. This shows in particular that $\lim_{r \uparrow 1} \phi_r(e_k) = \hat{a}_k$ for all $k \in \mathbb{Z}$, and since $\|\phi_r\| = 1$ for all $0 \leq r < 1$, this implies that there exists $0 \leq \phi \in C(\mathbb{T})^*$ such that $\lim_{r \uparrow 1} \phi_r(g) = \phi(g)$ for all $g \in C(\mathbb{T})$. Let μ be the probability measure on $\mathbb{T}$ which represents the functional ϕ, i.e., $\phi(g) = \int_0^{2\pi} g(\theta)\, d\mu(\theta)$ for all $g \in C(\mathbb{T})$. In particular we have

$$a_n = \int_0^{2\pi} e^{-in\theta}\, d\mu(\theta)$$

for all $n \geq 0$. We thus have shown that the a_n's are the Fourier coefficients of a unique probability measure μ on $\mathbb{T}$. If the function $f(z)$ has an analytic extension over an arc $J \subseteq \mathbb{T}$, then the measure μ is absolutely continuous on J with respect to $d\theta$, and for all measurable $A \subseteq J$ we have

$$\mu(A) = \frac{1}{\pi} \int_A \mathrm{Re}\left\{ f(e^{i\theta}) - \frac{1}{2} \right\} d\theta.$$

Observe that the sequence $\{\hat{a}_n\}_{n=-\infty}^{\infty}$ in the above, being the Fourier coefficients of a positive measure, is positive definite. Using Toeplitz determinants this can also be formulated as follows. For all $n \geq 1$ we have

$$\begin{vmatrix} 1 & a_1 & a_2 & \cdots & a_n \\ \bar{a}_1 & 1 & a_1 & \cdots & a_{n-1} \\ \bar{a}_2 & \bar{a}_1 & 1 & \cdots & a_{n-2} \\ \vdots & \vdots & \vdots & \ddots & \vdots \\ \bar{a}_n & \bar{a}_{n-1} & \bar{a}_{n-2} & \cdots & 1 \end{vmatrix} \geq 0.$$

Furthermore we note that if, conversely, a probability measure μ on $\mathbb{T}$ is given, and we define $a_n = \int_0^{2\pi} e^{-in\theta}\, d\mu(\theta)$ $(n \in \mathbb{N})$ and $f(z) = \sum_{n=0}^{\infty} a_n z^n$ for $|z| < 1$, then the function $f(z)$ satisfies $f(0) = 1$ and $\mathrm{Re}\, f(z) \geq 1/2$. The function f can also be represented as

$$f(z) = \int_0^{2\pi} \frac{1}{1 - e^{-it}z}\, d\mu(t)$$

for all $|z| < 1$.

Let μ be a probability measure on $\mathbb{T} = [0, 2\pi]$ and consider the Hilbert space $L_2(\mathbb{T}, \mu)$. We define the unitary operator U on $L_2(\mathbb{T}, \mu)$ by $Uf(\theta) = e^{-i\theta}f(\theta)$. The Fourier coefficients $\{c_n\}$ of μ can now be written as $a_n = \langle U^n \chi_{\mathbb{T}}, \chi_{\mathbb{T}} \rangle$ for all $n \in \mathbb{Z}$. By the mean ergodic theorem we know that

$$\lim_{n \to \infty} \frac{I + U + \cdots + U^{n-1}}{n} = P,$$

strongly on $L_2(\mathbb{T}, \mu)$, where P is the orthogonal projection onto the kernel $\ker (I - U)$ of $I - U$. This implies in particular that

$$\lim_{n \to \infty} \frac{a_0 + a_1 + \cdots + a_{n-1}}{n} = \langle P \chi_{\mathbb{T}}, \chi_{\mathbb{T}} \rangle.$$

It is easy to see that $\ker (I - U) = \{\alpha \chi_{\{0\}} : \alpha \in \mathbb{C}\}$, which is non-zero if and only if $\mu(\{0\}) > 0$. Moreover, the orthogonal projection P is given by $Ph = h(0)\mu(\{0\})\chi_{\{0\}}$ for $h \in L_2(\mathbb{T}, \mu)$. This yields the well-known result (see e.g. [12], p. 42) that

$$\lim_{n \to \infty} \frac{a_0 + a_1 + \cdots + a_{n-1}}{n} = \mu(\{0\}).$$

Now we return again to the situation where E is a Dedekind complete Banach lattice and $T \in \mathcal{L}_r(E)$ with $r(|T|) = 1$. Define the function U for $|z| < 1$ by

$$U(z) = \sum_{n=0}^{\infty} z^n \mathcal{D}(T^n).$$

Then $U(z) = \mathcal{D}[\frac{1}{z}R(\frac{1}{z}, T)]$ for all $0 < |z| < 1$, and so it follows from Proposition 3.3 that $\operatorname{Re} U(z) \geq \frac{1}{2}I$ for all $|z| < 1$. To avoid development of a theory of vector valued positive definite sequences we shall use the isomorphism $Z(E) \cong C(K)$, where K is an extremally disconnected compact Hausdorff space, and consider the elements of $Z(E)$ as continuous functions on K. The value of $\mathcal{D}(T^n)$ at a point $\omega \in K$ is denoted by $\mathcal{D}(T^n)(\omega)$, whereas the value of $U(z)$ at $\omega \in K$ is denoted by $U(z, \omega)$. For fixed $\omega \in K$, the function $U(\cdot, \omega)$ has all the properties of the function f considered in Remark 3.4. The following corollaries are now immediate consequences of the results mentioned in Remark 3.4.

COROLLARY 3.5.   *Let E, T and K be as above. For every $\omega \in K$ there exists a positive Borel measure μ_ω on $[0, 2\pi]$ such that*

$$\mathcal{D}(T^n)(\omega) = \int_0^{2\pi} \epsilon^{-in\theta} d\mu_\omega(\theta)$$

for all $n = 0, 1, 2, \ldots$. Furthermore, if $J \subseteq [0, 2\pi]$ is an open interval such that $\sigma(T) \cap \{e^{i\theta} : \theta \in J\} = \emptyset$, then for each $\omega \in K$ the measure μ_ω is absolutely continuous on J with respect to $d\theta/2\pi$ (and $\mu_\omega = \frac{1}{\pi}\operatorname{Re}\{U(e^{i\theta}, \omega) - \frac{1}{2}\} d\theta$ on J).

COROLLARY 3.6.   *Let E and T be as above. Define $D_n = \mathcal{D}(T^n)$ and $D_{-n} = \overline{\mathcal{D}(T^n)}$ for all $n \in \mathbb{N}$. Then the sequence $\{D_n\}_{n=-\infty}^{\infty}$ is positive definite, which means in terms of Toeplitz determinants that*

$$\begin{vmatrix} I & \mathcal{D}(T) & \mathcal{D}(T^2) & \cdots & \mathcal{D}(T^n) \\ \overline{\mathcal{D}(T)} & I & \mathcal{D}(T) & \cdots & \mathcal{D}(T^{n-1}) \\ \overline{\mathcal{D}(T^2)} & \overline{\mathcal{D}(T)} & I & \cdots & \mathcal{D}(T^{n-2}) \\ \vdots & \vdots & \vdots & \ddots & \vdots \\ \overline{\mathcal{D}(T^n)} & \overline{\mathcal{D}(T^{n-1})} & \overline{\mathcal{D}(T^{n-2})} & \cdots & I \end{vmatrix} \geq 0.$$

for all $n \geq 1$.

The above results can be used to obtain some information about the asymptotic behavior of the sequence $\{\mathcal{D}(T^n)\}_{n=0}^{\infty}$. In the next remark we collect some known facts about the embedding of $C(K)$ into the space $\mathcal{B}(K)$ of all bounded Borel functions on an extremally disconnected compact Hausdorff space K.

REMARK 3.7. Let K be an extremally disconnected compact Hausdorff space, and let $\mathcal{B}(K)$ be the space of bounded Borel functions on K. For every $f \in \mathcal{B}(K)$ there exists a unique $g \in C(K)$ such that the set $\{\omega \in K : f(\omega) \neq g(\omega)\}$ is meager. Denoting $g = \mathcal{P}(f)$, the mapping $\mathcal{P} : \mathcal{B}(K) \to \mathcal{B}(K)$ is a positive projection onto $C(K)$. Moreover, $\mathcal{P}$ is a σ-order continuous lattice and algebra homomorphism. For a proof of these facts we refer to [10, p. 113] and [20]. We sketch the proof of the σ-order continuity of $\mathcal{P}$, since an explicit proof of this is hard to find in these references. Let $f_n \downarrow 0$ in $\mathcal{B}(K)$. Then $f_n(\omega) \downarrow 0$ for all $\omega \in K$, so $\inf_n f_n(\omega) = 0$ for all $\omega \in K$. From the positivity of $\mathcal{P}$ it follows that $0 \leq \mathcal{P}f_n \downarrow$ in $C(K)$. Let $g = \inf_n \mathcal{P}f_n$ in $C(K)$. We first show that $\{\omega \in K : g(\omega) \neq \inf_n \mathcal{P}f_n(\omega)\}$ is a meager set. To prove this, denote $K_m = \{\omega \in K : \inf_n \mathcal{P}f_n(\omega) \geq g(\omega) + \frac{1}{m}\}$. Then K_m is a closed subset of K and thus int K_m is a clopen subset of K. For all $n \in \mathbb{N}$ we have now that $\mathcal{P}f_n \geq g + \frac{1}{m}\chi_{\text{int } K_m}$ in $C(K)$, so that $\chi_{\text{int } K_m} = 0$ for all m. Hence K_m is nowhere dense in K for all m, from which we conclude that $\{\omega \in K : g(\omega) \neq \inf_n \mathcal{P}f_n(\omega)\} = \cup_{m=1}^{\infty} K_m$ is a meager set in K. Now for each n the set $\{\omega \in K : \mathcal{P}f_n(\omega) \neq f_n(\omega)\}$ is meager and thus the set $\{\omega \in K : \inf_n \mathcal{P}f_n(\omega) \neq 0\}$ is meager. From this we conclude that $\{\omega \in K : g(\omega) \neq 0\}$ is meager, which implies that $g = 0$. Thus the projection $\mathcal{P}$ is σ-order continuous.

It is easy to see that a bounded sequence $\{g_n\}_{n=1}^{\infty}$ in $\mathcal{B}(K)$ is order convergent to $g \in \mathcal{B}(K)$ if and only if $g_n(\omega) \to g(\omega)$ $(n \to \infty)$ for all $\omega \in K$. Combined with the above observations this implies that if $\{f_n\}_{n=1}^{\infty}$ is a sequence in $C(K)$ such that $\lim_{n\to\infty} f_n(\omega)$ exists for all $\omega \in K$, then there exists $f \in C(K)$ such that $f_n \xrightarrow{o} f$ $(n \to \infty)$ in $C(K)$. Indeed, let $g(\omega) = \lim_{n\to\infty} f_n(\omega)$. $\omega \in K$. Then $f_n \xrightarrow{o} g$ $(n \to \infty)$ in $\mathcal{B}(K)$. The σ-order continuity of the projection $\mathcal{P}$ implies that $f_n = \mathcal{P}f_n \xrightarrow{o} \mathcal{P}g$ $(n \to \infty)$ in $C(K)$.

The next result is an extension of Theorem 2.8.

THEOREM 3.8. *Let E be a Dedekind complete Banach lattice and let $T \in \mathcal{L}_r(E)$ with $r(|T|) = 1$. Then there exists $0 \leq Q \in Z(E)$ such that*

$$(8) \qquad o- \lim_{n\to\infty} \frac{I + \mathcal{D}(T) + \cdots + \mathcal{D}(T^{n-1})}{n} = Q$$

in $Z(E)$.

PROOF. As before, we identify $Z(E)$ with a space $C(K)$, where K is an extremally disconnected compact Hausdorff space. By Corollary 3.5, for each $\omega \in K$ there exists a positive Borel measure μ_ω on $[0, 2\pi]$ such that

$$\mathcal{D}(T^n)(\omega) = \int_0^{2\pi} e^{-in\theta} d\mu_\omega(\theta)$$

for all $n = 0, 1, 2, \ldots$. As observed in Remark 3.4, for each $\omega \in K$ the limit

$$\lim_{n \to \infty} \frac{1 + \mathcal{D}(T)(\omega) + \cdots + \mathcal{D}(T^{n-1})(\omega)}{n} = \mu_\omega(\{0\})$$

exists. Now it follows via Remark 3.7 that there exists $Q \in C(K) \cong Z(E)$ satisfying (8). Since $\mu_\omega(\{0\}) \geq 0$ for all $\omega \in K$, it is clear that $Q \geq 0$. $\square$

THEOREM 3.9. *Let E be a Dedekind complete Banach lattice and let $T \in \mathcal{L}_r(E)$ with $r(|T|) = 1$. Assume moreover that $\sigma(T) \cap \{|\lambda| = 1\} = \{1\}$. Then there exists $0 \leq Q \in Z(E)$ such that*

$$o- \lim_{n \to \infty} \mathcal{D}(T^n) = Q$$

in $Z(E)$.

PROOF. We use the same notation as in the proof of Theorem 3.8. Since $\sigma(T) \cap \{|\lambda| = 1\} = \{1\}$, it follows that for each $\omega \in K$ the measure μ_ω is absolutely continuous on $(0, 2\pi)$ with respect to $d\theta/2\pi$. It is now a direct consequence of the Riemann-Lebesgue theorem that

$$\lim_{n \to \infty} \mathcal{D}(T^n)(\omega) = \lim_{n \to \infty} \int_{(0,2\pi)} e^{-in\theta} d\mu_\omega(\theta) + \mu_\omega(\{0\}) = \mu_\omega(\{0\})$$

for all $\omega \in K$. The conclusion of the theorem follows in the same way as in the proof of the previous theorem. $\square$

REMARK 3.10. (i) If the Banach lattice E has order continuous norm, then the order limits in Theorems 3.7 and 3.8 are strong limits as well (cf. Corollary 2.10).

(ii) The above theorems are only of interest if $r(T) = r(|T|) = 1$. Indeed, if $r(T) < r(|T|) = 1$, then it is clear from Proposition 1.2 (v) that $\mathcal{D}(T^n) \to 0$ in operator norm as $n \to \infty$.

4. RENEWAL SEQUENCES

In the preceding section we have seen that for any $T \in \mathcal{L}_r(E)$ with $r(|T|) = 1$, the sequence $\{\mathcal{D}(T^n)\}_{n=0}^\infty$ is positive definite. In this section we show that for a positive operator T this sequence $\{\mathcal{D}(T^n)\}_{n=0}^\infty$ has actually a much stronger property. To this end we first recall some facts about Markov chains (for details see [13]).

We consider a discrete parameter Markov chain, with state space $\{0, 1, 2, \ldots\}$ and with stationary transition probabilities $(p_{ij} : i, j = 0, 1, 2, \ldots)$, i.e., p_{ij} is the one step transition probability from state i to j. If we denote by $A = (p_{ij})$ the transition matrix, and write $A^n = (p_{ij}^{(n)})$, then $p_{ij}^{(n)}$ is the n–step transition probability from state i to j (put $A^0 = (\delta_{ij})$). For a fixed state k we consider the sequence $u_n = p_{kk}^{(n)} (n = 0, 1, 2, \ldots)$ of return probabilities (diagonal transition probabilities). For this fixed state k we can also consider the sequence $\{f_n\}_{n=1}^{\infty}$ of recurrence time probabilities, i.e., f_n is the probability of first return in n steps to state k. This sequence $\{f_n\}_{n=1}^{\infty}$ satisfies $f_n \geq 0$ and $\sum_{n=1}^{\infty} f_n \leq 1$ (as $\sum_1^{\infty} f_n$ is the probability of return to state k in finite time). The sequence $\{u_n\}_{n=0}^{\infty}$ is completely determined by the f_n's via the recurrence relation

$$(9) \qquad u_0 = 1, \qquad u_n = f_n + f_{n-1}u_1 + \cdots + f_1 u_{n-1} \qquad (n \geq 1).$$

Any sequence $\{u_n\}_{n=0}^{\infty}$ which is defined via the recurrence relation (9), where the f_n's satisfy $f_n \geq 0 \, (n = 1, 2, \ldots)$ and $\sum_{n=1}^{\infty} f_n \leq 1$, is called a *renewal sequence*. It is an important result (see [13], Theorem 1.1) that the class of renewal sequences coincides with the class of diagonal transition probabilities corresponding to a state in a discrete parameter Markov chain as described above.

The recurrence relation (9) can also be formulated in terms of generating functions. Indeed, let the sequence $\{f_n\}_{n=1}^{\infty}$ be given, satisfying $f_n \geq 0$ and $\sum_{n=1}^{\infty} f_n \leq 1$, and let $F(z) = \sum_{n=1}^{\infty} f_n z^n$ for $|z| < 1$. If we now define

$$(10) \qquad\qquad U(z) = \frac{1}{1 - F(z)}, \qquad |z| < 1$$

and write $U(z) = \sum_{n=0}^{\infty} u_n z^n$, then (10) is equivalent to (9).

Since $|F(z)| \leq 1$ for $|z| < 1$, it follows from (10) that $\operatorname{Re} U(z) \geq \frac{1}{2}$. This implies (see Remark 3.4) that the sequence $\{u_n\}_{n=0}^{\infty}$ is positive definite. It should be observed that, if a sequence $\{u_n\}_{n=0}^{\infty}$, with $u_0 = 1$, is positive definite, then the function $U(z) = \sum_{n=0}^{\infty} u_n z^n$, $|z| < 1$, can be written as in (10) with $|F(z)| \leq 1$ for all $|z| < 1$. Writing again $F(z) = \sum_{n=1}^{\infty} f_n z^n$, we see that renewal sequences are precisely those positive definite sequences, with first term equal to one, for which $f_n \geq 0$ for $n = 1, 2, \ldots$.

Let E be a Dedekind complete Banach lattice. As before we identify the center $Z(E)$ with a space $C(K)$ for some extremally disconnected compact Hausdorff space K. Our first objective is to show that for a positive operator $0 \leq T \in \mathcal{L}_r(E)$, with $r(T) \leq 1$, the sequence $\{\mathcal{D}(T^n)(\omega)\}_{n=0}^{\infty}$ is a renewal sequence for any $\omega \in K$. Actually we will prove this result under the slightly weaker condition that $\mathcal{D}(T^n) \leq I$ for all $n \in \mathbb{N}$.

REMARK 4.1. Let E be as above, and $0 \le T \in \mathcal{L}_r(E)$. By Proposition 1.2 (ii), the sequence $\{\mathcal{D}(T^n)\}_{n=0}^{\infty}$ satisfies $\mathcal{D}(T^{n+m}) \ge \mathcal{D}(T^n)\mathcal{D}(T^m)$ for all $n, m \ge 0$. From this property it is easy to see that the condition that $\mathcal{D}(T^n) \le I$ for all $n \in \mathbb{N}$ is equivalent to the seemingly weaker condition that $\limsup_{n \to \infty} \mathcal{D}(T^n) \le I$. Indeed, suppose the latter condition is fulfilled, then for fixed n we have

$$\bigwedge_{k=1}^{\infty} \bigvee_{j=kn}^{\infty} \mathcal{D}(T^j) \le I,$$

and hence $\bigwedge_{k=1}^{\infty} \mathcal{D}(T^{kn}) \le I$, so $\bigwedge_{k=1}^{\infty} \mathcal{D}(T^n)^k \le I$, which implies that $\mathcal{D}(T^n) \le I$.

Now it is easy to see that the condition that $\mathcal{D}(T^n) \le I$ for all $n \in \mathbb{N}$ is equivalent to saying that the analytic function $\lambda \mapsto \mathcal{D}[R(\lambda, T)]$, which is defined for $|\lambda| > r(T)$, has, in case $r(T) > 1$, an analytic extension, via the Laurent expansion $\sum_{n=0}^{\infty} \mathcal{D}(T^{n+1})\lambda^{-n-1}$, to the domain $\{|\lambda| > 1\}$.

THEOREM 4.2. Let E be a Dedekind complete Banach lattice and let $0 \le T \in \mathcal{L}_r(E)$. Assume that $\mathcal{D}(T^n) \le I$ for all $n \in \mathbb{N}$ (which is in particular satisfied if $r(T) \le 1$). Then there exists $0 \le F_n \in Z(E)\,(n = 1, 2, \ldots)$ with $\sum_{n=1}^{\infty} F_n \le I$ such that

(11) $$\mathcal{D}(T^n) = F_n + F_{n-1}\mathcal{D}(T) + \cdots + F_1\mathcal{D}(T^{n-1}), \qquad n \ge 1$$

and hence the sequence $\{\mathcal{D}(T^n)(\omega)\}_{n=0}^{\infty}$ is a renewal sequence for each $\omega \in K$ (where $Z(E) \cong C(K)$ as before).

PROOF. First note that if $r(T) \le 1$, then the theorem follows immediately from Proposition 2.3 (with $S = T$). In the more general situation of the theorem we define the elements $0 \le F_n \in Z(E)$ as in the proof of the proposition. Note that it follows in particular from (11) that $0 \le F_n \le \mathcal{D}(T^n)$. It remains to show that $\sum_{n=1}^{\infty} F_n \le I$. To do so, we put

$$U(z) = \sum_{n=0}^{\infty} z^n \mathcal{D}(T^n), \qquad F(z) = \sum_{n=1}^{\infty} z^n F_n,$$

which are defined for $|z| < 1$, since $0 \le \mathcal{D}(T^n), F_n \le I$. Then it follows from (11) that $U(z) = U(z)F(z) + I$ for all $|z| < 1$. For $0 \le x < 1$ we have $U(x) \ge I$, which implies that $0 \le F(x) = I - U(x)^{-1} \le I$, and hence $\sum_{n=1}^{\infty} F_n = o\text{-}\lim_{x \uparrow 1} \sum_{n=1}^{\infty} x^n F_n \le I$. $\square$

In order to formulate the next corollary we recall the definition of the order spectrum of an operator $T \in \mathcal{L}_r(E)$ (see e.g. [15]). The order resolvent set of $T \in \mathcal{L}_r(E)$ is defined by

$$\rho_o(T) = \{\lambda \in \rho(T) : R(\lambda, T) \in \mathcal{L}_r(E)\},$$

and the *order spectrum* of T is $\sigma_o(T) = \mathbb{C} \setminus \rho_o(T)$. In general, even for positive operators, $\sigma(T)$ is a proper subset of $\sigma_o(T)$. However, if $0 \leq T \in \mathcal{L}_r(E)$, then it follows via the Neumann series of $R(\lambda, T)$ that $\{\lambda \in \mathbb{C} : |\lambda| > r(T)\} \subseteq \rho_o(T)$. Using Theorem 4.2 we obtain the following extension of Andô's inequality (Theorem 3.1).

COROLLARY 4.3. *Let E be a Dedekind complete Banach lattice and let $0 \leq T \in \mathcal{L}_r(E)$. Assume that $\mathcal{D}(T^n) \leq I$ for all $n \in \mathbb{N}$, and denote by $\rho_{o,\infty}(T)$ the unbounded component of the order resolvent set. Then $|TR(\lambda, T)| \leq |\lambda R(\lambda, T)|$ for all $\lambda \in \rho_{o,\infty}(T)$ with $|\lambda| > 1$.*

PROOF. Let $U(z)$ and $F(z)$ be defined for $|z| < 1$ as in the proof of Theorem 4.2. Then $U(z) = (I - F(z))^{-1}$ for all $|z| < 1$, and since $|F(z)| \leq I$, this implies that $\operatorname{Re} U(z) \geq \frac{1}{2} I$ for all $|z| < 1$, i.e., $\operatorname{Re} U(\frac{1}{\lambda}) \geq \frac{1}{2} I$ for all $|\lambda| > 1$. If $|\lambda| > \max(1, r(T))$, then $U(\frac{1}{\lambda}) = \mathcal{D}[\lambda R(\lambda, T)]$, which implies by uniqueness of analytic continuation that $U(\frac{1}{\lambda}) = \mathcal{D}[\lambda R(\lambda, T)]$ for all $\lambda \in \rho_{o,\infty}(T)$ with $|\lambda| > 1$. Since $\operatorname{Re} \mathcal{D}[\lambda R(\lambda, T)] \geq \frac{1}{2} I$ is equivalent to $|\mathcal{D}[\lambda R(\lambda, T)] - I| \leq |\mathcal{D}[\lambda R(\lambda, T)]|$ (see the proof of Proposition 3.3), it follows that $|\mathcal{D}[TR(\lambda, T)]| \leq |\mathcal{D}[\lambda R(\lambda, T)]|$ for all $\lambda \in \rho_{o,\infty}(T)$ with $|\lambda| > 1$. Since $|\mathcal{D}_\perp[TR(\lambda, T)]| = |\mathcal{D}_\perp[\lambda R(\lambda, T)]|$, we may conclude that $|TR(\lambda, T)| \leq |\lambda R(\lambda, T)|$ holds for all such λ. $\square$

Via a simple normalization we obtain the following result.

COROLLARY 4.4. *Let E be a Dedekind complete Banach lattice and let $0 \leq T \in \mathcal{L}_r(E)$. Let $r = \limsup_{n \to \infty} \|\mathcal{D}(T^n)\|^{\frac{1}{n}}$. Then $|TR(\lambda, T)| \leq |\lambda R(\lambda, T)|$ for all $\lambda \in \rho_{o,\infty}(T)$ with $|\lambda| > r$.*

Note that the above provides in particular an alternative proof of Andô's inequality for positive operators. It should also be observed that the above result does not hold for operators $T \in \mathcal{L}_r(E)$ which are not positive, as can be seen in the example of Remark 3.2(i). We now show another extension of Andô's inequality.

PROPOSITION 4.5. *Let E be a Dedekind complete Banach lattice and let $0 \leq S, T \in \mathcal{L}_r(E)$ such that S and T commute. i.e., $ST = TS$, and $r(S), r(T) \leq 1$. Then $|\mathcal{D}(SR(\lambda, T))| \leq |\mathcal{D}(\lambda R(\lambda, T))|$ for all $|\lambda| > 1$.*

PROOF. By Proposition 2.3 there exist $0 \leq F_n \in Z(E)$ with $\sum_{n=1}^{\infty} F_n \leq I$

such that $\mathcal{D}(ST^{n-1}) = \sum_{k=0}^{n-1} F_{n-k}\mathcal{D}(T^k)$ for all $n \geq 1$. From this it follows that

$$\left(\sum_{k=1}^{\infty} F_k z^k\right)\left(\sum_{m=0}^{\infty} \mathcal{D}(T^m)z^m\right) = \sum_{n=1}^{\infty}\left(\sum_{k=0}^{n-1} F_{n-k}\mathcal{D}(T^k)\right)z^n$$

$$= \sum_{n=1}^{\infty} \mathcal{D}(ST^{n-1})z^n$$

$$= \mathcal{D}\left(S\sum_{n=1}^{\infty} T^{n-1}z^n\right)$$

for $|z| < 1$. Substituting $z = \frac{1}{\lambda}$ we obtain

$$\mathcal{D}(SR(\lambda, T)) = \left(\sum_{k=1}^{\infty} F_k \lambda^{-k}\right)\mathcal{D}(\lambda R(\lambda, T))$$

for all $|\lambda| > 1$, from which the desired inequality immediately follows, since $\sum_{k=1}^{\infty} F_k \leq I$. $\square$

Theorem 4.2 allows us to give another description of the Cesàro order limit of the sequence $\{\mathcal{D}(T^n)\}_{n=0}^{\infty}$ for positive operators.

PROPOSITION 4.6. *Let E be a Dedekind complete Banach lattice and let $0 \leq T \in \mathcal{L}_r(E)$ with $r(T) = 1$. Then*

$$(\lambda - 1)\mathcal{D}[R(\lambda, T)] \downarrow o- \lim_{n\to\infty} \frac{I + \mathcal{D}(T) + \cdots + \mathcal{D}(T^{n-1})}{n}$$

in $Z(E)$ as $\lambda \downarrow 1$.

PROOF. Let $U(z)$ and $F(z)$ be defined for $|z| < 1$ as in the proof of Theorem 4.2. If we put $\lambda = \frac{1}{z}$, then we have the identity $(\lambda - 1)\mathcal{D}[R(\lambda, T)] = (1 - z)U(z)$. For $0 < x < 1$ we find that

$$\frac{U(x)^{-1}}{1 - x} = \frac{I - F(x)}{1 - x} = (1 + x + \ldots)(I - xF_1 - x^2F_2 - \ldots)$$

$$= I + x(I - F_1) + x^2(I - F_1 - F_2) + \ldots,$$

and since $\sum_{1}^{\infty} F_n \leq I$, this implies that

$$I \leq \frac{U(x)^{-1}}{1 - x} \uparrow \quad \text{as} \quad x \uparrow 1.$$

This shows $I \geq (1 - x)U(x) \downarrow \geq 0$ in $Z(E)$ as $x \uparrow 1$. We know already that

$$\lim_{n\to\infty} \frac{1 + \mathcal{D}(T)(\omega) + \cdots + \mathcal{D}(T^{n-1})(\omega)}{n}$$

exists for all $\omega \in K$ (where as before $Z(E) \cong C(K)$). Via a classical Abelian limit theorem it now follows that

$$(1 - x)U(x, \omega) \downarrow \lim_{n \to \infty} \frac{1 + \mathcal{D}(T)(\omega) + \cdots + \mathcal{D}(T^{n-1})(\omega)}{n}$$

as $x \uparrow 1$ for all $\omega \in K$. Using the σ–order continuity of the projection of $\mathcal{B}(K)$ onto $C(K)$ (see Remark 3.7), we obtain the conclusion of the proposition. $\square$

REMARK 4.7. From the above proposition one can see that it is essential to distinguish between order convergence in $Z(E) \cong C(K)$ and pointwise convergence on K. Indeed, let $0 \leq T \in \mathcal{L}_r(E)$ with $r(T) = 1$, where, as before, E is a Dedekind complete Banach lattice. We denote by Q the Cesàro order limit of the sequence $\{\mathcal{D}(T^n)\}_{n=0}^{\infty}$. Now assume that

$$\lim_{n \to \infty} \frac{1 + \mathcal{D}(T)(\omega) + \cdots + \mathcal{D}(T^{n-1})(\omega)}{n} = Q(\omega)$$

for *all* $\omega \in K$. From the proof of the preceding proposition we see that this implies

$$(\lambda - 1)\mathcal{D}[R(\lambda, T)](\omega) \downarrow Q(\omega) \qquad (\lambda \downarrow 1)$$

for all $\omega \in K$. Now it follows from Dini's theorem that $\|(\lambda - 1)\mathcal{D}[R(\lambda, T)] - Q\| \to 0$ as $\lambda \downarrow 1$. Using a Banach lattice valued version of a Tauberian theorem of Hardy and Littlewood (see [11], Theorem A in section 4), we may conclude that

$$\frac{I + \mathcal{D}(T) + \cdots + \mathcal{D}(T^{n-1})}{n} \to Q$$

in *norm* in $Z(E)$. It is easily seen that this last statement is not true in general.

We note that it follows in particular from the above that, if $\mathcal{D}(T^n)(\omega) \to 0$ for all $\omega \in K$ as $n \to \infty$ (which is the case e.g. if the series $\sum_{n=0}^{\infty} \mathcal{D}(T^n)$ is order convergent in $Z(E)$), then the Cesàro means of $\{\mathcal{D}(T^n)\}_{n=0}^{\infty}$ are norm convergent to 0. However, in this situation one cannot conclude that the sequence $\{\mathcal{D}(T^n)\}_{n=0}^{\infty}$ is norm convergent to 0, as is shown by the following example.

EXAMPLE 4.8. For $n = 2, 3, \ldots$ we denote by $\ell_2^{(n)}$ the n–dimensional Euclidean space and let $T_n : \ell_2^{(n)} \to \ell_2^{(n)}$ be the linear operator with matrix

$$\sqrt[n]{\frac{1}{2}} \begin{pmatrix} 0 & 0 & \cdots & 0 & 1 \\ 1 & 0 & \cdots & 0 & 0 \\ 0 & 1 & \cdots & 0 & 0 \\ \vdots & \vdots & \ddots & \vdots & \vdots \\ 0 & 0 & \cdots & 1 & 0 \end{pmatrix}$$

with respect to the standard basis. Define the Banach lattice E by $E = \bigoplus\limits_{n=2}^{\infty} {}_{(2)}\ell_2^{(n)}$ and
$T : E \to E$ by $T = \bigoplus\limits_{n=2}^{\infty} T_n$. It is easy to see that $\sum_{n=0}^{\infty} \mathcal{D}(T^n) \leq 2I$ in $Z(E)$ and

$$\left\| \frac{I + \mathcal{D}(T) + \cdots + \mathcal{D}(T^{n-1})}{n} \right\| \leq \frac{3}{2n},$$

but $\|\mathcal{D}(T^n)\| = \frac{1}{2}$ for all $n = 2, 3, \ldots$.

Considering the above example, we see that the space E has a band decomposition $E = \{\bigoplus\limits_{n=2}^{\infty} \ell_2^{(n)}\}^{dd}$, such that on each band in the decomposition the sequence $\{\mathcal{D}(T^n)\}_{n=0}^{\infty}$ has a periodic character. Next we will show that such a decomposition actually exists for any positive operator. First we recall some facts about renewal sequences.

Let $\{u_n\}_{n=0}^{\infty}$ be a renewal sequence. We define the natural number d by

$$d = \gcd\{n \geq 1 : u_n > 0\},$$

where we put $d = \infty$ in case $u_n = 0$ for all $n \geq 1$. From [13], Theorem 1.4 we have, if $d < \infty$, that

(1) if d does not divide n, then $u_n = 0$;

(2) there exists k_0 such that $u_n > 0$ whenever $n = kd$ for all $k \geq k_0$.

This number d is called the *period* of the sequence $\{u_n\}_{n=0}^{\infty}$. Moreover, if $d < \infty$, then it follows from the Erdös–Feller–Pollard theorem (see [13], Theorem 1.6) that $\lim_{n\to\infty} u_{nd}$ exists. Renewal sequences with period $d = 1$ are called a-periodic.

THEOREM 4.9. *Let E be a Dedekind complete Banach lattice and let $0 \leq T \in \mathcal{L}_r(E)$ with $\mathcal{D}(T^n) \leq I$ for all $n \in \mathbb{N}$. For all $p \in \mathbb{N} \cup \{\infty\}$ there exist bands B_p in E, with corresponding band projections P_p, such that the following hold.*

(a) *The bands B_p are mutually disjoint and $((\bigoplus\limits_{p=1}^{\infty} B_p) \oplus B_\infty)^{dd} = E$.*

(b) *$P_\infty \mathcal{D}(T^n) = 0$ for all $n \in \mathbb{N}$ and if $B_p \neq \{0\}$, $p < \infty$, then*

(i) *$P_p \mathcal{D}(T^n) = 0$. whenever p does not divide n;*

(ii) *for every band projection P with $0 < P \leq P_p$ there exists $k(P) \in \mathbb{N}$ such that $P\mathcal{D}(T^n) > 0$ whenever $n = kp$ for all $k \geq k(P)$.*

(c) *The sequence $\{P_p \mathcal{D}(T^{kp})\}_{k=1}^{\infty}$ is order convergent in $Z(E)$ for all $p < \infty$.*

Moreover the bands B_p are uniquely determined by the properties (a) and (b).

PROOF. We identify $Z(E)$ with a space $C(K)$. By Theorem 4.2, the sequence $\{\mathcal{D}(T^n)(\omega)\}_{n=0}^{\infty}$ is a renewal sequence for all $\omega \in K$. For $\omega \in K$ we denote by $d(\omega)$ the period of the sequence $\{\mathcal{D}(T^n)(\omega)\}_{n=0}^{\infty}$, which defines a function $d : K \to \mathbb{N} \cup \{\infty\}$. First observe that the function d is upper–semicontinuous. Indeed, take $p \in \mathbb{N}$ and $\omega_0 \in K$ such that $d(\omega_0) < p$. From the definition of $d(\omega_0)$ it follows that there exist $1 \leq n_1 < n_2 < \cdots < n_k$ such that $\mathcal{D}(T^{n_j})(\omega_0) > 0$ and $d(\omega_0) = \gcd\{n_j : 1 \leq j \leq k\}$. By continuity there exists a neighborhood U of ω_0 such that $\mathcal{D}(T^{n_j})(\omega) > 0$ for all $\omega \in U$, and hence $d(\omega) \leq d(\omega_0) < p$ for all $\omega \in U$. This shows that the set $\{\omega \in K : d(\omega) < p\}$ is open.

Now define $A_p = \{\omega \in K : d(\omega) = p\}$ for $p \in \mathbb{N} \cup \{\infty\}$. By the above observation, each A_p is a Borel set. It is clear that the A_p's are mutually disjoint and $(\cup_{p=1}^{\infty} A_p) \cup A_\infty = K$, hence

$$\sum_{p=1}^{\infty} \chi_{A_p} + \chi_{A_\infty} = \chi_K$$

pointwise in $\mathcal{B}(K)$. Let $\mathcal{P}$ denote the projection of $\mathcal{B}(K)$ onto $C(K)$ (see Remark 3.7). From the σ–order continuity of $\mathcal{P}$ it follows that

$$(12) \qquad \sum_{p=1}^{\infty} \mathcal{P}(\chi_{A_p}) + \mathcal{P}(\chi_{A_\infty}) = \chi_K,$$

as an order convergent series in $C(K)$. The fact that $\mathcal{P}$ is a lattice homomorphism implies that $\mathcal{P}(\chi_{A_p}) = \chi_{B_p}$ for $p \in \mathbb{N} \cup \{\infty\}$, where the sets B_p are mutually disjoint clopen subsets of K such that $K \setminus \{(\cup_{p=1}^{\infty} B_p) \cup B_\infty\}$ has empty interior. Let P_p denote the band projection in E corresponding to $\chi_{B_p} \in C(K) \cong Z(E)$, and let B_p denote the corresponding band in E. Then (12) implies that

$$\sum_{p=1}^{\infty} P_p + P_\infty = I$$

as an order convergent series in $Z(E)$, by which (a) is proved.

To prove (b), we first note $\chi_{A_\infty} \mathcal{D}(T^n) = 0$, and so $P_\infty \mathcal{D}(T^n) = 0$ for all $n \geq 1$. Now assume that $p < \infty$ and $B_p \neq \{0\}$, i.e., A_p is nonmeager. Then $\mathcal{D}(T^n)(\omega) = 0$ for all $\omega \in A_p$ whenever p does not divide n, and so $\chi_{A_p} \mathcal{D}(T^n) = 0$ in $\mathcal{B}(K)$ for all such n. This implies that $\chi_{B_p} \mathcal{D}(T^n) = \mathcal{P}(\chi_{A_p} \mathcal{D}(T^n)) = 0$, i.e., $P_p \mathcal{D}(T^n) = 0$ in $Z(E)$ whenever p does not divide n. Now take a band projection P such that $0 < P \leq P_p$. Then P corresponds to a non–empty clopen set $O \subseteq B_p$. Since $B_p \triangle A_p$ is meager, we have $O \cap A_p \neq \emptyset$. Take $\omega \in O \cap A_p$. Then there exists $k(\omega) \in \mathbb{N}$ such that $\mathcal{D}(T^{kp})(\omega) > 0$ for all $k \geq k(\omega)$, and thus also $P\mathcal{D}(T^n) > 0$ whenever $n = kp$ for all $k \geq k(\omega)$. Hence $\chi_O \mathcal{D}(T^{kp}) > 0$, i.e., $P\mathcal{D}(T^{kp}) > 0$ for all $k \geq k(\omega)$. Therefore (ii) of (b) holds if we take $k(P) = k(\omega)$.

For the proof of (c), take $p < \infty$ such that $B_p \neq \{0\}$. It follows from the above mentioned theorem of Erdös–Feller–Pollard that the sequence $\{\mathcal{D}(T^{kp})(\omega)\}_{k=1}^{\infty}$ is convergent For all $\omega \in A_p$, i.e., the sequence $\{\chi_{A_p} \mathcal{D}(T^{kp})(\omega)\}_{k=1}^{\infty}$ is pointwise convergent in $\mathcal{B}(K)$. Since the projection $\mathcal{P}$ is σ-order continuous, this implies that the sequence $\{\chi_{B_p}\mathcal{D}(T^{kp})\}_{k=1}^{\infty}$ is order convergent in $C(K)$, which shows that (c) holds.

Finally we show the uniqueness of the above band decomposition. For this purpose, suppose that $\{C_j : j \in \mathbb{N} \cup \{\infty\}\}$ are bands in E satisfying (a) and (b). Denote the corresponding band projections by Q_j. We claim that if $C_j \cap B_p \neq \{0\}$, then $j = p$. Indeed, let P be the band projection onto $C_j \cap B_p$. If $p < \infty$, then it follows from $0 < P \leq P_p$ and (b) that there exists $k(P) \in \mathbb{N}$ such that $P\mathcal{D}(T^{kp}) > 0$ for all $k \geq k(P)$, which implies that $j < \infty$ and $j|p$, so $j \leq p$. A similar argument shows that $p \leq j$, and therefore $j = p$. Since (a) holds for the C_j's and the B_p's it follows now easily that $C_p = B_p$ for all $p \in \mathbb{N} \cup \{\infty\}$. $\square$

The above theorem can be used to define local periods for $\{\mathcal{D}(T^n)\}_{n=0}^{\infty}$; we say that $\{\mathcal{D}(T^n)\}_{n=0}^{\infty}$ has period p on the band B_p. We note that in part (ii) of (b) we do not get strict positivity of $P\mathcal{D}(T^{kp})$ on B_p in general.

EXAMPLE 4.10. Let $E = \ell_{\infty}$ and define $0 \leq T : \ell_{\infty} \to \ell_{\infty}$ by means of the matrix

$$\begin{pmatrix} \frac{1}{2} & \frac{1}{2} & 0 & 0 & 0 & \cdots \\ \frac{1}{2} & 0 & \frac{1}{2} & 0 & 0 & \cdots \\ \frac{1}{2} & 0 & 0 & \frac{1}{2} & 0 & \cdots \\ \vdots & \vdots & & & \ddots & \ddots \end{pmatrix}$$

Identifying $\mathcal{D}(T^n)$ with the ℓ_{∞}-sequence along the diagonal we have that $\mathcal{D}(T^n)(m) > 0$ for $1 \leq m \leq n$ and $\sum_{n=0}^{\infty} \mathcal{D}(T^n)(m)z^n = 1 + \frac{1}{2^m}\frac{z^m}{1-z}(|z| < 1)$, and so $\{\mathcal{D}(T^n)(m)\}$ is a-periodic for all m. For each n, however, $\mathcal{D}(T^n)(m)$ is nonzero for only finitely many values of m. This example also illustrates that the period function $d(\omega)$ is in general not continuous. Indeed, in this case, $Z(E) \cong C(\beta\mathbb{N})$ and so for all $\omega \in \mathbb{N}$, $d(\omega) = 1$ for $\omega \in \mathbb{N}$ and $d(\omega) = \infty$ for $\omega \in \beta\mathbb{N} \setminus \mathbb{N}$. Hence $\mathcal{A}_1 = \mathbb{N}$ and $\mathcal{A}_{\infty} = \beta\mathbb{N} \setminus \mathbb{N}$. However $B_1 = \ell_{\infty}$ and $B_{\infty} = \{0\}$. Finally we observe that the operator T satisfies $r(T) = 1$ and that its peripheral spectrum Per $\sigma(T) = \sigma(T) \cap \{z : |z| = 1\} = \{1\}$. In fact, it is easy to see that $\sigma(T) = \{\frac{1}{2}, 1\}$ and $\sigma_p(T) = \{1\}$.

It may be of some interest to point out that in general if $0 \leq T \in \mathcal{L}_r(E), r(T) = 1$ and Per $\sigma(T) = \{1\}$, then the period function of T is two-valued, namely it is either equal to 1 or to ∞. To see this observe that for all $\omega \in K$, where $Z(E) \cong C(K)$, using Voigt's theorem (Proposition 1.2 (v)) we have for all $|z| < 1$

$$\left| \sum_{n=0}^{\infty} \mathcal{D}(T^n)(\omega)z^n \right| \leq \left\| \mathcal{D}\left(\frac{1}{z}R(\frac{1}{z}, T)\right) \right\| \leq \left\| \left(\frac{1}{z}R(\frac{1}{z}, T)\right) \right\|,$$

and so if z_0 is a singularity of the power series $\sum_{n=0}^{\infty} \mathcal{D}(T^n)(\omega)z^n$ satisfying $|z_0| = 1$, then $z_0 = 1$. Then, if $d(\omega) \neq \infty$, observe that if $d(\omega) = d > 1$ then, by a well-known result of

Pólya concerning singularities of power series, the power series $\sum_{n=0}^{\infty} \mathcal{D}(T^n)(\omega)z^n$ having a d-periodic sequence of coefficients has a singularity on each arc on $|z| = 1$ of length $\geq \frac{2\pi}{d}$. Hence, since $z = 1$ is the only singularity of $\sum_{n=0}^{\infty} \mathcal{D}(T^n)(\omega)z^n$ on $|z| = 1$ we obtain that $d = 1$.

5. DISJOINTNESS PRESERVING OPERATORS

In this section we will study in detail the diagonals of lattice homomorphisms and, more generally, of disjointness preserving operators. We will show how these results can be used to obtain decompositions of such operators. Furthermore applications to the spectral theory of disjointness preserving operators will be given. Disjointness preserving operators, and in particular the properties of their spectrum, have been studied in the last decade by several authors (see e.g. [1], [5], [22], [8], and the references in these papers). Some of our results extend the results of the above mentioned authors, and some of our results are similar, but obtained via different methods.

Let E be a Dedekind complete Banach lattice. We recall that the operator $0 \leq T \in \mathcal{L}_r(E)$ is called a *lattice homomorphism* (= Riesz homomorphism) if $|Tx| = T|x|$ for all $x \in E$. In the next proposition we collect some special properties of the diagonal of a lattice homomorphism. In the proof of this proposition the following remark will be used.

REMARK 5.1. Let E be a Dedekind complete Banach lattice. For $\pi \in Z(E)$ define the mapping $L_\pi : \mathcal{L}_r(E) \to \mathcal{L}_r(E)$ by $L_\pi(T) = \pi T$. It is easy to see that $L_\pi \in Z(\mathcal{L}_r(E))$. The mapping $\pi \mapsto L_\pi$ is a unital algebra homomorphism from $Z(E)$ into $Z(\mathcal{L}_r(E))$, and hence this mapping is a lattice homomorphism. This implies in particular that $(\pi_1 \wedge \pi_2)T = (\pi_1 T) \wedge (\pi_2 T)$ holds for all $0 \leq T \in \mathcal{L}_r(E)$ and all $\pi_1, \pi_2 \in ReZ(E)$ (and similarly $T(\pi_1 \wedge \pi_2) = (T\pi_1) \wedge (T\pi_2)$).

PROPOSITION 5.2. *Let* $0 \leq T \in \mathcal{L}_r(E)$ *be a lattice homomorphism. If* P *denotes the band projection onto* $\{ran\ \mathcal{D}(T)\}^{dd}$, *then* $\mathcal{D}(T) = PT$.

Moreover $PT = PTP$, $T(ker\ \mathcal{D}(T)) \subseteq ker\ \mathcal{D}(T)$, $\mathcal{D}(T)\mathcal{D}_\perp(T) = 0$ *and* $\mathcal{D}(T)^n = PT^n$ *for all* $n \geq 1$.

PROOF. Since $0 \leq \mathcal{D}(T) \leq T$ and $0 \leq \mathcal{D}_\perp(T) \leq T$, it follows from a theorem of Kutateladze (see e.g. [15], Corollary 3.1.19) that there exist $0 \leq \pi_1, \pi_2 \leq I$ in $Z(E)$ such that $\mathcal{D}(T) = \pi_1 T$ and $\mathcal{D}_\perp(T) = \pi_2 T$. Using Remark 5.1 it follows that $(\pi_1 \wedge \pi_2)T = (\pi_1 T) \wedge (\pi_2 T) = \mathcal{D}(T) \wedge \mathcal{D}_\perp(T) = 0$. Therefore, replacing π_1 and π_2 by $\pi_1 - \pi_1 \wedge \pi_2$ and $\pi_2 - \pi_1 \wedge \pi_2$ respectively, we may assume that $\pi_1 \wedge \pi_2 = 0$, i.e., that $\pi_1 \pi_2 = 0$. Since $\mathcal{D}(T) = \mathcal{D}(\pi_1 T) = \pi_1 \mathcal{D}(T) = \mathcal{D}(T)\pi_1$, we see already that $\mathcal{D}(T)\mathcal{D}_\perp(T) = \mathcal{D}(T)\pi_1\pi_2 T = 0$. Now define $P = \sup_n \{n\ \mathcal{D}(T) \wedge I\}$, i.e., P is the component of I in the band

$\{\mathcal{D}(T)\}^{dd}$ in $Z(E)$, which is a band projection in E. Since $(I - P) \wedge \mathcal{D}(T) = 0$, we have $P\mathcal{D}(T) = \mathcal{D}(T)$ and hence $\{\text{ran } \mathcal{D}(T)\}^{dd} \subseteq \text{ran } P$. From the definition of P it follows that $\ker \mathcal{D}(T) \subseteq \ker P$, so $\text{ran } P \subseteq \{\ker \mathcal{D}(T)\}^d = \{\text{ran } \mathcal{D}(T)\}^{dd}$. This shows that $\text{ran } P = \{\text{ran } \mathcal{D}(T)\}^{dd}$, i.e., P is the band projection onto $\{\text{ran } \mathcal{D}(T)\}^{dd}$. Using Remark 5.1 once more we find that

$$PT = \sup_{n}\{n\,\mathcal{D}(T) \wedge I\}T = \sup_{n}\{(n\,\mathcal{D}(T) \wedge I)T\} =$$
$$= \sup_{n}\{n\,\mathcal{D}(T)T \wedge T\}.$$

It follows from $\mathcal{D}(T)\mathcal{D}_{\perp}(T) = 0$ that

$$n\,\mathcal{D}(T)T \wedge T = n\,\mathcal{D}(T)\{\mathcal{D}(T) + \mathcal{D}_{\perp}(T)\} \wedge T =$$
$$= n\,\mathcal{D}(T)^2 \wedge T = n\,\mathcal{D}(T)^2 \wedge \mathcal{D}(T) = (n\,\mathcal{D}(T) \wedge I)\mathcal{D}(T)$$

and hence

$$PT = \sup_{n}\{(n\mathcal{D}(T) \wedge I)\mathcal{D}(T)\} =$$
$$= \sup_{n}\{n\mathcal{D}(T) \wedge I\} \cdot \mathcal{D}(T) = P\mathcal{D}(T) = \mathcal{D}(T).$$

Furthermore, $PT = P\mathcal{D}(T) = P\mathcal{D}(T)P = PTP$, and so $T(I - P) = (I - P)T(I - P)$, which shows that $\text{ran}(I - P)$ is T-invariant. Since $\text{ran}(I - P) = \ker \mathcal{D}(T)$, this shows that $T(\ker \mathcal{D}(T)) \subseteq \ker \mathcal{D}(T)$. Finally we show that $\mathcal{D}(T)^n = PT^n (n = 1, 2, \dots)$ via induction on n. For $n = 1$ this was already proved above. Now suppose that $\mathcal{D}(T)^n = PT^n$ holds for some $n \geq 1$. Then

$$\mathcal{D}(T)^{n+1} = \mathcal{D}(T)\mathcal{D}(T)^n = \mathcal{D}(T)PT^n = \mathcal{D}(T)T^n = PT^{n+1}.$$

By this all the statements of the proposition have been proved. $\square$

Using the same notation as in the above proof, observe that $n\,\mathcal{D}(T) \wedge I = nT \wedge I$ for all $n \geq 1$, and so $P = \sup_n\{nT \wedge I\}$. In this form the band projection P was already used, in the case that T is a lattice isomorphism in [4]. The following result was proved in [9] under the additional hypothesis that T is order continuous.

COROLLARY 5.3. *Let E be a Dedekind complete Banach lattice and $0 \leq T \in \mathcal{L}_r(E)$ a lattice homomorphism. If $\mathcal{D}(T)$ is injective, then $T = \mathcal{D}(T)$.*

PROOF. Let P be the band projection onto $\{\text{ran } D(T)\}^{dd}$. If $\mathcal{D}(T)$ is injective, then $\{\text{ran } \mathcal{D}(T)\}^{dd} = \{\ker \mathcal{D}(T)\}^d = E$, and so $P = I$. Now it follows from the above proposition that $T = \mathcal{D}(T)$. $\square$

In general the band projection P of Proposition 5.2 does not commute with T, as is shown by the following example. Note already that, since $PT = \mathcal{D}(T) = P\mathcal{D}(T)$, P commutes with T if and only if $\mathcal{D}_{\perp}(T)P = 0$.

EXAMPLE 5.4. Consider the interval $[0,1]$ with Lebesgue measure and define the null-preserving map $\psi : [0,1] \to [0,1]$ by

$$\psi(\omega) = \begin{cases} \omega & \text{if } 0 \leq \omega \leq \frac{1}{2} \\ 1 - \omega & \text{if } \frac{1}{2} \leq \omega \leq 1. \end{cases}$$

Let $E = L_p[0,1]$ $(1 \leq p \leq \infty)$, and define the lattice homomorphism T on E by $Tf(\omega) = f(\psi(\omega))$. Observe that this operator is of the type considered in Example 1.3(iii). Since $\text{Fix}(\psi) = [0,\frac{1}{2}]$, it follows from the results obtained in Example 1.3 that $\mathcal{D}(T) = \chi_{[0,\frac{1}{2}]}T$, and hence $\mathcal{D}(T)f = \chi_{[0,\frac{1}{2}]}f$ for all $f \in E$. Therefore, the band projection P of Proposition 5.2 is multiplication by $\chi_{[0,\frac{1}{2}]}$. Since $TP = T$ in this case, it is clear that P does not commute with T.

In the following proposition we will give some sufficient conditions for $PT = TP$ to hold.

PROPOSITION 5.5. *Let E be a Dedekind complete Banach lattice and $0 \leq T \in \mathcal{L}_r(E)$ a lattice homomorphism. Let P be the band projection onto $\{\text{ran } \mathcal{D}(T)\}^{dd}$, and suppose that T is interval preserving. Then $TP = PT$.*

PROOF. First note that it follows from $PT = \mathcal{D}(T) = P\mathcal{D}(T)$ that $P\mathcal{D}_\perp(T) = 0$. Now take $0 \leq x \in E$ and put $y = \mathcal{D}_\perp(T)Px$. Then $0 \leq y \leq T(Px)$, and since T is interval preserving, this implies that there exists $0 \leq z \leq Px$ such that $y = Tz$. Now $\mathcal{D}(T)z = PTz = Py = P\mathcal{D}_\perp(T)Px = 0$, so $z \in \ker \mathcal{D}(T)$ and hence $Pz = 0$. From $0 \leq z \leq Px$ it follows that $Pz = z$, so $z = 0$. We thus have shown that $\mathcal{D}_\perp(T)P = 0$. As observed before, this implies that $PT = TP$. $\square$

In connection with the above proposition we note that if $0 \leq T \in \mathcal{L}_r(E)$ is an order continuous lattice homomorphism such that T^* is a lattice homomorphism as well, then T is interval preserving (cf. [7], Theorem 3.13). Indeed, take $0 \leq u$. Then $T([0,u])$ is norm dense in $[0,Tu]$ (see e.g. [15], Theorem 1.4.19). Hence, if $0 \leq y \leq Tu$, then there exists a sequence $\{x_n\}_{n=1}^\infty$ such that $0 \leq x_n \leq u$ $(n = 1, 2, \dots)$ and $Tx_n \to y$ in norm as $n \to \infty$. By passing to a subsequence we can assume that $Tx_n \to y$ in order as well (see [21], Theorem 100.6). Let $x = \limsup_{n \to \infty} x_n$. Then $0 \leq x \leq u$, and since T is an order continuous lattice homomorphism we have $Tx = \limsup_{n \to \infty} Tx_n = y$. It follows from this observation that the above proposition generalizes Proposition 2.3(ii) of [22].

Next we will study the sequence $\{\mathcal{D}(T^n)\}_{n=0}^\infty$ of diagonals of a lattice homomorphism $0 \leq T \in \mathcal{L}_r(E)$. As before, E is a Dedekind complete Banach lattice and we identify $Z(E)$ with a $C(K)$ space. As we have seen in Theorem 4.2, if $r(T) \leq 1$, then for each $\omega \in K$ the sequence $\{\mathcal{D}(T^n)(\omega)\}_{n=0}^\infty$ is a renewal sequence. For such sequences the period was defined in the remarks preceding Theorem 4.9.

LEMMA 5.6. *Let E be a Dedekind complete Banach lattice and let $0 \leq T \in$
$\mathcal{L}_r(E)$ be a lattice homomorphism with $r(T) \leq 1$. Assume that $\mathcal{D}(T^p)(\omega) = 0$ for some
$p \geq 1$ and $\omega \in K$ (where $Z(E) \cong C(K)$). Then the sequence $\{\mathcal{D}(T^n)(\omega)\}_{n=0}^\infty$ does not
have period p.*

PROOF. By the properties of the period of the sequence $\{\mathcal{D}(T^n)(\omega)\}_{n=0}^\infty$ it
is sufficient to show that $\mathcal{D}(T^{np})(\omega)\mathcal{D}(T^{(n+1)p})(\omega) = 0$ for all $n \geq 1$. For $n \geq 1$ we
have $T^{np} = \mathcal{D}(T^{np}) + \mathcal{D}_\perp(T^{np})$, and so $T^{(n+1)p} = \mathcal{D}(T^{np})T^p + \mathcal{D}_\perp(T^{np})T^p$. Let Q_{np}
be the band projection in E onto $\{\text{ran } \mathcal{D}(T^{np})\}^{dd}$. It follows from Proposition 5.2 that
$\mathcal{D}_\perp(T^{np}) = (I - Q_{np})T^{np}$, and hence

$$T^{(n+1)p} = \mathcal{D}(T^{np})T^p + (I - Q_{np})T^{(n+1)p}.$$

This implies that

$$\mathcal{D}(T^{(n+1)p}) = \mathcal{D}(T^{np})\mathcal{D}(T^p) + (I - Q_{np})\mathcal{D}(T^{(n+1)p}),$$

and so

$$\mathcal{D}(T^{np})\mathcal{D}(T^{(n+1)p}) = \mathcal{D}(T^{np})^2\mathcal{D}(T^p) + \mathcal{D}(T^{np})(I - Q_{np})\mathcal{D}(T^{(n+1)p}).$$

Since $\mathcal{D}(T^{np})(I - Q_{np}) = 0$, it follows that $\mathcal{D}(T^{np})\mathcal{D}(T^{(n+1)p}) = \mathcal{D}(T^{np})^2\mathcal{D}(T^p)$, and
therefore $\mathcal{D}(T^{np})(\omega)\mathcal{D}(T^{(n+1)p})(\omega) = 0$. $\square$

PROPOSITION 5.7. *Let E be a Dedekind complete Banach lattice, and let
$0 \leq T \in \mathcal{L}_r(E)$ be a lattice homomorphism with $r(T) \leq 1$. As before $Z(E) \cong C(K)$. If
$\omega \in K$ and d is the period of $\{\mathcal{D}(T^n)(\omega)\}_{n=0}^\infty$, then $d = \min\{n \geq 1 : \mathcal{D}(T^n)(\omega) \neq 0\}$.
Moreover, if $d < \infty$, then $\mathcal{D}(T^{nd})(\omega) = (\mathcal{D}(T^d)(\omega))^n$ for all $n \geq 0$.*

PROOF. We may assume that $d < \infty$. Let $m = \min\{n \geq 1 : \mathcal{D}(T^n)(\omega) \neq$
$0\}$. From the definition of the period it follows that $d \leq m$, and from Lemma 5.6 it
follows that $m \leq d$. Hence $d = m$. Let Q_d be the band projection onto $\{\text{ran } \mathcal{D}(T^d)\}^{dd}$.
Then by Proposition 5.2, $\mathcal{D}(T^d)^n = Q_d T^{nd}$, and so $\mathcal{D}(T^d)^n = Q_d \mathcal{D}(T^{nd})$ for all $n \geq 1$.
Hence $Q_d(\omega)\mathcal{D}(T^{nd})(\omega) = (\mathcal{D}(T^d)(\omega))^n$. In particular $Q_d(\omega)\mathcal{D}(T^d)(\omega) = \mathcal{D}(T^d)(\omega)$, and
since $\mathcal{D}(T^d)(\omega) > 0$, this implies that $Q_d(\omega) = 1$. Therefore, we may conclude that
$\mathcal{D}(T^{nd})(\omega) = (\mathcal{D}(T^d)(\omega))^n$ for all $n \geq 1$. $\square$

We note that the above proposition is essentially a combination of Lemma 3.1.b
and the proof of Lemma 3.1.c of [22]. For a lattice homomorphism T the bands B_p in
the decomposition given in Theorem 4.9 have an alternative and simpler description, as is
shown in the following proposition.

PROPOSITION 5.8. *Let E be a Dedekind complete Banach lattice and let $0 \leq T \in \mathcal{L}_r(E)$ be a lattice homomorphism with $r(T) \leq 1$. Then the bands B_p ($p \in \mathbb{N} \cup \{\infty\}$) in the decomposition in Theorem 4.9 are given by*

$$(13) \qquad B_p = \{\operatorname{ran} \mathcal{D}(T^p)\}^{dd} \cap \left\{ \bigcap_{k=1}^{p-1} \ker \mathcal{D}(T^k) \right\}$$

for $p \in \mathbb{N}$, and $B_\infty = \bigcap_{k=1}^{\infty} \ker \mathcal{D}(T^k)$.

PROOF. We identify $Z(E)$ with a space $C(K)$. The projection $\mathcal{P}$ from $\mathcal{B}(K)$ onto $C(K)$ (see Remark 3.7) induces a Boolean σ-homomorphism from the Borel sets in K into the clopen sets of K (see the proof of Theorem 4.9). This homomorphism can also be considered as a Boolean σ-homomorphism which we will call h, from the Borel sets of K into the Boolean algebra of all bands in E. For $0 \leq \pi \in Z(E)$, it is easy to see that $h(\{\omega \in K : \pi(\omega) = 0\}) = \ker(\pi)$ and $h(\{\omega \in K : \pi(\omega) > 0\}) = \{\operatorname{ran}(\pi)\}^{dd}$. As in the proof of Theorem 4.9, we consider the sets $A_p = \{\omega \in K : \{\mathcal{D}(T^n)(\omega)\}_{n=0}^{\infty} \text{ has period } p\}$ ($p \in \mathbb{N} \cup \{\infty\}$). From Proposition 5.7 it follows that

$$(14) \qquad A_p = \{\omega \in K : \mathcal{D}(T^p)(\omega) > 0\} \cap \left\{ \bigcap_{k=1}^{p-1} \{\omega \in K : \mathcal{D}(T^k)(\omega) = 0\} \right\}$$

for all $p \in \mathbb{N}$. In the proof of Theorem 4.9, the bands B_p are defined by $B_p = h(A_p)$. From the above remarks it follows that, if we apply h to the right hand side of (14), then we get precisely (13).

Since $A_\infty = \bigcap_{k=1}^{\infty} \{\omega \in K : \mathcal{D}(T^k)(\omega) = 0\}$, the formula for B_∞ follows similarly. $\square$

Before we give some applications of the above results, we first recall some facts about disjointness preserving operators. Let E be a Dedekind complete Banach lattice. A linear operator $T \in \mathcal{L}(E)$ is called *disjointness preserving* if $x \perp y$ in E implies that $Tx \perp Ty$. Such operators are automatically order bounded, i.e., $T \in \mathcal{L}_r(E)$ (see e.g. [15], Theorem 3.15), and the modulus of T satisfies $|Tx| = |T|(|x|)$ for all $x \in E$ (see e.g. [15], Theorem 3.14) and $|T|$ is a lattice homomorphism. Moreover, there exists $\sigma \in Z(E)$ with $|\sigma| = I$ such that $T = \sigma|T|$ (see e.g. [15], Corollary 3.1.20). Since $|Tx| = |T|(|x|)$ for all $x \in E$, it is clear that $|T^n| = |T|^n$ for all $n \geq 1$. It is now easy to see that $r(|T|) = r(T)$. It also follows that $|\mathcal{D}(T^n)| = \mathcal{D}(|T^n|) = \mathcal{D}(|T|^n)$ for all $n \geq 1$. Furthermore, if we denote by P the band projection in E onto $\{\operatorname{ran} \mathcal{D}(|T|)\}^{dd} = \{\operatorname{ran} \mathcal{D}(T)\}^{dd}$, then $P|T| = \mathcal{D}(|T|)$ by Proposition 5.2. We claim that $PT = \mathcal{D}(T)$ holds as well. Indeed, writing $T = \sigma|T|$ as above, we have $PT = P\sigma|T| = \sigma P|T| = \sigma\mathcal{D}(|T|) = \mathcal{D}(\sigma|T|) = \mathcal{D}(T)$. As in the proof of Proposition 5.2 it now follows that $PT^n = D(T)^n$, and hence $PD(T^n) = D(T)^n$ for all $n \geq 1$.

After these preliminary remarks we come to our next result.

PROPOSITION 5.9. *Let E be a Dedekind complete Banach lattice and $T \in \mathcal{L}_r(E)$ a disjointness preserving operator. Suppose that there exists an open connected subset $U \subseteq \rho(T)$ with the property that for each $0 < r \leq r(T)$ there exists θ_r such that the circular arc*

(15) $$C_r = \{\lambda \in \mathbb{C} : |\lambda| = r, \ \theta_r \leq \arg \lambda \leq \theta_r + \pi\}$$

is contained in U. Then $T = \pi + S$, where $\pi \in Z(E)$ and $S^n \perp I$ for all $n \geq 1$.

PROOF. Without loss of generality we may assume that $r(T) = r(|T|) = 1$. Identifying $Z(E)$ with $C(K)$, we define the sets

$$A_p = \{\omega \in K : \{D(|T|^n)(\omega)\}_{n=0}^{\infty} \text{ has period } p\}$$

for $p \in \mathbb{N} \cup \{\infty\}$. From the above remarks and from Proposition 5.7 it follows that

$$A_p = \{\omega \in K : D(T^p)(\omega) \neq 0 \text{ and } D(T^k)(\omega) = 0 \text{ for } 1 \leq k < p\}$$

for all $p \in \mathbb{N}$, and $A_\infty = \{\omega \in K : D(T^k)(\omega) = 0 \text{ for all } k \geq 1\}$.

Now suppose that $\omega \in A_p$ for some $p \in \mathbb{N}$. Then $D(T^n)(\omega) = 0$ whenever n is not a multiple of p. We claim that $D(T^{kp})(\omega) = (D(T^p)(\omega))^k$ for all $k \geq 1$. Indeed, let Q_p be the band projection in E onto $\{\text{ran } D(T^p)\}^{dd}$. From the observations above we know that $Q_p D(T^{kp}) = D(T^p)^k$ for all $k \geq 1$. Since $D(T^p)(\omega) \neq 0$ and $Q_p(\omega)D(T^p)(\omega) = D(T^p)(\omega)$, we have $Q_p(\omega) = 1$, and from this the claim follows. Consequently, for $|\lambda| > 1$ we find that

$$D[R(\lambda, T)](\omega) = \sum_{n=0}^{\infty} \frac{D(T^n)(\omega)}{\lambda^{n+1}}$$

$$= \sum_{k=0}^{\infty} \frac{(D(T^p)(\omega))^k}{\lambda^{kp+1}} = \frac{\lambda^p - 1}{\lambda^p - D(T^p)(\omega)}.$$

The given open connected set $U \subseteq \rho(T)$ has a non-empty intersection with $\{\lambda \in \mathbb{C} : |\lambda| > 1\}$. By analytic continuation we therefore have

$$D[R(\lambda, T)](\omega) = \frac{\lambda^{p-1}}{\lambda^p - D(T^p)(\omega)}$$

for all $\lambda \in U$. If $p > 1$, then the right hand side has a pole on the circular arc C_r with $r = |D(T^p)(\omega)|^{\frac{1}{p}}$, which is impossible. We thus have shown that $A_p = \emptyset$ for all $1 < p < \infty$.

Hence $K = \mathcal{A}_1 \cup \mathcal{A}_\infty$. This implies that $\mathcal{D}(|T|^n) = \mathcal{D}(|T|)^n$ for all $n \geq 1$, since the equality holds on $\mathcal{A}_1$ and both sides are zero on $\mathcal{A}_\infty$. Let $S = \mathcal{D}_\perp(T)$. If we can show that $|S|^n \wedge I = 0$ for all $n \geq 1$, then the conclusion of the proposition follows. To this end, first observe that $|T| = \mathcal{D}(|T|) + |S|$, and so $|T|^n \geq \mathcal{D}(|T|)^n + |S|^n$ for all $n \geq 1$. However, by the above we also have $|T|^n = \mathcal{D}(|T|^n) + \mathcal{D}_\perp(|T|^n) = \mathcal{D}(|T|)^n + \mathcal{D}_\perp(|T|^n)$, and combined with the previous inequality, this implies that $|S|^n \leq \mathcal{D}_\perp(|T|^n)$. Hence $|S|^n \wedge I = 0$ for all $n \geq 1$. $\square$

The following lemma will enable us to obtain the result of the above proposition under seemingly weaker conditions on the spectrum of T.

LEMMA 5.10. *Let K be a closed subset of $\{z \in \mathbb{C} : |z| \leq 1\}$. Assume that for all $0 < r \leq 1$ there exists θ_r such that the circular arc C_r as defined by (15), is contained in K^c. Then there exists an open connected set $U \subseteq K^c$ such that $C_r \subseteq U$ for all $0 < r \leq 1$.*

PROOF. For $0 < \epsilon < r$ and θ we define the open sets

$$U(r, \theta, \epsilon) = \{z \in \mathbb{C} : \theta - \epsilon < \arg z < \theta + \pi + \epsilon, r - \epsilon < |z| < r + \epsilon\}$$

and we make the following observations.

(i) $C_r \subseteq U(r, \theta_r, \epsilon)$ for all $0 < \epsilon < r \leq 1$.

(ii) $U(r, \theta, \epsilon) \cap U(r', \theta', \epsilon') \neq \emptyset$ whenever $0 < \epsilon < r, 0 < \epsilon' < r'$ such that $r - \epsilon < r' < r + \epsilon$.

(iii) For each $0 < r \leq 1$ there exists $0 < \epsilon_r < r$ such that $U(r, \theta_r, \epsilon_r) \subseteq K^c$ (since C_r and K are disjoint closed sets).

Now define $U = \bigcup_{0 < r \leq 1} U(r, \theta_r, \epsilon_r)$, where θ_r is as in the definition of C_r, and ϵ_r is as in (iii). Then clearly $C_r \subseteq U$ for all $0 < r \leq 1$ and $U \subseteq K^c$.

It remains to show that U is connected. Assume that $U = O_1 \cup O_2$, with O_1, O_2 open and $O_1 \cap O_2 = \emptyset$. Let $A_j = \{r \in (0, 1] : U(r, \theta_r, \epsilon_r) \cap O_j \neq \emptyset\}$ for $j = 1, 2$. Clearly $(0, 1] = A_1 \cup A_2$. Furthermore $A_1 \cap A_2 = \emptyset$, since $O_j \cap U(r, \theta_r, \epsilon_r) \neq \emptyset$ implies that $U(r, \theta_r, \epsilon_r) \subseteq O_j$ as $U(r, \theta_r, \epsilon_r)$ is connected. From (ii) it follows that each A_j is an open subset of $(0, 1]$. Hence $A_1 = \emptyset$ or $A_2 = \emptyset$, i.e., $O_1 = \emptyset$ or $O_2 = \emptyset$, which shows that U is connected. $\square$

A combination of Proposition 5.9 with Lemma 5.10 yields the following result.

THEOREM 5.11. *Let E be a Dedekind complete Banach lattice and let $T \in \mathcal{L}_r(E)$ be a disjointness preserving operator. Suppose that for each $0 < r \leq r(T)$ there exists θ_r such that $C_r \subseteq \rho(T)$, where C_r is as defined in (15). Then $T = \pi + S$, where $\pi \in Z(E)$ and $S^n \perp I$ for all $n \geq 1$.*

The above result can be considered as an extension of a result of W. Arendt and D. Hart ([5], Theorem 5.3), who considered the more restrictive class of quasi-invertible disjointness preserving operators (recall that a disjointness preserving operator T is called quasi-invertible if T is order continuous, injective, $\{\mathrm{ran}(T)\}^{dd} = E$ and T^* is disjointness preserving as well). As another illustration of our techniques we mention the following corollary, which is also contained in [5].

COROLLARY 5.12. *Let E be a Dedekind complete Banach lattice and $T \in \mathcal{L}_r(E)$ a disjoint preserving operator, whose spectrum satisfies the same condition as in Theorem 5.11. Assume in addition that $0 \notin \sigma(T)$. Then $T \in Z(E)$.*

PROOF. We may assume that $r(T) = 1$. Using the same notation as in the proof of Proposition 5.9, we have $K = \mathcal{A}_1 \cup \mathcal{A}_\infty$. For $\omega \in \mathcal{A}_\infty$ and $\lambda \in U$ it is clear that $\mathcal{D}[R(\lambda, T)](\omega) = \frac{1}{\lambda}$. Since, however $0 \in U$, this is impossible, which shows that $\mathcal{A}_\infty = \emptyset$, and hence $K = \mathcal{A}_1$. This implies that $\mathcal{D}(|T|)$ is injective, and so, by Corollary 5.3, it follows that $|T| \in Z(E)$, and hence $T \in Z(E)$. $\square$

Now we will analyze the situation of Theorem 5.11 in more detail, in particular the relation between the spectra of T, π and S. For this we need the following lemma.

LEMMA 5.13. *Let E be a Dedekind complete Banach lattice and let $T \in \mathcal{L}_r(E)$ be disjointness preserving. Suppose that $T = \pi + S$ with $\pi \in Z(E)$ and $S \perp I$. Then $\sigma(T) \subseteq \sigma(\pi) \cup \sigma(S) \subseteq \sigma(T) \cup \{0\}$.*

PROOF. It follows from Proposition 5.2 that $\pi S = 0$, and hence for all $\lambda \in \mathbb{C}$ we have

$$(16) \qquad (\lambda I - \pi)(\lambda I - S) = \lambda(\lambda I - T).$$

Clearly, this implies that $\rho(\pi) \cap \rho(S) \subseteq \rho(T)$, i.e., $\sigma(T) \subseteq \sigma(\pi) \cup \sigma(S)$. Now assume that $\lambda \in \rho(T)$ and $\lambda \neq 0$. It follows from (16) that $\mathrm{ran}(\lambda I - \pi) = E$, and so $\ker(\lambda I - \pi) = \{\mathrm{ran}(\lambda I - \pi)\}^d = \{0\}$, hence $\lambda \in \rho(\pi)$. Using (16) once again we find that $\lambda \in \rho(S)$. This shows that $\rho(T) \backslash \{0\} \subseteq \rho(\pi) \cap \rho(S)$ and therefore $\sigma(\pi) \cup \sigma(S) \subseteq \sigma(T) \cup \{0\}$. $\square$

If $S \in \mathcal{L}_r(E)$ is a disjointness preserving operator such that $S^n \perp I$ for all $n \geq 1$, then there are several conditions, on S or on E, implying that S has rotationally invariant spectrum (i.e., $\alpha\sigma(S) = \sigma(S)$ for all $\alpha \in \mathbb{C}$ with $|\alpha| = 1$). For the case that S is quasi-invertible, the rotational invariance of $\sigma(S)$ has been proved by W. Arendt and D. Hart ([5], Corollary 4.7). More general results were recently obtained by Y. Abramovich, E. Arenson and A. Kitover in [1], Chapter 13. In order to formulate one of their results, which we will use below, we need some notation. Given a Dedekind complete Banach lattice E, the Lorentz seminorm on E is defined by

$$\ell_E(x) = \inf \left\{ \sup_\alpha \|x_\alpha\| : 0 \leq x_\alpha \uparrow |x| \text{ in } E \right\}.$$

In general ℓ_E is a Riesz semi-norm on E. Following [1], we will say that E has property (*) if ℓ_E is a norm on E. This is for example the case if E has a weak Fatou norm, or if the order continuous functionals E_n^* separate the points of E (which holds in particular, if E is any Banach function space; see e.g. [21], Theorem 112.1). The following result is proved in [1] (Theorem 13.4). Suppose that E has property (*) and that $S \in \mathcal{L}_r(E)$ is a disjointness preserving operator satisfying $S^n \perp S^m$ whenever $0 \leq n < m$, then $\sigma(S)$ is rotationally invariant. Observe that if $S \in \mathcal{L}_r(E)$ is an order continuous disjointness preserving operator such that $S^n \perp I$ for all $n \geq 1$, then $S^n \perp S^m$ whenever $0 \leq n < m$.

COROLLARY 5.14. *Let E be a Dedekind complete Banach lattice with property (*), and let $T \in \mathcal{L}_r(E)$ be an order continuous disjointness preserving operator. Assume that the spectrum of T satisfies the condition of Theorem 5.11. Then $T = \pi + S$ with $\pi \in Z(E), \sigma(|S|) = \{0\}$ and $\sigma(T) = \sigma(\pi)$.*

PROOF. By Theorem 5.11 we know that $T = \pi + S$ with $\pi \in Z(E)$ and $S^n \perp I$ for all $n \geq 1$. Since T, and hence S is order continuous, this implies that $S^n \perp S^m$ whenever $0 \leq n < m$. Therefore $\sigma(S)$ is rotationally invariant. From Lemma 5.13 we know that $\sigma(S) \subseteq \sigma(T) \cup \{0\}$, and from the condition on $\sigma(T)$ it now follows that $\sigma(S) = \{0\}$. Since S is disjointness preserving this implies that $\sigma(|S|) = \{0\}$. Therefore, again by Lemma 5.13, $\sigma(T) \subseteq \sigma(\pi) \cup \{0\} \subseteq \sigma(T) \cup \{0\}$. Finally observe that if $0 \notin \sigma(T)$ then $T = \pi$ by Corollary 5.12, and if $0 \notin \sigma(\pi)$ then $T = \pi$ by Corollary 5.3. Hence we may conclude that $\sigma(T) = \sigma(\pi)$. □

We note that under slightly stronger assumptions on E a similar result was obtained by X. Zhang ([22], Theorem 3.1). Our proof of Theorem 5.11 is similar to the proof of his Theorem 3.1

COROLLARY 5.15. *Let E be a Dedekind complete Banach lattice and let $T \in \mathcal{L}_r(E)$ be a disjointness preserving operator. Assume that the spectrum of T satisfies the condition of Theorem 5.11. Then $T^{**} = \pi + S$ with $\pi \in Z(E^{**}), \sigma(|S|) = \{0\}$ and $\sigma(T) = \sigma(\pi)$.*

PROOF. Since E^{**} has Fatou norm, and hence property (*), and since T^{**} is an order continuous disjointness preserving operator with $\sigma(T^{**}) = \sigma(T)$, this is an immediate consequence of the previous result. □

The order continuity of T in Corollary 5.14 cannot be omitted, as is shown by the following example.

EXAMPLE 5.16. Let $E = \ell_\infty \cong C(\beta\mathbb{N})$. Take $\omega \in \beta\,\mathbb{N}\backslash\mathbb{N}$ and define $\phi \in \ell_\infty^*$ by $< x, \phi >= x(\omega)$ for all $x \in \ell_\infty$. Let the lattice homomorphism $T : \ell_\infty \to \ell_\infty$ be given by $Tx = < x, \phi > \mathbf{1}$ for all $x \in \ell_\infty$ (where $\mathbf{1} = (1, 1, ...)$). It is easy to see that $T \wedge I = 0$ and $T^n = T$ for all $n \geq 1$. Hence $T^n \wedge I = 0$ for all $n \geq 1$, but also $\sigma(T) = \{0, 1\}$. The operator T cannot be decomposed as in Corollary 5.14. Using that $\phi \in \ell_\infty^*$ is an atom, it is readily verified that the band projection P_ϕ in ℓ_∞^* onto $\{\phi\}^{dd}$ is given by $P_\phi x^* =< P_\phi^* \mathbf{1}, x^* > \phi$ for al $x^* \in \ell_\infty^*$. Now T^{**} can be decomposed as $T^{**} = \pi + S$, where π is given by $\pi x^{**} =< x^{**}, \phi > P_\phi^* \mathbf{1}$ for all $x^{**} \in \ell_\infty^{**}$, and $S x^{**} =< x^{**}, \phi > (\mathbf{1} - P_\phi^* \mathbf{1})$ for all $x^{**} \in \ell_\infty^{**}$. It now follows easily that $\pi \in Z(\ell_\infty^{**})$ is actually a band projection, and $S^2 = 0$. This is in accordance with Corollary 5.15.

As observed before, if $T \in \mathcal{L}_r(E)$ is an order continuous disjointness preserving operator with $T^n \perp I$ for all $n \geq 1$, then $T^n \perp T^m$ whenever $0 \leq n < m$. The above example shows that this is false in general if T is not order continuous. The following lemma shows that by assuming a different hypothesis we can make the same conclusion.

LEMMA 5.17. Let E be a Dedekind complete Banach lattice and let $T \in \mathcal{L}_r(E)$ be such that T and T^* are both disjointness preserving. Assume that $T^n \perp I$ for all $n \geq 1$. Then $T^n \perp T^m$ whenever $0 \leq n < m$.

PROOF. First we assume that $0 \leq T \in \mathcal{L}_r(E)$ such that T^* is a lattice homomorphism. We claim that the mapping $S \mapsto ST$ is a lattice homomorphism on $\mathcal{L}_r(E)$. Indeed, it is sufficient to show that $(ST)^+ = S^+T$ for any $S \in \mathrm{Re}\,\mathcal{L}_r(E)$. The inequality $(ST)^+ \leq S^+T$ is clear. To prove the reverse inequality, take $0 \leq u \in E$ and $0 \leq y \leq Tu$. Since T^* is a lattice homomorphism, there exists a sequence $\{x_n\}_{n=1}^\infty$ in E such that $0 \leq x_n \leq u$ $(n = 1, 2, ...)$ and $Tx_n \to y$ in norm as $n \to \infty$ (see [15], Theorem 1.4.19). Now

$$Sy = STx_n + (Sy - STx_n) \leq (ST)^+u + (Sy - STx_n),$$

and so $(Sy - (ST)^+u)^+ \leq |Sy - STx_n|$ for all n. Since $\|Sy - STx_n\| \to 0$ $(n \to \infty)$, this implies that $(Sy - (ST)^+u)^+ = 0$, i.e., $Sy \leq (ST)^+u$. From this it follows that

$$S^+Tu = \sup\{Sy : 0 \leq y \leq Tu\} \leq (ST)^+u,$$

by which the claim is proved. Now assume that $T \in \mathcal{L}_r(E)$ such that T and T^* are disjointness preserving, and $T^n \perp I$ for all $n \geq 1$. Since $|T^*| = |T|^*$ (see [4], Lemma 2.6), it follows from the above that

$$|T^{m+n}| \wedge |T^n| = |T|^{m+n} \wedge |T|^n = (|T|^m \wedge I)|T|^n = 0$$

for all $m \geq 1$ and $n \geq 0$. $\Box$

Using this lemma we are able to prove the following consequence of Theorem 5.11.

COROLLARY 5.18. *Let E be a Dedekind complete Banach lattice and let $T \in \mathcal{L}_r(E)$ be disjointness preserving. Assume that T^* is disjointness preserving as well and that the spectrum of T satisfies the condition of Theorem 5.11. Then $T = \pi + S$ with $\pi \in Z(E), \sigma(|S|) = \{0\}$ and $\sigma(T) = \sigma(\pi)$.*

PROOF. By Theorem 5.11 we know that $T = \pi + S$ with $\pi \in Z(E)$ and $S^n \perp I$ for all $n \geq 1$. Clearly, S is disjointness preserving and since

$$|S^*| = |S|^* \leq (|\pi| + |S|)^* = |T|^* = |T^*|,$$

it follows that S^* is disjointness preserving. Hence, by Lemma 5.17, $S^n \perp S^m$ for all $0 \leq n < m$. Now Theorem 13.14 in [1] states that under these conditions $\sigma(S)$ is rotationally invariant. Since, by Lemma 5.13, $\sigma(S) \subseteq \sigma(T) \cup \{0\}$ it follows from the condition on $\sigma(T)$ that $\sigma(S) = \{0\}$, and hence $\sigma(|S|) = \{0\}$, as S is disjointness preserving. Finally, the proof that $\sigma(T) = \sigma(\pi)$ is the same as in the proof of Corollary 5.14. $\Box$

EXAMPLE 5.19. In connection with the above results it is of interest to know that there exist disjointness preserving operators T, for which T^* is disjointness preserving as well, but which are not order continuous. Consider the space $E = \ell_\infty \cong C(\beta\mathbb{N})$, and choose a sequence $\{p_n\}_{n=1}^\infty$ of mutually different points in $\beta\mathbb{N}\backslash\mathbb{N}$. Define the operator $T : \ell_\infty \to \ell_\infty$ by $Tx = \sum_{n=1}^\infty x(p_n)e_n$ for all $x \in \ell_\infty$. Then T is a surjective lattice homomorphism which is not order continuous. Since T is interval preserving, T^* is disjointness preserving as well. Moreover, it is not difficult to see that $T^n \wedge T^m = 0$ whenever $0 \leq n < m$, and that $\sigma(T) = \sigma_p(T) = \{\lambda \in \mathbb{C} : |\lambda| \leq 1\}$.

The last part of this section will be devoted to the study of order continuous disjointness preserving operators. We will in particular be interested in the properties of the diagonals, the structure, and the spectral properties of such operators. We emphasize the fact that we do *not* assume that T^* is disjointness preserving as well, as is the case in most of [7], [5] and [22].

We recall from Proposition 5.2 that if $0 \leq T \in \mathcal{L}_r(E)$ is a lattice homomorphism, then $T(\ker \mathcal{D}(T)) \subseteq \ker \mathcal{D}(T)$. Our next result shows that $\ker \mathcal{D}(T^n)$ is T-invariant for all $n \geq 1$, provided T is order continuous.

PROPOSITION 5.20. *Let E be a Dedekind complete Banach lattice and let $0 \leq T \in \mathcal{L}_r(E)$ be an order continuous lattice homomorphism. For $n = 1, 2, \ldots$ we* denote by Q_n the band projection in E onto $\{\text{ran } \mathcal{D}(T^n)\}^{dd}$. Then $Q_nT(I - Q_n) = 0$, i.e., $T(\ker \mathcal{D}(T^n)) \subseteq \ker \mathcal{D}(T^n)$ for all $n \geq 1$.

PROOF. Since $I - Q_n$ is the band projection onto $\ker \mathcal{D}(T^n)$, it is clear that $\mathcal{D}(T^n)(I - Q_n) = 0$. Now we have

$$0 \leq T^n(I - Q_n) \wedge (I - Q_n) = \mathcal{D}(T^n)(I - Q_n) \wedge (I - Q_n) = 0,$$

which shows that $T^n(I - Q_n) \wedge (I - Q_n) = 0$. Since T is an order continuous lattice homomorphism, this implies that $T^{n+1}(I - Q_n) \wedge T(I - Q_n) = 0$. It follows from $0 \leq \mathcal{D}(T^n) \leq T^n$ that

$$(17) \qquad\qquad \mathcal{D}(T^n)T(I - Q_n) \wedge T(I - Q_n) = 0,$$

and since $0 \leq \mathcal{D}(T^n)T(I-Q_n) \leq \|\mathcal{D}(T^n)\|\|T(I-Q_n)\|$, we conclude that $\mathcal{D}(T^n)T(I-Q_n) = 0$. This shows that $T(I - Q_n)(E) \subseteq \ker \mathcal{D}(T^n)$, and hence $Q_nT(I - Q_n) = 0$. $\square$

The following example shows that the order continuity of T in Proposition 5.20 cannot be omitted.

EXAMPLE 5.21. Consider the space $E = \ell_\infty \cong C(\beta\mathbb{N})$. Let $u = (0,1,0,1,\dots)$ and $v = (1,0,1,0,\dots)$. Take a point $p \in \beta\mathbb{N}\backslash\mathbb{N}$ which is in the closure of the set $\{2n : n \in \mathbb{N}\}$. Define $\phi, \psi \in \ell_\infty^*$ by $< x, \phi > = x(1)$ and $< x, \psi > = x(p)$ for all $x \in \ell_\infty$.

Let the operator $T : \ell_\infty \to \ell_\infty$ be defined by $Tx = < x, \phi > u + < x, \psi > v$ for all $x \in \ell_\infty$. Then T is a lattice homomorphism which satisfies $T \wedge I = 0$. A simple computation shows that $T^2x = < x, \psi > u + < x, \phi > v$ and $\mathcal{D}(T^2)x = < x, \phi > e_1$ for all $x \in \ell_\infty$ (where $e_1 = (1,0,0,\dots)$). The element $z = (0,1,1,1,\dots)$ satisfies $\mathcal{D}(T^2)z = 0$ and $Tz = v$, so $\mathcal{D}(T^2)Tz = < v, \phi > e_1 = e_1$. Hence $z \in \ker \mathcal{D}(T^2)$, but $Tz \notin \ker \mathcal{D}(T^2)$.

COROLLARY 5.22. Let E be a Dedekind complete Banach lattice and let $T \in \mathcal{L}_r(E)$ be an order continuous disjointness preserving operator. For $n = 1, 2, \dots$ we denote by Q_n the band projection in E onto $\{\text{ran } \mathcal{D}(T^n)\}^{dd}$. Then $Q_nT(I - Q_n) = 0$, i.e., $T(\ker \mathcal{D}(T^n)) \subseteq \ker \mathcal{D}(T^n)$ for all $n \geq 1$.

PROOF. Since $\{\text{ran } \mathcal{D}(T^n)\}^{dd} = \{\text{ran } \mathcal{D}(|T|^n)\}^{dd}$, the above proposition implies that $Q_n|T|(I - Q_n) = 0$. The result now follows from the observation that $|Q_nT(I - Q_n)| = Q_n|T|(I - Q_n)$. $\square$

Now we introduce some notation which will be used from now on. Let E be a Dedekind complete Banach lattice and $T \in \mathcal{L}_r(E)$ an order continuous disjointness preserving operator. We denote by $Q_n(n = 1, 2, \dots)$ the band projection onto $\{\text{ran } \mathcal{D}(T^n)\}^{dd}$.

So by the above we have $Q_n T(I - Q_n) = 0$, i.e., $Q_n T Q_n = Q_n T$ for all $n \geq 1$. Observe that this implies that $Q_n T^k(I - Q_n) = 0$ for all $n \geq 1$ and $k \geq 1$. Furthermore it should be noted that $Q_n T^n = \mathcal{D}(T^n)$ for all $n \geq 1$ (see the remarks preceding Proposition 5.9). For $n = 1, 2, \ldots$ we define the bands

$$B_n = \{\operatorname{ran} \mathcal{D}(T^n)\}^{dd} \cap \left\{ \bigcap_{k=1}^{n-1} \ker \mathcal{D}(T^k) \right\}$$

and $B_\infty = \bigcap_{k=1}^{\infty} \ker \mathcal{D}(T^k)$ (cf. Proposition 5.8). The bands B_n ($n \in \mathbb{N} \cup \{\infty\}$) are mutually disjoint and

(18)
$$\left\{ \bigoplus_{n=1}^{\infty} B_n \right\}^{dd} \oplus B_\infty = E.$$

Let P_n be the band projection onto $B_n (n \in \mathbb{N} \cup \{\infty\})$. Then

(19) $P_1 = Q_1; \ P_n = Q_n(I - Q_1) \ldots (I - Q_{n-1})$ $(n = 2, 3, \ldots)$.

and from (18) it is clear that

$$\sum_{n=1}^{\infty} P_n + P_\infty = I.$$

A less trivial property of the projections P_n is given in the following proposition.

PROPOSITION 5.23. *With the notation introduced above we have* $P_n T P_k = 0$ *for all* $n, k \in \mathbb{N}$ *with* $n \neq k$.

PROOF. First assume that $k > n$. From (19) it follows that $|P_n T P_k| = P_n|T|P_k \leq Q_n|T|(I - Q_n) = |Q_n T(I - Q_n)|$, and by Corollary 5.22 we have $Q_n T(I - Q_n) = 0$. This implies that $P_n T P_k = 0$.

Now assume that $1 \leq k < n$, and take $x, y \in E$. By (19) we have $P_n \leq I - Q_k$ and so

$$|P_k T^{n-1} P_n| = P_k|T^{n-1}|P_n \leq Q_k|T^{n-1}|(I - Q_k) = |Q_k T^{n-1}(I - Q_k)|.$$

Via Corollary 5.22 it follows that $Q_k T^{n-1}(I - Q_k) = 0$, and hence $P_k T^{n-1} P_n = 0$. This implies that

$$T^{n-1} P_n = P_k T^{n-1} P_n + (I - P_k) T^{n-1} P_n = (I - P_k) T^{n-1} P_n,$$

and so $P_k x \perp T^{n-1} P_n y$. Using that T is disjointness preserving we get $T P_k x \perp T^n P_n y$ and consequently $T P_k x \perp Q_n T^n P_n y = \mathcal{D}(T^n) P_n y$. From the definition of Q_n it now follows that $T P_k x \perp Q_n P_n y = P_n y$. Since this holds for all $y \in E$, we may conclude that $P_n T P_k x = 0$. □

Since the bands $\ker \mathcal{D}(T^n)$ ($n = 1, 2, \ldots$) are all T-invariant, it is clear that B_∞ is T-invariant as well, i.e., $P_\infty T P_\infty = T P_\infty$. This implies in particular that $P_n T P_\infty = P_n P_\infty T P_\infty = 0$ for all $n = 1, 2, \ldots$.

COROLLARY 5.24. *Same situation as above.*

(i) $P_n T P_n = P_n T$ for all $n = 1, 2, \ldots$.

(ii) For each $n \in \mathbb{N}$ we have $(P_n T P_n)^k \perp I$ for $k = 1, 2, \ldots, n - 1$, and $(P_n T P_n)^n = P_n \mathcal{D}(T^n) \in Z(E)$.

PROOF. (i) Using the properties of the P_n's obtained above, and the order continuity of T, we get

$$P_n T = P_n T \left(\sum_{k=1}^{\infty} P_k + P_\infty \right) = \sum_{k=1}^{\infty} P_n T P_k + P_n T P_\infty = P_n T P_n.$$

(ii) It follows from (i) that $(P_n T P_n)^k = P_n T^k$ for all $k \in \mathbb{N}$. Hence $\mathcal{D}[(P_n T P_n)^k] = P_n \mathcal{D}(T^k) = P_n Q_k T^k = 0$ for $k = 1, 2, \ldots, n - 1$. Furthermore

$$(P_n T P_n)^n = P_n T^n = P_n Q_n T^n = P_n \mathcal{D}(T^n). \quad \square$$

From the above results it follows that the 'matrix-representation' of T with respect to the band decomposition $E = (\bigoplus_{n=1}^{\infty} B_n)^{dd} \oplus B_\infty$ has the following simple form

(20)
$$\begin{pmatrix} P_1 T P_1 & & \vdots & 0 \\ & P_2 T P_2 & \vdots & 0 \\ \emptyset & & \ddots & \vdots \\ \cdots\cdots\cdots\cdots\cdots\cdots\cdots\cdots\cdots\cdots\cdots & & & \\ P_\infty T P_1 & P_\infty T P_2 & \cdots\cdots & \vdots & P_\infty T P_\infty \end{pmatrix}.$$

REMARK 5.25. Statement (ii) in Corollary 5.24 shows that $P_n T P_n (n \in \mathbb{N})$ has strict period n in the sense of Arendt and Hart ([5], Proposition 3.8(d)). Moreover, the operator $P_n T P_n : B_n \to B_n$ $(n \in \mathbb{N})$ is quasi-invertible. Indeed, to show that $P_n T P_n$ is injective on B_n suppose that $x \in B_n$ such that $P_n T P_n x = 0$. Then $\mathcal{D}(T^n) P_n x = (P_n T P_n)^n x = 0$, so $P_n x \in \ker \mathcal{D}(T^n)$, and hence $x = P_n x = Q_n P_n x = 0$. Now assume that $x \in B_n$ is such that $x \perp P_n T P_n y$ for all $y \in B_n$. Then $(P_n T P_n)^{n-1} x \perp (P_n T P_n)^n y = P_n \mathcal{D}(T^n) y$ for all $y \in B_n$, and since $\{P_n \mathcal{D}(T^n)(B_n)\}^{dd} = B_n$, it follows that $(P_n T P_n)^{n-1} x = 0$, and so $x = 0$. This shows that $\{(P_n T P_n)(B_n)\}^{dd} = B_n$. It remains to prove that $(P_n T P_n)^*$ is disjointness preserving. Since $|(P_n T P_n)^*| = |P_n T P_n|^*$ (see [4], Lemma 2.6), it is sufficient to show that $|P_n T P_n|$ is interval preserving (see e.g., [15], Theorem 1.4.19(ii)). To this end, suppose that $0 \leq x, y \in B_n$ such that $0 \leq y \leq |P_n T P_n| x$. Then $0 \leq |P_n T P_n|^{n-1} y \leq |P_n T P_n|^n x$, and $|P_n T P_n|^n = |(P_n T P_n)^n| \in Z(E)$ is interval preserving (as B_n is Dedekind complete).

Therefore, there exists $u \in B_n$ such that $0 \le u \le x$ and $|P_n T P_n|^n u = |P_n T P_n|^{n-1} y$, and so

$$|(P_n T P_n)^{n-1}(|P_n T P_n|)|u - y)| = |P_n T P_n|^{n-1}||P_n T P_n|u - y| = 0.$$

As observed above, $P_n T P_n$ is injective, and we may conclude that $y = |P_n T P_n|u$, which shows that $|P_n T P_n|$ is inverval preserving. By this we have proved that $P_n T P_n : B_n \to B_n$ is a quasi-invertible operator.

Next we will analyze the relation between the spectrum of T and the spectra of the entries in the matrix representation (20) of T. For this purpose we need the following preliminary lemmas.

LEMMA 5.26. _Let E be a Dedekind complete Banach lattice, $\pi \in Z(E)$ and $T \in \mathcal{L}_r(E)$ an order continuous operator such that $\pi T = T \pi$. Let P be the band projection onto $\{ran(\pi)\}^{dd}$. Then $PT = TP$._

PROOF. Denote by $\mathcal{L}_n(E)$ the band in $\mathcal{L}_r(E)$ consisting of all order continuous operators, and define

$$\mathcal{C}_n(\pi) = \{ S \in \mathcal{L}_n(E) : S\pi = \pi S \}.$$

First we observe that $\mathcal{C}_n(\pi)$ is a band in $\mathcal{L}_n(E)$. Indeed, defining $L_\pi, R_\pi \in Z(\mathcal{L}_n(E))$ by $L_\pi(S) = \pi S$ and $R_\pi(S) = S\pi$ respectively for all $S \in \mathcal{L}_n(E)$, we have $\mathcal{C}_n(\pi) = \ker(L_\pi - R_\pi)$, which is a band in $\mathcal{L}_n(E)$. Consequently, for the proof of the lemma we may assume that $T \ge 0$. Then $|\pi T| = |\pi| T$ and $|T \pi| = T |\pi|$ (see Remark 5.1) and hence $|\pi| T = T |\pi|$. The band projection P onto $\{ran(\pi)\}^{dd}$ is given by $P = \sup_n n|\pi| \wedge I$ (see e.g. the proof of Proposition 5.2). Using remark 5.1 once more, in combination with the order continuity of T, we find that

$$PT = \left(\sup_n n|\pi| \wedge I \right) T = \sup_n (n|\pi| \wedge I) T$$
$$= \sup_n n|\pi|T \wedge T = \sup_n nT|\pi| \wedge T$$
$$= \sup_n T(n|\pi| \wedge I) = T\left(\sup_n n|\pi| \wedge I \right) = TP,$$

and the proof is complete. □

LEMMA 5.27. _Let E be a Dedekind complete Banach lattice and $T \epsilon \mathcal{L}_r(E)$ an order continuous operator. Suppose that there exists $n \in \mathbb{N}$ such that $T^k \perp I$ for $k = 1, \ldots, n-1$ and $T^n \in Z(E)$. If $\lambda \in \mathbb{C}$ such that $\lambda I - T$ is surjective, then $\lambda \in \rho(T)$._

PROOF. First assume that $\lambda = 0$. Then T is surjective, and hence T^n is surjective. Since $T^n \in Z(E)$ this implies that T^n is injective, and so T is injective.

Therefore $0 \in \rho(T)$. Now assume that $\lambda \neq 0$. Let Q be the band projection onto $\ker(\lambda^n I - T^n)$. Then $(\lambda^n I - T^n)Q = 0$. i.e.,

$$(\lambda^{n-1}I + \lambda^{n-2}T + \cdots + T^{n-1})(\lambda I - T)Q = 0.$$

Since $\lambda I - T$ commutes with $\lambda^n I - T^n \in Z(E)$, it follows from Lemma 5.26 that $(\lambda I - T)Q = Q(\lambda I - T)$ and hence

$$(\lambda^{n-1}I + \lambda^{n-2}T + \cdots + T^{n-1})Q(\lambda I - T) = 0.$$

By hypothesis $\lambda I - T$ is surjective, which now implies that

(21) $$(\lambda^{n-1}I + \lambda^{n-2}T + \cdots + T^{n-1})Q = 0.$$

Applying the diagonal map $\mathcal{D}$ to (21), and using that $\mathcal{D}(T^k) = 0$ for $k = 1, 2, \ldots, n - 1$, we find that $\lambda^{n-1}Q = 0$. and so $Q = 0$. This shows that $\lambda^n I - T^n$ is injective, and consequently $\lambda I - T$ is injective as well. We have thus shown that $\lambda I - T$ is a bijection, hence $\lambda \in \rho(T)$. $\square$

The proof of the following simple lemma is left to the reader.

LEMMA 5.28. *Let X be a Banach space and $X_1, X_2 \subseteq X$ closed linear subspaces such that $X = X_1 \oplus X_2$. Furthermore, let $T : X \to X$ be a bounded linear operator with matrix representation*

$$T = \begin{pmatrix} A & 0 \\ B & C \end{pmatrix}$$

with respect to $X = X_1 \oplus X_2$. where $A : X_1 \to X_1$. $B : X_1 \to X_2$ and $C : X_2 \to X_2$ are bounded linear operators. Assume that A has the property that $\lambda \in \rho(A)$ whenever $\lambda I - A$ is surjective. Then $\sigma(T) = \sigma(A) \cup \sigma(C)$.

Now we return to the situation where E is a Dedekind complete Banach lattice and $T \in \mathcal{L}_r(E)$ is an order continuous disjointness preserving operator. Let the bands B_n, and the corresponding band projections P_n ($n \in \mathbb{N} \cup \{\infty\}$) be as introduced before. In particular, T has the matrix representation (20) with respect to the band decomposition $E = \{\oplus_{n=1}^{\infty} B_n\}^{dd} \oplus B_\infty$. For $n \in \mathbb{N}$ we define

$$S_n = \left(\sum_{k=n}^{\infty} P_k\right) T \left(\sum_{k=n}^{\infty} P_k\right),$$

considered as an operator $S_n : (\oplus_{k=n}^{\infty} B_k)^{dd} \to (\oplus_{k=n}^{\infty} B_k)^{dd}$. Moreover, $\sigma(P_n T P_n)$ will denote the spectrum of $P_n T P_n$ considered as an operator on the band B_n ($n \in \mathbb{N} \cup \{\infty\}$).

THEOREM 5.29. *Let E be a Dedekind complete Banach lattice and $T \in \mathcal{L}_r(E)$ an order continuous disjointness preserving operator. With the notation introduced above we have $\sigma(S_{n+1}) \subseteq \sigma(S_n)$ for all $n \in \mathbb{N}$, and*

$$\sigma(T) = \sigma(P_\infty T P_\infty) \cup \left\{ \bigcup_{n=1}^{\infty} \sigma(P_n T P_n) \right\} \cup \left\{ \bigcap_{n=1}^{\infty} \sigma(S_n) \right\}.$$

PROOF. Let $B_f = (\oplus_{n=1}^{\infty} B_n)^{dd}$, and denote by $P_f = \sum_{n=1}^{\infty} P_n$ the corresponding band projection. The operator T has the matrix representation

$$T = \begin{pmatrix} S_1 & 0 \\ P_\infty T P_f & P_\infty T P_\infty \end{pmatrix}$$

with respect to the band decomposition $E = B_f \oplus B_\infty$. First we will show that $S_1 : B_f \to B_f$ has the property of the operator A in Lemma 5.28. Clearly, with respect to the decomposition $B_f = (\oplus_{n=1}^{\infty} B_n)^{dd}$, the matrix of S_1 is the diagonal matrix

$$S_1 = \begin{pmatrix} P_1 T P_1 & & & \emptyset \\ & P_2 T P_2 & & \\ & & \ddots & \\ \emptyset & & & \ddots \end{pmatrix}.$$

Now suppose that $\lambda \in \mathbb{C}$ is such that $\lambda I - S_1 : B_f \to B_f$ is surjective. It is easy to see that this implies that $\lambda I - P_n T P_n$ is surjective on B_n for all $n \in \mathbb{N}$. By Corollary 5.24, $P_n T P_n$ satisfies the hypothesis of Lemma 5.27, and hence $\lambda I - P_n T P_n$ is injective on B_n for all $n \in \mathbb{N}$. Therefore $\lambda I - S_1$ is injective on B_f, which shows that $\lambda \in \rho(S_1)$. An application of Lemma 5.28 now yields that $\sigma(T) = \sigma(S_1) \cup \sigma(P_\infty T P_\infty)$. For each $n \in \mathbb{N}$ the operator S_1 is reduced by the band decomposition $B_f = B_1 \oplus \cdots \oplus B_n \oplus \left\{ \bigoplus_{k=n+1}^{\infty} B_k \right\}^{dd}$. This implies that

$$\sigma(S_1) = \sigma(P_1 T P_1) \cup \cdots \cup \sigma(P_n T P_n) \cup \sigma(S_{n+1})$$

for all $n \in \mathbb{N}$, and hence

$$\sigma(S_1) = \left\{ \bigcup_{n=1}^{\infty} \sigma(P_n T P_n) \right\} \cup \left\{ \bigcap_{n=2}^{\infty} \sigma(S_n) \right\}.$$

Finally, for each $n \in \mathbb{N}$, the operator S_n is reduced by the decomposition

$$\left\{ \bigoplus_{k=n}^{\infty} B_k \right\}^{dd} = B_n \oplus \left\{ \bigoplus_{k=n+1}^{\infty} B_k \right\}^{dd},$$

implying that $\sigma(S_n) = \sigma(P_n T P_n) \cup \sigma(S_{n+1})$. By this the theorem is completely proved. □

REMARK 5.30. (1) In general, the band decomposition $E = \{\oplus_{n=1}^\infty B_n\}^{dd} \oplus B_\infty$ does not reduce the operator T, as is shown by the operator considered in Example 5.4. However, if we assume in addition that T^* is disjointness preserving as well, then T is reduced by this decomposition (as was already observed by X. Zhang in [22] Theorem 4.1). Indeed, take $x \in E$ and put $y = (I - Q_n)T Q_n x$. Since, under the present assumptions, $|T|$ is interval preserving (see the remarks following Proposition 5.5), it follows from $|y| \leq |T|Q_n|x|$ that there exists $0 \leq u \leq Q_n|x|$ such that $|y| = |T|u$. Then

$$\begin{aligned}|\mathcal{D}(T^n)u| &= |Q_n T^n u| = Q_n|T|^{n-1}|y| \\ &= Q_n|T|^{n-1}(I - Q_n)|T|Q_n|x| \\ &= |Q_n T^{n-1}(I - Q_n)| \, |T|Q_n|x| = 0,\end{aligned}$$

since $Q_n T^{n-1}(I - Q_n) = 0$. This shows that $(I - Q_n)T Q_n = 0$, and hence $Q_n T = T Q_n$. Our results above extend [22] Theorem 4.1 (where it is assumed that both T and T^* are disjointness preserving), and [5], Theorem 3.10 (where it is assumed that T is quasi-invertible).

(2) As observed in Remark 5.25, the operator $P_n T P_n : B_n \to B_n$ ($n \in \mathbb{N}$) is quasi-invertible and (by Corollary 5.24 (2)) has strict period n. Therefore, it follows from [5], Theorem 4.3 that the spectrum of $P_n T P_n$ is fully cyclic of order n, i.e., $\sigma(P_n T P_n) = \alpha \sigma(P_n T P_n)$ for all n-th roots of unity α. Note that this is equivalent to saying that $\sigma(P_n T P_n) = \{\lambda \in \mathbb{C} : \lambda^n \in \sigma((P_n T P_n)^n)\}$.

(3) Using the observations in Remark 5.25 it is not difficult to see that the operator $S_1 : \{\oplus_{n=1}^\infty B_n\}^{dd} \to \{\oplus_{n=1}^\infty B_n\}^{dd}$ is actually quasi-invertible. Therefore it follows from [5], Theorem 4.6 that the set $\bigcap_{n=1}^\infty \sigma(S_n)$ is rotationally invariant (i.e., $\alpha \bigcap_{n=1}^\infty \sigma(S_n) = \bigcap_{n=1}^\infty \sigma(S_n)$ for all $\alpha \in \mathbb{C}$ with $|\alpha| = 1$).

As observed above, the band decomposition $E = \{\bigoplus_{n=1}^\infty B_n\}^{dd} \oplus B_\infty$ does not reduce T in general. However, by enlarging B_n ($n \in \mathbb{N}$) in an appropriate way, we can obtain a band decomposition of E which reduces T, as will be shown next.

Again, E will be a Dedekind complete Banach lattice and $T \in \mathcal{L}_r(E)$ an order continuous disjointness preserving operator, and we will use the B_n's and P_n's as before. We first prove a simple lemma.

LEMMA 5.31. For all $n \in \mathbb{N}$ and $k = 0, 1, 2, \ldots$ we have

$$\{T^k(B_n)\}^{dd} \subseteq \{T^{k+1}(B_n)\}^{dd}.$$

PROOF. First we show that $B_n \subseteq \{T(B_n)\}^{dd}$. From Corollary 5.24 we know that $(P_n T P_n)^n = P_n \mathcal{D}(T^n)$, and from the definition of B_n it follows that

$$B_n = P_n \{\mathcal{D}(T^n)(E)\}^{dd} = \{P_n \mathcal{D}(T^n)(B_n)\}^{dd}.$$

Hence

$$B_n = \{P_n \mathcal{D}(T^n)(B_n)\}^{dd} = \{(P_n T P_n)^n (B_n)\}^{dd}$$
$$\subseteq \{P_n T P_n(B_n)\}^{dd} = P_n \{T(B_n)\}^{dd} \subseteq \{T(B_n)\}^{dd}.$$

Using that T is order continuous, we now find that

$$T^k(B_n) \subseteq T^k(\{T(B_n)\}^{dd}) \subseteq \{T^{k+1}(B_n)\}^{dd}$$

for all $k = 1, 2, \ldots$. $\square$

For $n = 1, 2, \ldots$ we define the band B_n^ω by

$$B_n^\omega = \left\{ \bigcup_{k=0}^\infty T^k(B_n) \right\}^{dd}.$$

Since $\{T^k(B_n)\}^{dd} \subseteq B_n^\omega$, it is obvious that

$$B_n^\omega = \left\{ \bigcup_{k=0}^\infty \{T^k(B_n)\}^{dd} \right\}^{dd},$$

so, by the above lemma, B_n^ω is the band generated by an increasing sequence of bands. We note that the definition of the band B_n^ω is related to, and actually motivated by the definition of the set $F^\#$ in [1] (Appendix A). In the following lemma we collect some properties of the bands B_n^ω.

LEMMA 5.32. *Same situation as above.*

(i) $B_n^\omega \perp B_m^\omega$ whenever $n, m \in \mathbb{N}$ with $n \neq m$;

(ii) *For each $n \in \mathbb{N}$ the band decomposition $E = B_n^\omega \oplus (B_n^\omega)^d$ reduces the operator T.*

PROOF.(i) Since T is disjointness preserving, $B_n \perp B_m$ implies that $T^k(B_n) \perp T^k(B_m)$, and hence $\{T^k(B_n)\}^{dd} \perp \{T^k(B_m)\}^{dd}$ for all $k = 0, 1, \ldots$. Via Lemma 5.31 it now follows that

$$\bigcup_{k=0}^\infty \{T^k(B_n)\}^{dd} \perp \bigcup_{k=0}^\infty \{T^k(B_m)\}^{dd},$$

and hence $B_n^\omega \perp B_m^\omega$.

(ii) Clearly $\cup_{k=0}^{\infty} T^k(B_n)$ is T-invariant, and from the order continuity of T it follows that $T(B_n^{\omega}) \subseteq B_n^{\omega}$. To show that $(B_n^{\omega})^d$ is T-invariant as well, take $x \in (B_n^{\omega})^d$. Then $x \perp T^k(B_n)$, and hence $Tx \perp T^{k+1}(B_n)$ for all $k = 0, 1, \ldots$. From Lemma 5.31 it follows that $Tx \perp \cup_{k=0}^{\infty} T^k(B_n)$ and so $Tx \in (B_n^{\omega})^d$. $\square$

Observe that B_n^{ω} is clearly the smallest T-invariant band containing B_n. The band projection onto B_n^{ω} will be denoted by P_n^{ω} $(n \in \mathbb{N})$. By (ii) in the lemma above we have $P_n^{\omega} T = T P_n^{\omega}$ for all $n \in \mathbb{N}$. Furthermore we define

$$B_{\infty}^{\omega} = \left\{ \bigoplus_{n=1}^{\infty} B_n^{\omega} \right\}^d.$$

Denoting the band projection onto B_{∞}^{ω} by P_{∞}^{ω}, it is clear that $P_{\infty}^{\omega} = I - \sum_{n=1}^{\infty} P_n^{\omega}$. The order continuity of T implies that $P_{\infty}^{\omega} T = T P_{\infty}^{\omega}$, and this yields the following corollary.

COROLLARY 5.33. *The band decomposition* $E = \left\{ \bigoplus_{n=1}^{\infty} B_n^{\omega} \right\}^{dd} \oplus B_{\infty}^{\omega}$ *reduces the operator* T.

From the above result we conclude immediately that

$$\sigma(T) = \sigma(P_{\infty}^{\omega} T P_{\infty}^{\omega}) \cup \left\{ \bigcup_{n=1}^{\infty} \sigma(P_n^{\omega} T P_n^{\omega}) \right\} \cup \left\{ \bigcap_{n=1}^{\infty} \sigma(S_n^{\omega}) \right\},$$

where $S_n^{\omega} = (\sum_{k=n}^{\infty} P_k^{\omega}) T (\sum_{k=n}^{\infty} P_k^{\omega})$, considered as operator on the space $\{\oplus_{k=n}^{\infty} B_k^{\omega}\}^{dd}$. In order to analyze further the operator $P_n^{\omega} T P_n^{\omega}$ on B_n^{ω} $(n \in \mathbb{N})$, we first observe that, since $B_n^{\omega} \perp B_m$ for all $m \in \mathbb{N}$ with $m \neq n$, we have

$$(22) \qquad B_n^{\omega} = B_n \oplus (B_n^{\omega} \cap B_{\infty}).$$

where, as before $B_{\infty} = \cap_{k=1}^{\infty} B_k^d$. Hence $P_n^{\omega} = P_n + P_n^{\omega} P_{\infty}$. We denote the band projection $P_n^{\omega} P_{\infty}$ by R_n. It follows from $P_n T R_n = P_n T P_{\infty} P_n^{\omega} = 0$, that the matrix of the operator $P_n^{\omega} T P_n^{\omega} : B_n^{\omega} \to B_n^{\omega}$ with respect to the band decomposition (22) is given by

$$(23) \qquad \begin{pmatrix} P_n T P_n & 0 \\ R_n T P_n & R_n T R_n \end{pmatrix}.$$

As before, from Corollary 5.24, and Lemmas 5.27 and 5.28 we can conclude that $\sigma(P_n^{\omega} T P_n^{\omega}) = \sigma(P_n T P_n) \cup \sigma(R_n T R_n)$. Note that $R_n T P_n = P_{\infty} T P_n$ (i.e., the block in the lower left corner of (23) is essentially equal to the block $P_{\infty} T P_n$ in the matrix (20)). Indeed, since

$$P_{\infty} = I - \sum_{k=1}^{\infty} P_k = I - \sum_{k=1}^{\infty} (P_k^{\omega} - R_k) = \sum_{k=1}^{\infty} R_k + \left(I - \sum_{k=1}^{\infty} P_k^{\omega} \right) = \sum_{k=1}^{\infty} R_k + P_{\infty}^{\omega},$$

and $P_k^\omega P_n = 0$ for all $k \in \mathbb{N} \cup \{\infty\}, k \neq n$, we have

$$P_\infty T P_n = \sum_{k=1}^\infty R_k T P_n + P_\infty^\omega T P_n = \sum_{k=1}^\infty P_\infty P_k^\omega T P_n + P_\infty^\omega T P_\infty^\omega P_n$$

$$= \sum_{k=1}^\infty P_\infty P_k^\omega T P_k^\omega P_n = P_\infty P_n^\omega T P_n^\omega P_n = R_n T P_n.$$

Moreover, observe that a similar calculation shows that $P_\infty T P_\infty = \sum_{k=1}^\infty R_k T R_k + P_\infty^\omega T P_\infty^\omega$.

By remark 5.30(2), the spectrum $\sigma(P_n T P_n)$ is fully cyclic of order n $(n \in \mathbb{N})$. The spectrum of $R_n T R_n$ is rather special as well. In fact we will show that $\sigma(R_n T R_n)$ is always a disc around 0. For this purpose it is convenient to introduce the following terminology.

DEFINITION 5.34. *Let E be a Dedekind complete Banach lattice. An order continuous operator $T \in \mathcal{L}_r(E)$ is called a generalized unilateral shift if there exist band* .*projections $\{W_k\}_{k=1}^\infty$ in E satisfying:*

(i) *$W_j W_k = 0$ whenever $j \neq k$, and $\sum_{k=1}^\infty W_k = I$;*
(ii) *$W_{k+1} T W_k = T W_k$ for all $k \in \mathbb{N}$.*

Examples of generalized unilateral shifts are, of course, the weighted unilateral shifts on the space ℓ_2, which are discussed in detail by A. Shields in [18]. In [7], Definition 4.6 the notion 'forward shift type operator' is defined for (bi)-disjointness preserving operators. In a Dedekind complete Banach lattice every disjointness preserving operator of forward shift type is a generalized unilateral shift (cf. [7], Lemma 5.5).

LEMMA 5.35. *Let E be a Dedekind complete Banach lattice and $T \in \mathcal{L}_r(E)$ a generalized unilateral shift. Then $\sigma(T)$ is rotationally invariant.*

PROOF. Let the band projections $\{W_k\}_{k=1}^\infty$ be as in Definition 5.34. First we observe that $W_1 T = 0$ and $W_{k+1} T = W_{k+1} T W_k$ for all $k \geq 1$. Indeed, using that T is order continuous, we have

$$W_1 T = W_1 T \left(\sum_{j=1}^\infty W_j \right) = \sum_{j=1}^\infty W_1 T W_j = \sum_{j=1}^\infty W_1 W_{j+1} T W_j = 0,$$

and

$$W_{k+1} T = W_{k+1} T \left(\sum_{j=1}^\infty W_j \right) = \sum_{j=1}^\infty W_{k+1} T W_j = \sum_{j=1}^\infty W_{k+1} W_{j+1} T W_j = W_{k+1} T W_k.$$

Now take $\alpha \in \mathbb{C}$, $|\alpha| = 1$, and define $\pi \in Z(E)$ by $\pi = \sum_{k=1}^{\infty} \alpha^{-k} W_k$ (order convergent series in $Z(E)$). Note that π is an invertible operator, with $\pi^{-1} = \sum_{k=1}^{\infty} \alpha^k W_k$. Using the above observation we find that

$$T\pi = \sum_{k=1}^{\infty} \alpha^{-k} TW_k = \sum_{k=1}^{\infty} \alpha^{-k} W_{k+1} TW_k$$

$$= \sum_{k=1}^{\infty} \alpha^{-k} W_{k+1} T = \alpha \left(\sum_{k=2}^{\infty} \alpha^{-k} W_k \right) T,$$

and also that

$$\pi T = \sum_{k=1}^{\infty} \alpha^{-k} W_k T = \sum_{k=2}^{\infty} \alpha^{-k} W_k T = \left(\sum_{k=2}^{\infty} \alpha^{-k} W_k \right) T.$$

This shows that $T\pi = \alpha \pi T$, i.e., $\pi^{-1} T\pi = \alpha T$, and consequently $\sigma(T) = \sigma(\pi^{-1} T\pi) = \sigma(\alpha T) = \alpha \sigma(T)$. We thus have proved that $\sigma(T) = \alpha \sigma(T)$ for all $\alpha \in \mathbb{C}$ with $|\alpha| = 1$, i.e., $\sigma(T)$ is rotationally invariant. $\square$

THEOREM 5.36. *Let E be a Dedekind complete Banach lattice and $T \in \mathcal{L}_r(E)$ a generalized unilateral shift. Then $\sigma(T) = \{\lambda \in \mathbb{C} : |\lambda| \leq r(T)\}$.*

PROOF. For the proof we may, of course, assume that $r(T) > 0$. Suppose that $\lambda_0 \in \mathbb{C}$ with $0 < |\lambda_0| < r(T)$ is such that $\lambda_0 \notin \sigma(T)$. Since, by the previous lemma, $\sigma(T)$ is rotationally invariant, this implies that $\{\lambda \in \mathbb{C} : |\lambda| = |\lambda_0|\} \subseteq \rho(T)$. Put $\sigma = \sigma(T) \cap \{\lambda \in \mathbb{C} : |\lambda| > |\lambda_0|\}$, which is a non-empty spectral set of T. Let P_σ be the spectral projection of T corresponding to σ, and let $E_\sigma = P_\sigma(E)$. Then $T(E_\sigma) \subseteq E_\sigma$ and $T|_{E_\sigma} : E_\sigma \to E_\sigma$ is invertible as $0 \notin \sigma(T|_{E_\sigma}) = \sigma$. Let the band projections $\{W_k\}_{k=1}^{\infty}$ be as in Definition 5.34, and define

$$k = \min\{m \in \mathbb{N} : \text{there exists } y \in E_\sigma \text{ such that } W_m y \neq 0\}.$$

Take $y_0 \in E_\sigma$ such that $W_k y_0 \neq 0$. If $y \in E_\sigma$, then $y = \sum_{j=1}^{\infty} W_j y = \sum_{j=k}^{\infty} W_j y$, and so

$$Ty = \sum_{j=k}^{\infty} T W_j y = \sum_{j=k}^{\infty} W_{j+1} T W_j y.$$

This implies that $W_k Ty = 0$, which shows that $Ty \neq y_0$ for all $y \in E_\sigma$, and hence $T|_{E_\sigma} : E_\sigma \to E_\sigma$ is not surjective. This is clearly a contradiction since $T|_{E_\sigma}$ is invertible. We may conclude, therefore, that $\{\lambda \in \mathbb{C} : 0 < |\lambda| < r(T)\} \subseteq \sigma(T)$, from which the conclusion of the theorem follows. $\square$

We note that for weighted unilateral shifts the result of the above theorem can be found in [18], obtained via a different method of proof. The corresponding result for bi-disjointness preserving operators of forward shift type is contained in [7] Theorem 5.6. Our method of proof of Theorem 5.36 bears resemblance to the proof in [7].

We return again to the situation preceding Definition 5.34, where E is a Dedekind complete Banach lattice and $T \in \mathcal{L}_r(E)$ an order continuous disjointness preserving operator. We also use the notations which were introduced there.

PROPOSITION 5.37. *In the situation described above, the operator* $R_n T R_n :$ $B_n^\omega \cap B_\infty \to B_n^\omega \cap B_\infty$ *is a generalized unilateral shift for all* $n \in \mathbb{N}$.

PROOF. Let $n \in \mathbb{N}$ be fixed. From the definition of B_n^ω and from Lemma 5.31, it follows that

$$B_n \subseteq \{T(B_n)\}^{dd} \subseteq \{T^2(B_n)\}^{dd} \subseteq \ldots \uparrow B_n^\omega.$$

For $k \in \mathbb{N}$ let W_k be the band projection onto $A_k = \{T^k(B_n)\}^{dd} \cap \{T^{k-1}(B_n)\}^d$. Then it is clear that $W_j W_k = 0$ whenever $j \neq k$ and that $\sum_{k=1}^\infty W_k = P_n^\omega - P_n = R_n$. Since T is order continuous and disjointness preserving it follows that $T(A_k) \subseteq A_{k+1}$ for all $k \in \mathbb{N}$, i.e., $W_{k+1} T W_k = T W_k$ for all k, which is of course equivalent to saying that $W_{k+1}(R_n T R_n)W_k = (R_n T R_n)W_k$ for all k. This shows, formally by taking the restrictions of the W_k's to $B_n^\omega \cap B_\infty$, that the operator $R_n T R_n$ is a generalized unilateral shift on $B_n^\omega \cap B_\infty$. $\square$

Combining the above proposition with Theorem 5.36 we obtain the following result.

COROLLARY 5.38. *Same situation as above. For each* $n \in \mathbb{N}$ *the spectrum* $\sigma(R_n T R_n)$ *is a closed disc around 0.*

For the convenience of the reader we collect the main results of the last part of this section in the following theorem.

THEOREM 5.39. *Let* E *be a Dedekind complete Banach lattice and let* $T \in$ $\mathcal{L}_r(E)$ *be an order continuous disjointness preserving operator.*

A. *For* $n = 1, 2, \ldots$ *we define the mutually disjoint bands*

$$B_n = \{\operatorname{ran} \mathcal{D}(T^n)\}^{dd} \cap \left\{ \bigcap_{k=1}^{n-1} \ker \mathcal{D}(T^k) \right\}$$

and $B_\infty = \bigcap_{k=1}^\infty \ker \mathcal{D}(T^k)$. *Let* P_n *be the band projection onto* $B_n(n \in \mathbb{N} \cup \{\infty\})$. *With respect to the band decomposition* $E = (\oplus_{n=1}^\infty B_n)^{dd} \oplus B_\infty$ *the operator*

T has the matrix representation

$$
\begin{pmatrix}
P_1 T P_1 & & \emptyset & \vdots & 0 \\
& P_2 T P_2 & & \vdots & 0 \\
\emptyset & & \ddots & \vdots & \vdots \\
\cdots\cdots\cdots\cdots\cdots\cdots\cdots\cdots\cdots\cdots\cdots & & & & \\
P_\infty T P_1 & P_\infty T P_2 & \cdots\cdots & \vdots & P_\infty T P_\infty
\end{pmatrix}.
$$

Each of the operators $P_n T P_n$ $(n \in \mathbb{N})$ has strict period n and is quasi-invertible on B_n, whereas $(P_\infty T P_\infty)^n \perp I$ on B_∞ for all $n \in \mathbb{N}$. Furthermore,

$$
\sigma(T) = \sigma(P_\infty T P_\infty) \cup \left\{ \bigcup_{n=1}^{\infty} \sigma(P_n T P_n) \right\} \cup \left\{ \bigcap_{n=1}^{\infty} \sigma(S_n) \right\},
$$

where $S_n = (\sum_{k=n}^{\infty} P_k) T (\sum_{k=n}^{\infty} P_k)$ on $(\oplus_{k=n}^{\infty} B_k)^{dd}$. For $n \in \mathbb{N}$ the spectrum $\sigma(P_n T P_n)$ is fully cyclic of order n, and $\bigcap_{n=1}^{\infty} \sigma(S_n)$ is rotationally invariant.

B. For $n = 1, 2, \ldots$ we define the mutually disjoint bands

$$
B_n^\omega = \left\{ \bigcup_{k=0}^{\infty} T^k(B_n) \right\}^{dd}
$$

and $B_\infty^\omega = \left\{ \stackrel{\infty}{\underset{n=1}{\oplus}} B_n^\omega \right\}^d$. Let P_∞^ω be the band projection onto B_∞^ω $(n \in \mathbb{N} \cup \{\infty\})$.

The operator T is reduced by the band decomposition $E = \left\{ \stackrel{\infty}{\underset{n=1}{\oplus}} B_n^\omega \right\}^{dd} \oplus B_\infty^\omega$, and

$$
\sigma(T) = \sigma(P_\infty^\omega T P_\infty^\omega) \cup \left\{ \bigcup_{n=1}^{\infty} \sigma(P_n^\omega T P_n^\omega) \right\} \cup \left\{ \bigcap_{n=1}^{\infty} \sigma(S_n^\omega) \right\},
$$

where $S_n^\omega = (\sum_{k=n}^{\infty} P_k^\omega) T (\sum_{k=n}^{\infty} P_k^\omega)$ on $\{\oplus_{k=n}^{\infty} B_k^\omega\}^{dd}$. Moreover $(P_\infty^\omega T P_\infty^\omega)^n \perp I$ on B_∞^ω for all $n \in \mathbb{N}$.

C. For each $n \in \mathbb{N}$ we consider the band decomposition

$$
B_n^\omega = B_n \oplus (B_n^\omega \cap B_\infty),
$$

and let R_n be the band projection on $B_n^\omega \cap B_\infty$. With respect to this band decomposition the operator $P_n^\omega T P_n^\omega$ on B_n^ω has the matrix representation

$$
\begin{pmatrix}
P_n T P_n & 0 \\
R_n T P_n & R_n T R_n
\end{pmatrix}.
$$

Furthermore, $\sigma(P_n^\omega T P_n^\omega) = \sigma(P_n T P_n) \cup \sigma(R_n T R_n)$. The operator $R_n T R_n$ on $B_n^\omega \cap B_\infty$ is a generalized unilateral shift and $\sigma(R_n T R_n)$ is a disc around 0.

We conclude this paper with some simple examples illustrating the decompositions in the above theorem, in particular in connection with C.

EXAMPLES 5.40. (i) Take $1 \le p < \infty$, and consider the operator $T : \ell_p \to \ell_p$ defined by $T(x_1, x_2, \ldots) = (x_1, x_1, x_2, x_3, \ldots)$. We denote the unit basis vectors in ℓ_p by e_n ($n \in \mathbb{N}$). It is easy to see that $B_1 = \mathrm{span}\{e_1\}$ and $B_\infty = \overline{\mathrm{span}}\{e_n : n \ge 2\}$. Moreover $B_1^\omega = \ell_p$ and $B_1^\omega \cap B_\infty = B_\infty$. The operator $R_1 T R_1 : B_\infty \to B_\infty$ is given by $R_1 T R_1 (x_2, x_3, \ldots) = (0, x_2, x_3, \ldots)$, and is indeed a shift operator with $\sigma(R_1 T R_1) = \{\lambda \in \mathbb{C} : |\lambda| \le 1\}$.

(ii) Take $1 \le p < \infty$, and consider the Banach lattice $E = \ell_p(\mathbb{N} \times \mathbb{N})$. We define the disjointness preserving operator $T : E \to E$ by

$$(Tx)(n, m) = \begin{cases} x(n, 1) & \text{if } m = 1 \\ x(n, m-1) & \text{if } 2 \le m \le n+1 \\ 0 & \text{if } m > n+1 \end{cases}$$

for all $(m, n) \in \mathbb{N} \times \mathbb{N}$. We denote by $e_{n,m}$ ($n, m \in \mathbb{N}$) the unit basis vectors in E defined by $e_{n,m}(k, l) = 1$ if $(k, l) = (n, m)$ and 0 otherwise. In this case we find that $B_1 = \overline{\mathrm{span}}\{e_{n,1} : n \in \mathbb{N}\}$ and $B_\infty = \overline{\mathrm{span}}\{e_{n,m} : n, m \in \mathbb{N}, m \ge 2\}$. Furthermore, $B_1^\omega = \overline{\mathrm{span}}\{e_{n,m} : n, m \in \mathbb{N}, 1 \le m \le n+1\}$ and $B_\infty^\omega = \overline{\mathrm{span}}\{e_{n,m} : n, m \in \mathbb{N}, m > n+1\}$. Clearly, the decomposition $E = B_1^\omega \oplus B_\infty^\omega$ reduces T. Now $B_1^\omega \cap B_\infty = \overline{\mathrm{span}}\{e_{n,m} : n, m \in \mathbb{N}, 2 \le m \le n+1\}$ and $R_1 T R_1$ is a generalized unilateral shift in $B_1^\omega \cap B_\infty$ satisfying $\|(R_1 T R_1)^n\| = 1$ for all $n \in \mathbb{N}$. This implies that $\sigma(R_1 T R_1) = \{\lambda \in \mathbb{C} : |\lambda| \le 1\}$, and hence $\sigma(T) = \sigma(P_1 T P_1) \cup \sigma(R_1 T R_1) = \{\lambda \in \mathbb{C} : |\lambda| \le 1\}$.

REFERENCES

1. Y.A. Abramovich, E.L. Arenson and A.K. Kitover, *Banach C(K)-modules and operators preserving disjointness*, Pitman Research Notes in Mathematics Series, vol. 277, Longman Scientific & Technical, Harlow, 1992.
2. Charalambos D. Aliprantis and Owen Burkinshaw, *Positive Operators*, Pure and Applied Mathematics, vol. 119, Academic Press, Orlando, 1985.
3. T. Andô, *Inequalities for M-matrices*, Lin. and Multilin. Alg. **8** (1979/80), 291–316.
4. W. Arendt, *Spectral properties of Lamperti operators*, Indiana Univ. Math. J. **32** (1983), 199–215.
5. W. Arendt and D.R. Hart, *The spectrum of quasi-invertible disjointness preserving operators*, J. of Funct. Anal. **68** (1986), 149–167.
6. Ulf Grenander and Gabor Szegö, *Toeplitz Forms and their applications*, University of California Press, Berkeley-Los Angeles, 1958.
7. D.R. Hart, *Disjointness preserving operators*, Ph.D. Thesis, California Institute of Technology, Pasadena, 1983.
8. C.B. Huijsmans and B. de Pagter, *Disjointness preserving and diffuse operators*, Compos. Math. **79** (1991), 351–374.

9. C.B. Huijsmans and B. de Pagter, *Some remarks on disjointness preserving operators*, Acta Appl. Math. **27** (1992), 73–78.
10. E. de Jonge and A.C.M. van Rooij, *Introduction to Riesz spaces*, Mathematical Centre Tracts 78, Mathematical Centre, Amsterdam, 1977.
11. S. Karlin, *Positive operators*, J. Math. Mech. **8** (1959), 907–937.
12. Yitzhak Katznelson, *An introduction to Harmonic Analysis*, Dover Publications, New York, 1976.
13. J.F.C. Kingman, *Regenerative Phenomena*, John Wiley & Sons Ltd., London-New York-Sydney-Toronto, 1972.
14. W.A.J. Luxemburg and A.C. Zaanen, *Riesz Spaces I*, North–Holland, Amsterdam, 1971.
15. Peter Meyer–Nieberg, *Banach Lattices*, Springer–Verlag, New York–Heidelberg–Berlin, 1991.
16. H.H. Schaefer, *Banach Lattices and Positive Operators*, Springer–Verlag, New York–Heidelberg–Berlin, 1974.
17. A.R. Schep, *Positive diagonal and triangular operators*, J. of Operator Theory **3** (1980), 165–178.
18. A.L. Shields, *Weighted shift operators and analytic function theory*, Topics in Operator Theory (C. Pearcy, ed.), Amer. Math. Soc., Providence, R.I., 1974.
19. J. Voigt, *The projection onto the center of operators in a Banach Lattice*, Math. Zeitschrift **199** (1988), 115–117.
20. J.D. Maitland Wright, *Stone-Algebra-Valued Measures and Integrals*, Proc. Lond. Math. Soc **19** (1969), 107–122.
21. A.C. Zaanen, *Riesz Spaces II*, North–Holland, Amsterdam, 1983.
22. X.D. Zhang, *Decomposition theorems for disjointness preserving operators*, J. of Funct. Anal. **116** (1993), 158–178.

Mathematics Subject Classifications (1991): 46B42, 47B65, 47B38

DEPARTMENT OF MATHEMATICS,
CALIFORNIA INSTITUTE OF TECHNOLOGY,
PASADENA, CA 91125

DEPARTMENT OF MATHEMATICS,
DELFT UNIVERSITY OF TECHNOLOGY,
P.O. BOX 356,
2600 AJ DELFT,
THE NETHERLANDS

DEPARTMENT OF MATHEMATICS,
UNIVERSITY OF SOUTH CAROLINA,
COLUMBIA, SC 29208

Operator Theory:
Advances and Applications, Vol. 75
© 1995 Birkhäuser Verlag Basel/Switzerland

A CHARACTERIZATION OF LIPSCHITZ
CONTINUOUS EVOLUTION FAMILIES
ON BANACH SPACES*

Rainer Nagel and Abdelaziz Rhandi**

Dedicated to Professor A.C. Zaanen in occasion of his 80th birthday.

We consider evolution families $(U(t,s))$ of bounded, linear operators on a Banach space X and associate to it an evolution semigroup $(T(t))_{t\geq 0}$ defined by

$$T(t)f(s) = U(s, s-t)f(s-t)$$

on the weighted function space $C_{v_0}(\mathbb{R}, X)$. In the case of Lipschitz continuity of $(U(t,s))$ we characterize the generator of this semigroup $(T(t))_{t\geq 0}$ and thus obtain a family of operators $A(t) \in \mathcal{L}(X)$ such that $(U(t,s))$ solves the non-autonomous Cauchy problem

$$\dot{u} = A(t)u(t), \ u(s) = u_0.$$

1. INTRODUCTION

If the nonautonomous linear Cauchy problem

$$(ACP) \qquad \dot{u}(t) = A(t)u(t), \quad u(s) = u_0, \quad s \leq t,$$

for functions $u(\cdot)$ on $\mathbb{R}$ with values in a Banach space X is well posed (see [Ac-Te], [Da-Ko], [Fa], [Go1], [Go2], [Ka1] or [Ka2] for sufficient conditions on the linear operators $A(t)$) then the solutions are given through a *strongly continuous evolution family* $(U(t,s))_{t\geq s}$. This means that for $r \geq t \geq s$ we have

(a) $U(t,s) \in \mathcal{L}(X)$,

(b) $U(r,t)U(t,s) = U(r,s)$, $U(t,t) = Id$,

(c) $(t,s) \mapsto U(t,s)$ is strongly continuous on $\{(t,s) : t \geq s\}$ and

*This paper is part of a research project supported by the Deutsche Forschungsgemeinschaft DFG.
**The support of DAAD is gratefully acknowledged.

(d) $\|U(t,s)\| \le M e^{\omega(t-s)}$ for some constants $M \ge 1$ and $\omega \in \mathbb{R}$.

In the autonomous case, i.e., if $A(\cdot) \equiv A$, then the problem (ACP) is well posed if and only if A generates a strongly continuous semigroup $(e^{tA})_{t \ge 0}$ and the evolution family is given by

$$U(t,s) := e^{(t-s)A} \quad (t \ge s).$$

In this situation it is quite elementary to show that the semigroup (resp., the corresponding evolution family) is (locally) Lipschitz continuous if and only if the operator A is bounded. We restate this basic fact (see [Go], Prop. I.2.5).

PROPOSITION 1.1 *The strongly continuous semigroup $(e^{tA})_{t \ge 0}$ on the Banach space X generated by some operator A satisfies*

$$\|e^{(t-s)A} - I\| \le M(t-s)e^{\omega(t-s)}$$

for some constants $M, \omega \ge 0$ and all $t \ge s$ if and only if A is bounded.

In the nonautonomous case (and for infinite dimensional spaces) a similar simple characterization of Lipschitz continuity seems to be unknown (see [Fa], 7.1.4). We propose to use semigroup theory in an X-valued function space in order to obtain an answer to this problem.

2. EVOLUTION SEMIGROUPS

Take a strongly continuous evolution family $(U(t,s))_{t \ge s}$ on a Banach space X. It is unclear in general how to find operators $A(t)$ such that $(U(t,s))_{t \ge s}$ represents the solution of (ACP). On the other hand there are various well known procedures how to associate a strongly continuous semigroup to the evolution family $(U(t,s))_{t \ge s}$ (see [Ev], [Ho], [Lu], [Ne] and [Pa]).
We briefly describe the construction of such an *evolution semigroup*.
As underlying function space we take

$$E := C_{v_0}(\mathbb{R}, X),$$

the space of all continuous functions f from $\mathbb{R}$ into X such that vf vanishes at infinity where $v(s) := e^{-|s|}, s \in \mathbb{R}$, endowed with the norm

$$\|f\|_v := \sup_{s \in \mathbb{R}} \|v(s)f(s)\|.$$

Note that E equipped with this norm is a Banach space.
On this space the translations

$$T_0(t)f(s) := f(s-t) \quad \text{for } s \in \mathbb{R} \text{ and } t \ge 0,$$

define a strongly continuous semigroup with generator $G_0 := -\frac{d}{ds}$ on the domain

$$D(G_0) = \{f \in E : f \text{ differentiable and } f' \in E\}.$$

Using this translation semigroup and the evolution family $(U(t,s))_{t \ge s}$ we define a new semigroup on E.

DEFINITION 2.1 Let $(U(t,s))_{t \geq s}$ be a strongly continuous evolution family on X. Then the corresponding *evolution semigroup* $(T(t))_{t \geq 0}$ on E is defined by

$$T(t)f(s) := U(s, s - t)f(s - t)$$

for $f \in E, s \in \mathbb{R}$ and $t \geq 0$.

It is now an easy excercise to show that $(T(t))_{t \geq 0}$ is a one-parameter semigroup and strongly continuous on E (see [Pa], Thm. 1.8).

Consequently this semigroup is uniquely determined by its generator which is a densely defined, closed operator on E and which will be denoted G.

Since the evolution operators are obtained from the action of the semigroup on the constant functions, i.e.,

$$U(t,s)x = (T(t-s)\mathbb{1} \otimes x)(t) \text{ for } x \in X \text{ and } t \geq s,$$

the generator G determines $(U(t,s))_{t \geq s}$. Therefore the following terminology seems to be justified.

DEFINITION 2.2 The generator G of the evolution semigroup $(T(t))_{t \geq 0}$ will be called the *generator of the evolution family* $(U(t,s))_{t \geq s}$.

For a qualitative study of the evolution family $(U(t,s))_{t \geq s}$ this generator G seems to be much more appropriate than the family of operators $A(t)$ from (ACP) (in case they exist). This has been made evident by R. Rau, Y. Latushkin and A. Stepin in case of the spectral characterization of hyperbolic evolution families (see [Ra], [La-St] and also [Na-Rh]). In a similar spirit we will solve the problem of characterizing Lipschitz continuous evolution families by determining precisely the form of its generator G on the function space $C_{v_0}(\mathbb{R}, X)$.

We start this investigation with a particular case.

3.   DIFFERENTIABLE EVOLUTION FAMILIES

Let $(U(t,s))_{t \geq s}$ be a strongly continuous evolution family on the Banach space X and consider its associated evolution semigroup $(T(t))_{t \geq 0}$ on $E := C_{v_0}(\mathbb{R}, X)$ with generator G. By $\mathcal{L}_s(X)$ we denote the space of all bounded, linear operators on X endowed with the strong operator topology. Then the space $C_b(\mathbb{R}, \mathcal{L}_s(X))$ of all bounded, continuous functions from $\mathbb{R}$ into $\mathcal{L}_s(X)$ can be identified canonically with a subspace of $\mathcal{L}(E)$. More precisely the following holds.

LEMMA 3.1 *Every operator valued function* $A(\cdot) \in C_b(\mathbb{R}, \mathcal{L}_s(X))$ *defines a bounded, linear operator on E by*

$$(A(\cdot)f)(s) := A(s)f(s) \quad for \ f \in E, s \in \mathbb{R}.$$

Moreover, one has

$$\|A(\cdot)\|_v = \sup_{s \in \mathbb{R}} \|A(s)\|.$$

only continuous with respect to the norms but also continuous with respect to the Mackey topology $\tau(L^\infty, L^1)$ on $L^\infty(\mu)$ and the norm topology on X. This observation leads us to investigate those linear operators which are continuous from $(Y^*, \tau(Y^*, Y))$ into $(X, \|\cdot\|)$ for some weakly sequentially complete Banach space Y.

Hence we obtain the vector valued Vitali-Hahn-Saks theorem for countably additive vector measures not only in a substantially more general form, but also by a considerably simpler proof than those that can be found in the standard literature such as [1]. It thus appears to the authors that the present paper is quite appropriately dedicated to Professor A. C. Zaanen, who not only devoted a great part of his research to questions of measure theory but also has always appreciated transparent and simple proofs.

In this paper, we use [1], [6], [7] and [5] as our general references. We use X, Y to denote locally convex topological vector spaces and use X^*, Y^* to denote their topological duals, i.e., the set of all continuous linear functionals. $L_s(Y, X)$ is the locally convex space of continuous linear mappings of Y into X, equipped with the topology of simple convergence (=strong operator topology if X is a normed space). $T_\alpha \to T$ under this topology if and only if $T_\alpha(y) \to T(y)$ under the topology of X for each $y \in Y$.

1. SOME PRELIMINARIES

To prove our main results, we first do some preparation. The results in this section are also independently interesting.

LEMMA 1 *Let S be any subset of $L_s(Y, X)$ and let A be an equicontinuous subset of X^*. Consider S under the topology of simple convergence, A under the topology $\sigma(X^*, X)$, and Y^* under the topology $\sigma(Y^*, Y)$. Then the mapping*

$$(T, x^*) \to T^* x^*$$

is continuous from $S \times A$ into Y^*.

Proof. It suffices to show that for each $y \in Y$, the function

$$(T, x^*) \to\ < y, T^* x^* >$$

is continuous. Let $T_0 \in S, x_0^* \in A$ and $\varepsilon > 0$ be given. There exists a 0-neighborhood $U \subset L_s(Y, X)$ such that for all $T - T_0 \in U$, we have $|\ < (T - T_0)y, x^* > | \leq \varepsilon/2$ for all $x^* \in A$ since A is an equicontinuous subset of X^*. Moreover, there exists a 0-neighborhood

for $t \geq s$, hence $S(t)$ is of the form stated above. If we use the semigroup property of $(S(t))_{t \geq 0}$ and evaluate at the constant functions $1 \hspace{-0.5em}\text{I} \otimes x$, $x \in X$, we obtain

$$V(t,s)V(s,r) = V(t,r)$$

and it follows from (1) that

$$\|V(t,s)\| \leq M e^{(MC+\omega)(t-s)}$$

for $r \leq s \leq t$. Hence $(V(t,s))_{t \geq s}$ is an evolution family. $\quad\square$

In the following theorem we now show that under an appropriate assumption on its domain the generator G (of our evolution semigroup) is obtained from the generator of the translation semigroup $(T_0(t))_{t \geq 0}$ by a bounded perturbation of the above type.

THEOREM 3.3 *Under the assumptions made in Definition 2.1 on the evolution family $(U(t,s))_{t \geq s}$ and its corresponding evolution semigroup $(T(t))_{t \geq 0}$ the following assertions are equivalent.*

(a) *For every $x \in X$ the constant function $1 \hspace{-0.5em}\text{I} \otimes x$ belongs to the domain of the generator G of $(T(t))_{t \geq 0}$ and $G(1 \hspace{-0.5em}\text{I} \otimes x)$ is a bounded function.*

(b) *There exists an operator valued function $A(\cdot) \in C_b(\mathbb{R}, \mathcal{L}_s(X))$ such that*

$$Gf = -f' + A(\cdot)f$$

for every $f \in D(G_0) = D(G)$.

PROOF. Since the implication $(b) \Rightarrow (a)$ is trivial we only show $(a) \Rightarrow (b)$. Define

$$A(s)x := G(1 \hspace{-0.5em}\text{I} \otimes x)(s)$$

for $s \in \mathbb{R}$ and $x \in X$. Since G is closed we obtain that $A(s)$ is a closed and everywhere defined, hence bounded, linear operator. Moreover $A(\cdot) \in C_b(\mathbb{R}, \mathcal{L}_s(X))$. Consider the perturbed operator

$$K := G - A(\cdot).$$

By Proposition 3.2 this operator generates an evolution semigroup $(S(t))_{t \geq 0}$ corresponding to an evolution family $(V(t,s))_{t \geq s}$ on X. Observe now that all constant functions $1 \hspace{-0.5em}\text{I} \otimes x$, $x \in X$, belong to $D(K) = D(G)$ and satisfy

$$K(1 \hspace{-0.5em}\text{I} \otimes x) = 0.$$

Consequently $S(t)(1 \hspace{-0.5em}\text{I} \otimes x) = 1 \hspace{-0.5em}\text{I} \otimes x$ for all $t \geq 0$, or equivalently

$$V(s, s-t)x = (1 \hspace{-0.5em}\text{I} \otimes x)(s) = x$$

for all $s \in \mathbb{R}, t \geq 0$. This show that $(S(t))_{t \geq 0}$ is just the translation semigroup and its generator K coincides with G_0. By adding $A(\cdot)$ we obtain $G = G_0 + A(\cdot)$. $\quad\square$

The above theorem deals with (evolution) semigroups on the function space E. We now restate it in terms of evolution families on the space X, thereby obtaining a first answer to the question raised by Fattorini in [Fa], p. 386.
As a preparation we show that the evolution families we are considering actually are invertible.

COROLLARY 3.4 *If the evolution family* $(U(t,s))_{t\geq s}$ *and its corresponding evolution semigroup* $(T(t))_{t\geq 0}$ *satisfy the conditions stated in Theorem 3.3, then each* $U(t,s)$ *is invertible and by defining*

$$U(s,t) := U(t,s)^{-1} \quad \text{for } t \geq s$$

we obtain an extension to a strongly continuous evolution family on $\mathbb{R}^2$.

PROOF. By the previous theorem the generator of $(T(t))_{t\geq 0}$ is $G = G_0 + A(\cdot)$. Now observe that G_0 generates a strongly continuous group and $A(\cdot)$ is a bounded perturbation. Therefore G is the generator of a strongly continuous group and each $T(t)$ is invertible. This implies the invertibility of $U(t,s)$. $\square$

THEOREM 3.5 *Let* $(U(t,s))_{t\geq s}$ *be a strongly continuous evolution family on* X. *Then the following assertions are equivalent.*

(a) *For every* $x \in X$ *the function* $\mathbb{R}_+ \ni t \mapsto e^{-|s|}U(s, s-t)x \in X$ *is right-differentiable in* $t = 0$, *uniformly in* $s \in \mathbb{R}$ *and the function* $\mathbb{R} \ni s \mapsto \frac{\partial^+}{\partial t}U(s, s-t)|_{t=0}x \in X$ *belongs to* $C_b(\mathbb{R}, X)$.

(b) *For every* $x \in X$ *the function*

$$\{(t,s) : t \geq s\} \ni (t,s) \mapsto U(t,s)x \in X$$

is continuously differentiable and there exists an operator function $A(\cdot) \in C_b(\mathbb{R}, \mathcal{L}_s(X))$ *such that*

$$\frac{\partial}{\partial t}U(t,s)x = A(t)U(t,s)x, \tag{2}$$

$$\frac{\partial}{\partial s}U(t,s)x = -U(t,s)A(s)x \tag{3}$$

for $t \geq s$ *and all* $x \in X$.

PROOF. For the evolution semigroup $(T(t))_{t\geq 0}$ corresponding to $(U(t,s))_{t\geq s}$ we have

$$T(t)(\mathbb{1} \otimes x)(s) = U(s, s-t)x$$

for $t \geq 0, s \in \mathbb{R}$ and $x \in X$. So (a) is just a reformulation of condition (a) in Theorem 3.3. By that theorem we find a function $A(\cdot) \in C_b(\mathbb{R}, \mathcal{L}_s(X))$ such that the generator of $(T(t))_{t\geq 0}$ is $G = G_0 + A(\cdot)$. Since $\mathbb{1} \otimes x \in D(G)$ we conclude that

$$t \mapsto T(t)(\mathbb{1} \otimes x) = U(\cdot, \cdot - t)x$$

is differentiable with derivative

$$\frac{\partial}{\partial t}U(s, s-t)x = T(t)G(\mathbb{1} \otimes x)(s) = U(s, s-t)A(s-t)x$$

or after an obvious substitution

$$\frac{\partial}{\partial s}U(t,s)x = -U(t,s)A(s)x,$$

which is Equation (3).

For the other equation recall that our evolution family is invertible by Corollary 3.4. Clearly $(U(t,0))_{t \in \mathbb{R}}$ is exponentially bounded and after a rescaling of the form $e^{-\omega(t-s)}U(t,s)$, for $t \geq s$, we may assume that the function $t \mapsto \|U(t,0)\|$ is bounded. Then $f \in E$ if we define $f(t) := U(t,0)x$. But for this function we have

$$T(t)f(s) = U(s, s-t)U(s-t,0)x = U(s,0)x = f(s)$$

for all $s \in \mathbb{R}$, hence $T(t)f = f$ for all $t \geq 0$. Therefore $Gf = 0$ or

$$-\tfrac{\partial}{\partial t}U(t,0)x + A(t)U(t,0)x = 0,$$

which is Equation (2) if we substitute x by $U(0,s)x$. This proves $(a) \Rightarrow (b)$. The other implication is trivial. $\square$

REMARK. It follows from condition (a) in Theorem 3.3 that

$$U(s, s-t)x = x + \int_0^t U(s, s-\tau)A(s-\tau)x\,d\tau \quad \text{for } x \in X \text{ and } s,t \in \mathbb{R}.$$

This implies that the family $(U(t,s))_{(t,s) \in \mathbb{R}^2}$ is norm continuous and

$$\|U(t,s) - Id\| \leq M|t-s|e^{\omega|t-s|}$$

for some $M \geq 1$, $\omega \geq 0$ and all $t, s \in \mathbb{R}$.

Remark that the contents of Theorem 3.3 and Theorem 3.5 are essentially the same, but the semigroup formulation in Theorem 3.3 is much more elegant. The fact that right-differentiability in 0 implies differentiability in both variables of $(U(t,s))$ is a consequence of the semigroup property of $(T(t))_{t \geq 0}$. Moreover, (2) shows that $(U(t,s))$, and hence the semigroup $(T(t))_{t \geq 0}$ yield the solution of the nonautonomous Cauchy problem (ACP). Therefore the existence of solutions of (ACP) for $A(\cdot) \in C_b(\mathbb{R}, \mathcal{L}_s(X))$ follows from the bounded perturbation theorem in semigroup theory. In fact, by more sophisticated tools from semigroup theory we even obtain an approximation formula for the evolution family.

COROLLARY 3.6 *Let $A(\cdot) \in C_b(\mathbb{R}, \mathcal{L}_s(X))$. Then the operator*

$$G = G_0 + A(\cdot) \quad \text{with } D(G) = D(G_0)$$

generates a strongly continuous evolution semigroup

$$T(t)f(s) = U(s, s-t)f(s-t), \quad s \in \mathbb{R}, t \geq 0, \tag{4}$$

on E and the evolution family $(U(t,s))_{t \geq s}$ is given by

$$U(s, s-t)x = \lim_{n \to \infty} \prod_{k=0}^{n-1} exp\left(\int_{\frac{kt}{n}}^{\frac{(k+1)t}{n}} A(s-\tau)d\tau\right)x \tag{5}$$

for all $x \in X$, where the product is defined as :

$$\prod_{k=0}^{n-1} exp\left(\int_{\frac{kt}{n}}^{\frac{(k+1)t}{n}} A(s-\tau)d\tau\right) := exp\left(\int_0^{\frac{t}{n}} A(s-\tau)d\tau\right)\cdots exp\left(\int_{\frac{(n-1)t}{n}}^t A(s-\tau)d\tau\right).$$

If the operators $A(s)$ mutually commute, then

$$U(t,s) = exp\left(\int_s^t A(\tau)d\tau\right) \quad for \ t \geq s. \tag{6}$$

PROOF. Since G is a bounded perturbation of the generator G_0 of the translation semigroup we obtain a semigroup $(T(t))_{t\geq0}$ which has to be of the form (4) by Proposition 3.2. In order to obtain the approximation formula (5) define operators $B_t(s) \in \mathcal{L}(X)$ by

$$B_t(s)x := \int_{s-t}^s A(\tau)x d\tau$$

for $s \in \mathbb{R}, t \geq 0$ and $x \in X$.

Since $A(\cdot)x \in C_b(\mathbb{R}, X)$ for every $x \in X$ we obtain that the function $B_t(\cdot)$ belongs to $C_b(\mathbb{R}, \mathcal{L}_s(X))$. Consequently the operators

$$(V(t)f)(s) := exp(B_t(s))f(s-t) \quad (f \in E, s \in \mathbb{R}, t \geq 0)$$

are linear bounded on E. It is easy to see that for $f \in D(G)$ one has

$$\frac{d}{dt}V(t)|_{t=0}f = -f' + A(\cdot)f = Gf.$$

Therefore $(V(t))_{t\geq0}$ satisfies the assumptions of the Chernoff Product Formula (see [Go], Thm. I.8.4), hence

$$T(t)f = \lim_{n\to\infty}[V(\frac{t}{n})]^n f$$

for $f \in E$ and $t \geq 0$. Taking $f = \mathbb{1} \otimes x$ we obtain

$$\begin{aligned} U(s, s-t)x &= T(t)(\mathbb{1} \otimes x)(s) \\ &= \lim_{n\to\infty}\prod_{k=0}^{n-1} exp\left(\int_{\frac{kt}{n}}^{\frac{(k+1)t}{n}} A(s-\tau)d\tau\right)x \end{aligned}$$

for all $s \in \mathbb{R}, t \geq 0$ and $x \in X$. In the case of commuting operators $A(s)$ the identity (6) is then a trivial consequence. $\square$

4. LIPSCHITZ CONTINUOUS EVOLUTION FAMILIES

Having characterized the differentiable evolution families we now turn to the problem posed by Fattorini [Fa], p. 386 : Given a Lipschitz continuous evolution family $(U(t,s))_{t \geq s}$ on X, i.e. satisfying

$$\|U(t,s) - Id\| \leq M(t-s)e^{\omega(t-s)}$$

for some $M \geq 1$, $\omega \geq 0$ and all $t \geq s$. Does there exist a family of operators $A(t) \in \mathcal{L}(X)$ such that $(U(t,s))_{t \geq s}$ solves the (ACP) corresponding to $A(\cdot)$?
We will use the same approach as in Section 3, i.e., we consider the evolution semigroup corresponding to $(U(t,s))_{t \geq s}$ and then study its generator. The following simple observation will be most helpful.

LEMMA 4.1 Let $(U(t,s))_{t \geq s}$ be a Lipschitz continuous evolution family on X. Then

$$(\tilde{T}(t)F)(s) := U(s, s-t)F(s-t), \quad s \in \mathbb{R}, t \geq 0,$$

defines a strongly continuous semigroup on $\tilde{E} := C_{UB}(\mathbb{R}, \mathcal{L}(X))$, the Banach space of all uniformly continuous, bounded, $\mathcal{L}(X)-$valued functions on $\mathbb{R}$.

This semigroup $(\tilde{T}(t))_{t \geq 0}$ will be used in order to convert the evolution semi-group corresponding to a Lipschitz continuous evolution family into an evolution semigroup corresponding to a differentiable evolution family.

THEOREM 4.2 Let $(U(t,s))_{t \geq s}$ be a strongly continuous evolution family on X. Consider its evolution semigroup $(T(t))_{t \geq 0}$ on $C_{v_0}(\mathbb{R}, X)$ and denote by G its generator. Then the following assertions are equivalent.

(a) $(U(t,s))_{t \geq s}$ is Lipschitz continuous.

(b) There exists an invertible operator valued function $\tilde{Q}(\cdot) \in C_{UB}(\mathbb{R}, \mathcal{L}(X)) \cap Lip(\mathbb{R}, \mathcal{L}(X))$ such that $\tilde{Q}(\cdot)^{-1} \in C_{UB}(\mathbb{R}, \mathcal{L}(X))$ and a function $B(\cdot) \in C_b(\mathbb{R}, \mathcal{L}_s(X))$ such that

$$G = \tilde{Q}G_0\tilde{Q}^{-1} + B(\cdot), \quad i.e.,$$

$D(G) = \{f \in C_{v_0}(\mathbb{R}, X) : \tilde{Q}^{-1}f \in D(G_0)\}$ and $Gf(s) = -\tilde{Q}(s)(\tilde{Q}^{-1}f)'(s) + B(s)f(s)$ for $f \in D(G)$ and $s \in \mathbb{R}$.

Before proving the theorem we recall that $Lip(\mathbb{R}, X)$ denotes the space of all Lipschitz continuous, $X-$valued functions on $\mathbb{R}$. Moreover, the Banach space X is said to have the *Radon-Nikodym* property if every $f \in Lip(\mathbb{R}, X)$ is differentiable a.e.. In that case $f' \in L^\infty(\mathbb{R}, X)$ and

$$f(t) = f(0) + \int_0^t f'(s)ds. \tag{7}$$

We refer to Arendt [Ar] for more information in this direction and only mention that separable dual Banach spaces and all reflexive Banach spaces have the Radon-Nikodym property. Another notion to be used in the sequel is the *Favard class* $Fav(S(t))$ of a strongly continuous semigroup $(S(t))_{t \geq 0}$. The relevant definitions can be found in [Bu-Be]. For us it will be important to observe that the Favard class of the translation semigroup on $C_{v_0}(\mathbb{R}, X)$ coincides with

$$Lip_{v_0}(\mathbb{R}, X) := \{f \in C_{v_0}(\mathbb{R}, X) : v \cdot f \in Lip(\mathbb{R}, X)\}.$$

Therefore every evolution semigroup $(T(t))_{t \geq 0}$ on $C_{v_0}(\mathbb{R}, X)$ induced by a Lipschitz continuous evolution family $(U(t, s))_{t \geq s}$ has this same space as its Favard class.

PROOF. $(a) \Rightarrow (b)$: Consider the evolution semigroup $(\tilde{T}(t))_{t \geq 0}$ on $\tilde{E}$ defined in Lemma 4.1 and denote by $\tilde{G}$ its generator. By standard semigroup theory one has $\lim_{\lambda \to \infty} \lambda R(\lambda, \tilde{G})\tilde{F} = \tilde{F}$ for every $\tilde{F} \in \tilde{E}$. Take now $\tilde{F}(s) := Id_X$ for all $s \in \mathbb{R}$ and choose $\mu \in \mathbb{R}_+$ such that

$$R(\mu, \tilde{G})\tilde{F} = \int_0^\infty e^{-\mu t}\tilde{T}(t)\tilde{F}dt \text{ and } \|\mu R(\mu, \tilde{G})\tilde{F} - \tilde{F}\| < \frac{1}{2}.$$

Then $\tilde{Q} := \mu R(\mu, \tilde{G})\tilde{F}$ is an invertible function in $\tilde{E}$ and since $\tilde{Q} \in D(\tilde{G}) \subset Fav(\tilde{T}(t))$ we obtain that $\tilde{Q}$ is Lipschitz continuous.
On the other hand we have

$$
\begin{aligned}
\tilde{Q}(\mathbb{1} \otimes x)(s) &= \tilde{Q}(s)x \\
&= \mu \int_0^\infty e^{-\mu t}U(s, s - t)x dt \\
&= \mu R(\mu, G)(\mathbb{1} \otimes x)(s),
\end{aligned}
$$

hence $\tilde{Q}(\mathbb{1} \otimes x) \in D(G)$ and $G(\tilde{Q}(\mathbb{1} \otimes x))$ is a bounded function.
Next we define a new evolution family

$$V(t, s) := \tilde{Q}(t)^{-1}U(t, s)\tilde{Q}(s) \quad \text{for } t \geq s,$$

which is again Lipschitz continuous. The evolution semigroup $(S(t))_{t \geq 0}$ on $C_{v_0}(\mathbb{R}, X]$ associated to $(V(t, s))_{t \geq s}$ satisfies

$$S(t) = \tilde{Q}^{-1}T(t)\tilde{Q} \quad \text{for } t \geq 0,$$

hence its generator is

$$K = \tilde{Q}^{-1}G\tilde{Q} \text{ with domain } D(K) = \{f \in C_{v_0}(\mathbb{R}, X) : \tilde{Q}f \in D(G)\}.$$

Therefore $\mathbb{1} \otimes x \in D(K)$ and $K(\mathbb{1} \otimes x)$ is a bounded function. So by Theorem 3.3 it follows that $(V(t, s))_{t \geq s}$ is a differentiable evolution family and its generator is of the form

$$K = G_0 + A(\cdot)$$

for some $A(\cdot) \in C_b(\mathbb{R}, \mathcal{L}_s(X))$. Applying the inverse similarity transformation yields

$$G = \tilde{Q}K\tilde{Q}^{-1},$$

which is (b).

(b) $\Rightarrow$ (a): By the results in Section 3 it is clear that $K = G_0 + \tilde{Q}^{-1}B\tilde{Q}$ is the generator of a differentiable, hence Lipschitz continuous evolution family $(V(t,s))_{t \geq s}$. Since $U(t,s) = \tilde{Q}(t)V(t,s)\tilde{Q}(s)^{-1}$ remains Lipschitz continuous we obtain (a).   □

In presence of the Radon-Nikodym property it will now be possible to differentiate a.e. the Lipschitz continuous evolution family $(U(t,s))_{t \geq s}$. But we have to be aware that $\mathcal{L}(X)$ will in general not inherit the Radon-Nikodym property from X. To overcome this obstacle we denote by $L^\infty(\mathbb{R}, \mathcal{L}_s(X))$ the space of all functions $F : \mathbb{R} \to \mathcal{L}(X)$ such that $F(\cdot)x \in L^\infty(\mathbb{R}, X)$ (see [Za] for basic information on vector valued integration) for every $x \in X$ and then show the following.

LEMMA 4.3 *Assume that X is separable and has the Radon-Nikodym property. If $\mathcal{G} \in Lip(\mathbb{R}, \mathcal{L}(X))$ then $\mathcal{G}$ is strongly differentiable a.e. and $\mathcal{G}' \in L^\infty(\mathbb{R}, \mathcal{L}_s(X))$.*
In particular, if f is differentiable a.e. then

$$(\mathcal{G}f)' = \mathcal{G}'f + \mathcal{G}f' \quad a.e.. \tag{8}$$

PROOF. For $\mathcal{G} \in Lip(\mathbb{R}, \mathcal{L}(X)$ it follows that $\mathcal{G}(\cdot)x \in Lip(\mathbb{R}, X)$ for each $x \in X$. Therefore $\mathcal{G}(\cdot)x$ is differentiable a.e. and $\mathcal{G}'(\cdot)x \in L^\infty(\mathbb{R}, X)$. The space X being separable there exists a countable dense subset $Z := \{x_n \in X : n \in \mathbb{N}\}$ in X. Hence there exists $N \subset \mathbb{R}$ with measure zero such that the functions $\mathcal{G}(\cdot)x_n$ are differentiable on $\mathbb{R} \backslash N$ and $\mathcal{G}'(\cdot)x_n \in L^\infty(\mathbb{R}, X)$ for each $n \in \mathbb{N}$.

It follows from the density of Z in X and (7) that $\mathcal{G}$ is strongly differentiable a.e. and $\mathcal{G}' \in L^\infty(\mathbb{R}, \mathcal{L}_s(X))$.

The proof of the second statement is trivial.   □

COROLLARY 4.4 *Let $(U(t,s))_{t \geq s}$ be a Lipschitz continuous evolution family on a separable Banach space X having the Radon-Nikodym property. Then there exists an operator valued function $A(\cdot) \in L^\infty(\mathbb{R}, \mathcal{L}_s(X))$ such that the generator G of the corresponding evolution semigroup on $E := C_{v_0}(\mathbb{R}, X)$ is given by*

$$D(G) = \{f \in E : f \ a.e. \ differentiable \ \ and \ - f' + A(\cdot)f \in E\},$$

$$Gf = -f' + A(\cdot)f$$

for every $f \in D(G)$.

PROOF. By Theorem 4.2 we know that the generator G is of the form

$$G = \tilde{Q}G_0\tilde{Q}^{-1} + B(\cdot)$$

for some $\tilde{Q} \in Lip(\mathbb{R}, \mathcal{L}(X)) \cap C_{UB}(\mathbb{R}, \mathcal{L}(X))$ and $B \in C_b(\mathbb{R}, \mathcal{L}_s(X))$.

For $f \in D(G)$ we are now allowed to differentiate a.e. applying the product rule (8), hence

$$\begin{aligned}
(Gf)(s) &= -\tilde{Q}(s)(\tilde{Q}^{-1}f)'(s) + B(s)f(s) \\
&= -f'(s) + (-\tilde{Q}(s)(\tilde{Q}^{-1})'(s) + B(s))f(s) \\
&= -f'(s) + A(s)f(s)
\end{aligned}$$

for $A(s) := -\tilde{Q}(s)(\tilde{Q}^{-1})'(s) + B(s)$ and a.e. $s \in \mathbb{R}$.

Conversely, let $f \in E$ such that f a.e. differentiable and $-f' + A(\cdot)f \in E$.

Set $g := -f' + A(\cdot)f$. It follows from Lemma 4.3 that $\tilde{Q}^{-1}f$ is differentiable a.e. and by (8) we have

$$(\tilde{Q}^{-1}f)'(s) = \tilde{Q}(s)^{-1}(B(s)f(s) - g(s)) \quad \text{for } a.e. \ s \in \mathbb{R}.$$

So by (7) we obtain

$$(\tilde{Q}^{-1}f)(t) = (\tilde{Q}^{-1}f)(0) + \int_0^t \tilde{Q}(s)^{-1}(B(s)f(s) - g(s))ds \quad (t \geq 0).$$

This implies that $\tilde{Q}^{-1}f \in D(G_0)$. Consequently, $f \in D(G)$ and its easy to see that $Gf = g$. $\square$

REMARK. It follows from the proof of Theorem 4.2 that

$$U(t,s) = \tilde{Q}(t)V(t,s)\tilde{Q}(s)^{-1} \quad \text{for } t \geq s,$$

where $(V(t,s))_{t \geq s}$ is the evolution family corresponding to $\tilde{Q}^{-1}B(\cdot)\tilde{Q}$. Since the function

$$\{(t,s) : t \geq s\} \ni (s,t) \mapsto V(t,s) \in \mathcal{L}(X)$$

is strongly differentiable, we conclude by Lemma 4.3 that the functions

$$t \mapsto U(t,s) \text{ and } s \mapsto U(t,s)$$

are differentiable a.e.. It follows immediately that for fixed s and a.e. $t \geq s$

$$\tfrac{\partial}{\partial t}U(t,s)x = A(t)U(t,s)x,$$

and for fixed $t \geq 0$ and a.e. $s \leq t$

$$\tfrac{\partial}{\partial s}U(t,s)x = -U(t,s)A(s)x$$

for all $x \in X$.

We conclude with an example showing that for the generators G_0 and $G = G_0 + A(\cdot)$ we can have $D(G_0) \cap D(G) = \{0\}$ while obviously the two Favard classes coincide.

EXAMPLE 4.5 It suffices to take $X = \mathbb{C}$ and the evolution family

$$U(t,s) := e^{(i\psi(t)-i\psi(s))}$$

for some $\psi \in E \cap Lip(\mathbb{R}, \mathbb{C})$ which is not differentiable on a dense subset of $\mathbb{R}$ (cf. [De-Sch]). For a complete discussion of evolution semigroups in this one-dimensional case we refer to [Na], B-II. 3.

References

[Ac-Te] Acquistapace, P. and Terreni, B.: A unified approch to abstract linear nonautonomous parabolic equations. Rend. Sem. Mat. Univ. Padova **78** (1987), 47-107.

[Ar] Arendt, W.: Vector-valued Laplace transform. Israël J. Math. **59** (1987), 327-352.

[Bu-Be] Butzer, P. and Behrens, H.: Semigroups of Operators and Approximation. Springer-Verlag 1967.

[Da-Ko] Daners, D. and Koch Medina, P.: Abstract Evolution Equations, Periodic Problems and Applications. Pitman Research Notes Math. Ser. **279**. Longman 1993.

[De-Sch] Desch, W. and Schappacher, W.: A note on the comparison of C_0-semigroups. Semigroup Forum **35** (1987), 237-243.

[Ev] Evans, D.E.: Time dependent perturbations and scattering of strongly continuous groups on Banach spaces. Math. Ann. **221** (1976), 275-290.

[Fa] Fattorini, H.O.: The Cauchy Problem. Addison-Wesley 1983.

[Go] Goldstein, J.A.: Semigroup of Linear Operators and Applications. Oxford 1985.

[Go1] Goldstein, J.A.: Abstract evolution equations. Trans. Amer. Math. Soc. **141** (1969), 159-186.

[Go2] Goldstein, J.A.: Time dependent hyperbolic equations. J. Functional Anal. **4** (1969), 31-49, and **6** (1970), 347.

[Ho] Howland, J.S.: Stationary scattering theory for time-dependent Hamiltonians. Math. Ann. **207** (1974), 315-335.

[Ka1] Kato, T.: Abstract evolution equations of parabolic type in Banach and Hilbert spaces. Nagoya Math. J. **19** (1961), 93-125.

[Ka2] Kato, T.: Linear evolution equations of hyperbolic type. J. Fac. Sci. Tokyo **17** (1970), 241-258.

[La-St] Latushkin, Y.D. and Stepin, A.M.: Weighted translation operators and linear extensions of dynamical systems. Russian Math. Surveys **46** (1991), 95-165.

[Lu] Lumer, G.: Equations de diffusion dans le domaine (x,t) non-cylindriques et semi-groupes "espace-temps". Séminaire de Théorie du Potentiel. Lecture Notes Math. **1393**, Springer-Verlag 1989, 161-179.

[Na] Nagel, R. (ed.): One-parameter Semigroups of Positive Operators. Lecture Notes Math. **1184**, Springer-Verlag 1986.

[Na-Rh] Nagel, R. and Rhandi, A.: Positivity and Lyapunov stability conditions for linear systems. Adv. Math. Sci. Appl. (to appear).

[Ne] Neidhardt, H.: On abstract linear evolution equations, I. Math. Nachr. **103** (1981), 283-298.

[Pa] Paquet, L.: Semi-groupes généralisés et équations d'évolution. Seminaire de Théorie du Potentiel. Lecture Notes Math. **713**, Springer-Verlag 1979, 243-263.

[Ra] Rau, R.T.: Hyperbolic Evolution Semigroups. Dissertation, Tübingen, 1992.

[Za] Zaanen, A.C.: Integration. North-Holland Publ. Comp. Amsterdam 1967.

Rainer Nagel
Abdelaziz Rhandi
Mathematisches Institut der Universität
Auf der Morgenstelle 10
D-72076 Tübingen
Federal Republic of Germany

Mathematic subject classification : Primary 47D05; Secondary 34G10.

Operator Theory:
Advances and Applications, Vol. 75
© 1995 Birkhäuser Verlag Basel/Switzerland

ON THE VITALI-HAHN-SAKS THEOREM

Helmut H. Schaefer and Xiao-Dong Zhang

Dedicated to Professor A. C. Zaanen on the occasion of his eightieth birthday

In this paper we generalize the classical Vitali-Hahn-Saks theorem to sets of countably additive vector measures which are compact in the strong operator topology. The main result asserts that a set of countably additive vector measures which is compact in the strong operator topology is uniformly countably additive. We accomplish this by first studying the properties of linear operators from Y^*, the dual of a Banach space Y, into a Banach space X which are continuous with respect to the Mackey topology $\tau(Y^*, Y)$ on Y^* and the norm topology on X, and then applying the results to the special case where $Y = L^1(\mu)$ and $Y^* = L^\infty(\mu)$. Other related results on vector measures are also included.

0. INTRODUCTION

The celebrated Vitali-Hahn-Saks theorem asserts that if a sequence of countably additive finite measures on a σ-algebra is pointwise convergent, then the sequence of measures is uniformly countably additive and the limit measure is also countably additive. This result has been generalized to the case of vector measures. In this paper, we prove that the Vitali-Hahn-Saks theorem for countably additive vector measures is in fact valid in a much more general context. In particular, we prove that a set of countably additive vector measures which is compact in the strong operator topology is uniformly countably additive. The essential ingredient in the proofs of these results is the following facts about the space $L^1(\mu)$: (1) the Banach space $L^1(\mu)$ is weakly sequentially complete, (2) the solid convex hull of any relatively weakly compact subset is also relatively weakly compact, and thus (3) any countably additive vector measure which is absolutely continuous with respect to μ and takes values in a Banach space X induces a linear operator from $L^\infty(\mu)$ into X which is not only continuous with respect to the norms but also continuous with respect to the Mackey topology $\tau(L^\infty, L^1)$ on $L^\infty(\mu)$ and the norm topology on X. This observation leads us to investigate those linear operators which are continuous from $(Y^*, \tau(Y^*, Y))$ into $(X, \|\cdot\|)$ for some weakly sequentially complete Banach space Y.

Hence we obtain the vector valued Vitali-Hahn-Saks theorem for countably additive vector measures not only in a substantially more general form, but also by a considerably simpler proof than those that can be found in the standard literature such as [1]. It thus appears to the authors that the present paper is quite appropriately dedicated to Professor A. C. Zaanen, who not only devoted a great part of his research to questions of measure theory but also has always appreciated transparent and simple proofs.

In this paper, we use [1], [6], [7] and [5] as our general references. We use X, Y to denote locally convex topological vector spaces and use X^*, Y^* to denote their topological duals, i.e., the set of all continuous linear functionals. $L_s(Y, X)$ is the locally convex space of continuous linear mappings of Y into X, equipped with the topology of simple convergence (=strong operator topology if X is a normed space). $T_\alpha \to T$ under this topology if and only if $T_\alpha(y) \to T(y)$ under the topology of X for each $y \in Y$.

1. SOME PRELIMINARIES

To prove our main results, we first do some preparation. The results in this section are also independently interesting.

LEMMA 1 *Let S be any subset of $L_s(Y, X)$ and let A be an equicontinuous subset of X^*. Consider S under the topology of simple convergence, A under the topology $\sigma(X^*, X)$, and Y^* under the topology $\sigma(Y^*, Y)$. Then the mapping*

$$(T, x^*) \to T^* x^*$$

is continuous from $S \times A$ into Y^*.

Proof. It suffices to show that for each $y \in Y$, the function

$$(T, x^*) \to < y, T^* x^* >$$

is continuous. Let $T_0 \in S, x_0^* \in A$ and $\varepsilon > 0$ be given. There exists a 0-neighborhood $U \subset L_s(Y, X)$ such that for all $T - T_0 \in U$, we have $| < (T - T_0)y, x^* > | \leq \varepsilon/2$ for all $x^* \in A$ since A is an equicontinuous subset of X^*. Moreover, there exists a 0-neighborhood

V in X^* such that for $x^* - x_0^* \in V$, we have

$$| < y, T_0^* x^* - T_0^* x_0^* > | \leq \varepsilon/2.$$

Therefore, $x^* \in x_0^* + V$ and $T \in T_0 + U$ implies that

$$| < y, T^* x^* - T_0^* x_0^* > | \leq | < y, (T^* - T_0^*) x^* > | + | < y, T_0^* (x^* - x_0^*) > | \leq \varepsilon$$

This completes the proof.

THEOREM 2 *Let S be a compact subset of $L_s(Y, X)$, and let A be a $\sigma(X^*, X)$-closed equicontinuous subset of X^*. Then*

$$\bigcup_{T \in S} T^*(A)$$

is $\sigma(Y^*, Y)$-compact in Y^*.

Proof. The result follows from the above lemma, since the union is a continuous image of the compact space $S \times A$.

Now let Y, X be Banach spaces. We use $\tau = \tau(Y^*, Y)$ to denote the Mackey topology on Y^*, i.e. the topology of uniform convergence on $\sigma(Y, Y^*)$-compact subsets of Y. It is well-known [6] that the Mackey topology is a consistent topology, i.e., the dual of (Y^*, τ) is Y. The following proposition is well-known and can be found in [6]. For the sake of completeness, we give a proof.

PROPOSITION 3 *Suppose that Y, X are Banach spaces. Let $T : Y^* \to X$ be a linear mapping. Then the following are equivalent:*

(1) *T is continuous with respect to the topology $\tau(Y^*, Y)$ and the norm topology on X.*

(2) *T is continuous with respect to the topology $\tau(Y^*, Y)$ and the topology $\sigma(X, X^*)$.*

(3) *$T^*(X^*) \subseteq Y$ and $T^* : X^* \to Y$ is continuous with respect to $\sigma(X^*, X)$ and $\sigma(Y, Y^*)$.*

Proof. (1) $\Rightarrow$ (2) is straightforward. (2) $\Rightarrow$ (3) is also easy to verify (see also [6]). We now show that (3) $\Rightarrow$ (1). Since T^* is continuous with respect to $\sigma(X^*, X)$ and $\sigma(Y, Y^*)$ and since the unit ball B of X^* is $\sigma(X^*, X)$-compact, its continuous image $T^*(B)$ is $\sigma(Y, Y^*)$-compact. Let y_α^* be a net in Y^* that converges to zero for the topology τ. Then it converges to zero uniformly on the $\sigma(Y, Y^*)$-compact subset $T^*(B)$, and this implies that $||T(y_\alpha^*)||$ converges to zero. So (1) holds.

2. THE MAIN RESULTS

In this section we present our main results. We first prove two theorems regarding linear operators from Y^*, the dual of a Banach space Y, into a Banach space X which are continuous with respect to the Mackey topology $\tau(Y^*, Y)$ on Y^* and the norm topology on X, and then apply the results to the special case where $Y = L^1(\mu)$ and $Y^* = L^\infty(\mu)$. We can regard Theorem 4 and Theorem 5 as the abstraction of the Vitali-Hahn-Saks theorem for countably additive vector measures.

THEOREM 4 *Let X, Y be Banach spaces. Equip Y^* and X with the Mackey topology and the norm topology respectively. Let S be a compact subset of $L_s(Y^*, X)$. Then S is uniformly continuous in the sense that:*

$$y_\alpha^* \to 0 \text{ in } Y^* \text{ for } \tau(Y^*, Y) \text{ implies that } \limsup_{\alpha \ T \in S} \|T(y_\alpha^*)\| = 0$$

Equivalently, every compact subset of $L_s(Y^*, X)$ is equicontinuous.

Proof. Since the unit ball B of X^* is equicontinuous and $\sigma(X^*, X)$-closed, it follows from Theorem 2 that the set

$$K = \bigcup_{T \in S} T^*(B)$$

is $\sigma(Y, Y^*)$-compact. So if $y_\alpha^* \to 0$ in Y^* for the topology τ, then the net converges to zero uniformly on the set K, and this implies that $\lim_\alpha \sup_{T \in S} \|T(y_\alpha^*)\| = 0$. The last statement is obvious.

THEOREM 5 *Let Y be a weakly sequentially complete Banach space and let X be any Banach space. Let T_n be a sequence of continuous linear mappings from (Y^*, τ) into $(X, \| \cdot \|)$. Suppose that there exists a continuous operator $T : (Y^*, \| \cdot \|) \to (X, \| \cdot \|)$ such that $\lim_n T_n(y^*) = T(y^*)$ weakly for any $y^* \in Y^*$. Then T is also continuous with respect to the topology τ on Y^* and the norm topology on X. If $\lim_n T_n(y^*) = T(y^*)$ in norm for any $y^* \in Y^*$, then the set $S = \{T_n\} \cup \{T\}$ is compact in $L_s(Y^*, X)$, and hence is uniformly continuous in the sense of Theorem 4.*

Proof. It follows from Proposition 3 that $T_n^*(X^*) \subseteq Y$. Fix $x^* \in X^*$. Then for any $y^* \in Y^*$, $T^*(x^*)(y^*) = x^*(Ty^*) = \lim_n x^*(T_n y^*) = \lim_n T_n^* x^*(y^*)$. This implies that $T^*(x^*)$ is the weak limit of the sequence $T_n^*(x^*) \in Y$. Since Y is weakly sequentially complete, we have $T^*(x^*) \in Y$. So $T^*(X^*) \subseteq Y$, and hence T^* is continuous with respect to the topology $\sigma(X^*, X)$ and the topology $\sigma(Y, Y^*)$. Now by Proposition 3, T is continuous with respect to τ on Y^* and the norm topology on X. The last statement follows from Theorem 4.

Remark. The first part of Theorem 5 signifies that the subspace of $(Y^*, \tau) - (X, \|\cdot\|)$ continuous operators is sequentially closed in $L(Y^*, X)$ for the weak operator topology.

Next we consider a special case of the above theorem. Let X be a Banach space and let Σ be a σ-algebra of subsets of a nonempty set Ω. We use $B(\Omega)$ to denote the Banach space of all bounded measurable functions on Ω. It is well-known that if F is a bounded finitely additive vector measure defined on Σ and taking values in X, then $T(f) = \int f dF$ defines a bounded linear operator from $B(\Omega)$ into X, and any bounded linear operator is given this way (see [1], page 6). It is also well-known that a countably additive vector measure is bounded, i.e., the range of the vector measure is bounded in X (see [1], page 9). If F is a countably additive vector measure, then there exists a countably additive finite nonnegative scalar measure μ such that $F \ll \mu$ (see [1], page 12). Consequently, if F_n is a sequence of countably additive vector measures, then there exists a countably additive finite nonnegative scalar measure μ such that $F_n \ll \mu$ for each n. If F is a finitely additive bounded vector measure which vanishes on μ-null sets for some nonnegative countably additive scalar measure μ, then $T(f) = \int f dF$ defines a norm bounded linear operator from $L^\infty(\mu)$ into X. For more information, see [1], page 6. Now Let $Y = L^\infty(\mu)$ be equipped with the Mackey topology $\tau = \tau(L^\infty, L^1)$. To apply Theorem 4 and Theorem 5, we need some preparation.

PROPOSITION 6 *Let μ be a nonnegative countably additive finite scalar measure on Σ. Then $\mu(E_\alpha) \to 0$, $E_\alpha \in \Sigma$ if and only if $\chi_{E_\alpha} \to 0$ in $L^\infty(\mu)$ for the topology τ.*

Proof. Since $L^1(\mu)$ is a KB-space (i.e., a band in its bidual, see [7] or [5]), it follows from [5], Proposition 2.5.12 (i) that the (weakly or norm) closed solid convex hull of a weakly compact subset of $L^1(\mu)$ is weakly compact. So $(L^\infty(\mu), \tau)$ is a topological Riesz space (see [6]). If $\chi_{A_n} \downarrow 0$ in $L^\infty(\mu)$, then $\chi_{A_n} \to 0$ in $L^\infty(\mu)$ for the topology $\sigma(L^\infty, L^1)$. Now it follows from Dini's Theorem ([6], Theorem 4.3, chapter 5) that $\chi_{A_n} \to 0$ for the Mackey topology τ.

Suppose that $\mu(E_\alpha) \to 0$, but χ_{E_α} does not converge to zero under the Mackey topology τ. Then there exists a solid convex $\sigma(L^1, L^\infty)$-compact set K such that $\sup\{\int_{E_\alpha} |f| d\mu$

$f \in K\}$ does not go to zero as $\mu(E_\alpha) \to 0$. So there exist a sequence E_n in the net E_α and a positive number c such that

(a) $\mu(E_n) \leq 1/2^n$ and

(b) $\sup\{\int_{E_n} |f|d\mu : f \in K\} \geq c$ for all n.

Form $A_n = \cup_{k=n}^\infty E_k$. Then $\chi_{A_n} \downarrow 0$ in $L^\infty(\mu)$. By the first paragraph of the proof, we have $\chi_{A_n} \to 0$ for the topology τ and so it converges to zero uniformly on the solid set K. But (b) shows that

$$\sup\{\int_{A_n} |f|d\mu : f \in K\} \geq \sup\{\int_{E_n} |f|d\mu : f \in K\} \geq c \text{ for all } n$$

This is a contradiction.

Conversely, suppose that $\chi_{E_\alpha} \to 0$ for the topology τ. Then it converges to zero uniformly on any $\sigma(L^1, L^\infty)$-compact subset; in particular on the singleton $\{\chi_\Omega\}$ (since μ is a finite measure, $\chi_\Omega \in L^1(\mu)$). So $\mu(E_\alpha) \to 0$.

Remark. Let μ be as before. The Dunford-Schwartz-metric ρ defined on the measure algebra Σ/N_μ is given by $\rho(E_1, E_2) = \mu(E_1 \Delta E_2)$, where $E_1 \Delta E_2$ is the symmetric difference of E_1 and E_2. By the above Proposition 6, $\rho(E_\alpha, E_0) \to 0$ if and only if $|\chi_{E_\alpha} - \chi_{E_0}| \to 0$ for the topology $\tau(L^\infty, L^1)$. Since this topology is a Riesz topology, we conclude that $\rho(E_\alpha, E_0) \to 0$ if and only if $\chi_{E_\alpha} \to \chi_{E_0}$ for the topology $\tau(L^\infty, L^1)$. Now the metric space $(\Sigma/N_\mu, \rho)$ is well known to be complete (see [2]), and since the correspondence $E \to \chi_E$ preserves Cauchy sequences, the extreme boundary of the order interval $[0, \chi_\Omega]$ in $L^\infty(\mu)$ is τ-complete. Thus the extreme boundary of the unit ball of $L^\infty(\mu)$ is complete (hence closed) and metrizable under $\tau(L^\infty, L^1)$.

PROPOSITION 7 *Let F be a finitely additive bounded vector measure. Let μ be a non-negative countably additive finite scalar measure μ on Σ such that F vanishes on μ-null sets. Define $T(f) = \int f dF$ for all $f \in L^\infty(\mu)$. Then the following are equivalent:*

(1) *F is countably additive;*

(2) *$T : (L^\infty(\mu), \tau) \to (X, \|\cdot\|)$ is continuous;*

(3) *$T : (L^\infty(\mu), \tau) \to (X, \sigma(X, X^*))$ is continuous.*

Proof. The equivalence of (2) and (3) follows from Proposition 3. We now show that (2) implies (1). Let $\chi_{E_n} \downarrow 0$ in L^∞. Then $\chi_{E_n} \to 0$ for the topology $\sigma(L^\infty, L^1)$. By Dini's Theorem ([6], Theorem 4.3, chapter 5), $\chi_{E_n} \to 0$ for the Mackey topology τ. So $\|T(\chi_{E_n})\| \to 0$, i.e., $\|F(E_n)\| \to 0$. Therefore, F is countably additive. Finally we show that (1) implies (2). Let F be countably additive. Then it is easy to verify that (see also [1], page 14) T is continuous with respect to the topology $\sigma(L^\infty, L^1)$ and the topology $\sigma(X, X^*)$. Hence $T^*(X^*) \subseteq L^1$ and T^* is continuous with respect to $\sigma(X^*, X)$ and $\sigma(L^1, L^\infty)$. By Proposition 3, (2) holds.

THEOREM 8 *Let S be a set of countably additive vector measures that are absolutely continuous with respect to a nonnegative countably additive finite scalar measure μ. Identify the vector measures with the linear operators they induce. If S is a compact subset of $L_s(L^\infty(\mu), X)$, then the vector measures $T \in S$ are uniformly countably additive, i.e.,*

$$\lim_{\mu(E_\alpha) \to 0} \sup_{T \in S} \|T\chi_{E_\alpha}\| = 0$$

Proof. By Proposition 7, each $F \in S$ induces a linear operator that is continuous with respect to the topology τ and the norm topology. Now the theorem follows from Theorem 4 and Proposition 6.

COROLLARY 9 (V.H.S. Theorem) *If F_n is a pointwise convergent sequence of countably additive vector measures $\Sigma \to X$, then the limit is also countably additive and the countable additivity of F_n is uniform with respect to n.*

Proof. It is well-known that there exists a nonnegative countably additive finite scalar measure μ such that all F_n are absolutely continuous with respect to μ (see the remark before Proposition 6). Define $T_n(f) = \int f dF_n$ and $T(f) = \int f dF$ for any $f \in L^\infty(\mu)$. By the Nikodým Boundedness Theorem (see [1], page 14), it is easy to see that $T_n(f)$ converges to $T(f)$ in norm for any $f \in L^\infty(\mu)$. By Proposition 7, all T_n are continuous with respect to the Mackey topology τ. Now by Theorem 5, T is also continuous with respect to the Mackey topology τ, and so F is countably additive by Proposition 7. Let $S = \{T\} \cup \{T_n\}$. Then S is compact in $L_s(L^\infty(\mu), X)$. By Theorem 8, the vector measures F_n are uniformly countably additive.

The following is another main result of this paper.

THEOREM 10 *Identify a bounded vector measure with the bounded operator it induces. If $C \subset L_s(B(\Omega), X)$ is a compact set consisting of countably additive vector measures, then the countable additivity of the measures in C is uniform, i.e.,*

$$E_n \downarrow \emptyset \text{ implies that } \sup_{F \in C} ||F(E_n)|| \to 0 \text{ as } n \to \infty$$

Proof. Suppose that the theorem is false. Then there exist a sequence of measurable subsets $E_n \downarrow \emptyset$, a sequence $F_n \in C$ and a positive number ε_0 such that $||F_n(E_n)|| > \varepsilon_0$ for all n. By an obvious generalization of a theorem of Bartle-Dunford-Schwartz ([1], page 14, corollary 6), there exists a nonnegative countably additive finite scalar measure μ such that each $F_n \ll \mu$. Let D be the closure of $\{F_n\}$ in C. Then D consists of countably additive vector measures. We claim that $F \ll \mu$ for all $F \in D$. By [1], Theorem 1, page 10, we only have to prove that $\mu(E) = 0$ implies $F(E) = 0$. Suppose that $\mu(E) = 0$. Then $F_n(E) = 0$ for all n. Since F is in the closure of $\{F_n\}$ under the strong operator topology, we must have $F(E) = 0$. So $F \ll \mu$. Now it is easy to see that D is a compact subset of $L_s(L^\infty(\mu), X)$. Since $\mu(E_n) \to 0$, it follows from Theorem 8 that $\lim_n \sup_{F \in D} ||F(E_n)|| = 0$. But this contradicts to the fact that $||F_n(E_n)|| > \varepsilon_0$ for all n.

Finally, we mention the weak convergence of vector measures. The following can be found in [1], page 38.

COROLLARY 11 *If F_n is a sequence of countably additive vector measures such that $\lim_n F_n(E) = F(E)$ weakly for all $E \in \Sigma$, then F is countably additive.*

Proof. First there exists a nonnegative countably additive finite scalar measure μ such that each $F_n \ll \mu$. Then the corollary follows from Proposition 7 and Theorem 5.

Concluding Remarks. Looking back, we see that the Vitali-Hahn-Saks theorem (Cor. 9) rests on two important properties of $L_s(L^\infty(\mu), X)$, where $L^\infty(\mu)$ is endowed with $\tau(L^\infty, L^1)$. First, this space is (even weakly) sequentially closed under the strong operator topology, in the space of all bounded linear maps from $L^\infty(\mu)$ into X (Theorem 5); and second, each of its compact subsets is equicontinuous with respect to $\tau(L^\infty, L^1)$ and the norm topology of X (Theorem 4). Thus the V.H.S. theorem is essentially an equicontinuity theorem, as are the Banach-Steinhaus theorem and the Principle of Uniform Boundedness. Another important equicontinuity theorem is the Nikodým Boundedness Theorem employed above: Considering the space $S(\Omega)$ of all Σ-step functions under its natural norm (Σ is a σ-algebra of subsets of Ω), it asserts that every bounded subset of $L_s(S(\Omega), X)$ is equicontinuous, or equivalently, that $S(\Omega)$ is a barreled space.

References

[1] Diestel, J. and Uhl, J. J. JR. – *Vector Measures*, Mathematical Surveys, Number **15**, American Mathematical Society, 1977.

[2] Dunford, N. and Schwartz, J. T., *Linear Operators*, Part I, Interscience, New York and London, 1958.

[3] Fremlin, D. H. – *Topological Riesz Spaces and Measure Theory*, Cambridge University Press, London–New York, 1974.

[4] Luxemburg, W. A. J. and Zaanen, A. C. – *Riesz Spaces I*, North-Holland, Amsterdam, 1971.

[5] Meyer-Nieberg, P. – *Banach Lattices*, Springer-Verlag, 1991.

[6] Schaefer, H. H. – *Topological Vector Spaces*, fifth printing, New York–Berlin–Heidelberg–Tokyo, 1986.

[7] Schaefer, H. H. – *Banach Lattices and Positive Operators*, Berlin–Heidelberg–New York, 1974.

[8] Zaanen, A. C.– *Riesz Spaces II*, North-Holland, Amsterdam, 1983.

1991 Mathematics Subject Classification: 28-02, 28A33, 28B05, 46-02, 46B20, 47-02, 47B38.

Department of Mathematics
Florida Atlantic University
Boca Raton, FL 33431
USA

Operator Theory:
Advances and Applications, Vol. 75
© 1995 Birkhäuser Verlag Basel/Switzerland

MINKOWSKI'S INTEGRAL INEQUALITY
FOR FUNCTION NORMS

ANTON R. SCHEP

Dedicated to Professor Dr. A.C. Zaanen on the occasion of his eightieth birthday

Let ρ and λ be Banach function norms with the Fatou property. Then the generalized Minkowski integral inequality $\rho(\lambda(f_x)) \leq M\lambda(\rho(f^y))$ holds for all measurable functions $f(x,y)$ and some fixed constant M if and only if there exists $1 \leq p \leq \infty$ such that λ is p–concave and ρ is p–convex.

INTRODUCTION

Let (X,μ) and (Y,ν) be σ-finite measure spaces. Let $0 < r \leq s < \infty$ and let $f(x,y)$ be a $\mu \times \nu$–measurable function. Then the classical integral inequality of Minkowski states that

$$\left(\int \left(\int |f(x,y)|^r d\nu(y)\right)^{\frac{s}{r}} d\mu(x)\right)^{\frac{1}{s}} \leq \left(\int \left(\int |f(x,y)|^s d\mu(x)\right)^{\frac{r}{s}} d\nu(y)\right)^{\frac{1}{r}}.$$

If we define for fixed $x \in X$ the function f_x by $f_x(y) = f(x,y)$ and for fixed $y \in Y$ the function f^y by $f^y(x) = f(x,y)$, then the above inequality is the same as

$$\||\|f_x\|_r\|_s \leq \||\|f^y\|_s\|_r.$$

The goal of this paper is to extend this inequality to function norms. For general information and terminology concerning function norms and Banach function spaces we refer to [10]. It is known that if ρ is a function norm that $\rho(f_x)$ need not be a measurable function on X (see [5]). To avoid this pathology we shall assume that all our function norms have the Fatou property, i.e., if $0 \leq f_k \uparrow f$, $f_k \in L_\rho$ and $\sup_k \rho(f_k) < \infty$, then

$f \in L_\rho$ and $\rho(f) = \sup_k \rho(f_k)$. A function norm ρ is said to have the weak Fatou property, if $0 \le f_k \uparrow f$, $f_k \in L_\rho$ and $\sup_k \rho(f_k) < \infty$ implies that $f \in L_\rho$. As was shown in [4] the Fatou property is sufficient to ensure the measurability of $\rho(f_x)$, but the weak Fatou property is not (see [5]). We note that the associate norm ρ' of a function norm always has the Fatou property. Let λ be a function norm on $L_0(Y, \nu)$ and ρ a function norm on $L_0(X, \mu)$, where $L_0(Y, \nu)$, respectively $L_0(X, \mu)$, denotes the space of (equivalence classes of) all measurable functions on Y, respectively X. The main result of this paper is that

$$(*) \qquad\qquad \rho(\lambda(f_x)) \le M\lambda(\rho(f^y))$$

holds for all measurable functions $f(x, y)$ and some fixed constant M if and only if there exists $1 \le p \le \infty$ such that λ is p–concave and ρ is p–convex. We note that equation $(*)$ generalizes Minkowski's integral inequality to arbitrary function norms. In section 2 of this paper the above mentioned result will be proved. The proof of this theorem depends crucially on the fundamental result of Krivine about the local structure of a Banach lattice ([3], see [8] for an expository account of this theorem). In section 3 we shall relate the main result to the theory of integral operators of finite double norm (the so called Hille-Tamarkin integral operators) and integral operators of finite inverse double norm.

1. PRELIMINARIES

Let us recall the notion of p–convexity and p–concavity. Let ρ be a Banach function norm and let $L_\rho = L_\rho(X, \mu)$ denote the corresponding Banach function space. Then L_ρ is called p–convex for $1 \le p \le \infty$ if there exists a constant M such that for all $f_1, \dots, f_n \in L_\rho$,

$$\rho\left(\left(\sum_{k=1}^{n} |f_k|^p\right)^{\frac{1}{p}}\right) \le M\left(\sum_{k=1}^{n} \rho(f_k)^p\right)^{\frac{1}{p}} \text{ if } 1 \le p < \infty$$

or $\rho(\sup|f_k|) \le M \max_{1 \le k \le n} \rho(f_k)$ if $p = \infty$. Similarly L_ρ is called p–concave for $1 \le p \le \infty$ if there exists a constant M such that for all $f_1, \dots, f_n \in L_\rho$,

$$\left(\sum_{k=1}^{n} \rho(f_k)^p\right)^{\frac{1}{p}} \le M\rho\left(\left(\sum_{k=1}^{n} |f_k|^p\right)^{\frac{1}{p}}\right) \text{ if } 1 \le p < \infty$$

or $\max_{1 \le k \le n} \rho(f_k) \le M\rho(\sup|f_k|)$ if $p = \infty$. The notions of p–convexity, respectively p–concavity are closely related to the notions of upper p–estimate (strong ℓ_p–composition property), respectively lower p–estimate (strong ℓ_p–decomposition property) as can be found in e.g. [6, Theorem 1.f.7]. In particular the lower index $s(L_\rho)$ of L_ρ is also equal to the supremum of $\{p \geq 1 : L_\rho$ is $p - \text{convex}\}$ and the upper index $\sigma(L_\rho)$ of L_ρ equals the infimum of $\{p \geq 1 : L_\rho$ is $p - \text{concave}\}$. We now recall the result of J.L. Krivine (see [8] for a discussion of this result as well as several relevant references).

THEOREM 1.1 (KRIVINE). *Let E be an infinite dimensional Banach lattice. Then for all integers n, all $\epsilon > 0$, $p = s(E)$ and $p = \sigma(E)$ there exist disjoint $x_1, \ldots, x_n$ in E such that for all n–tuples $\{a_i\}$ of real numbers we have*

$$\|\{a_i\}\|_p \leq \|\sum_{i=1}^n a_i x_i\| \leq (1 + \epsilon)\|\{a_i\}\|_p.$$

The following proposition describes p–convex function norms with the weak Fatou property. Its origin is in Pisier's work and the theory of p–concavication (see [6]).

PROPOSITION 1.2. *Let L_ρ be a p–convex Banach function space with the weak Fatou property. Then there exists a collection G of non-negative measurable functions on X such that ρ is equivalent to the function norm*

$$\rho_1(f) = \sup_{g \in G}\left(\int |f|^p g d\mu\right)^{\frac{1}{p}}.$$

PROOF. Define $\tau(f) = (\rho(|f|^{\frac{1}{p}}))^p$. Then τ has the weak Fatou property and all the properties of a norm, except that $\tau(f + g) \leq M(\tau(f) + \tau(g))$, where M is the convexity constant of ρ. Let

$$\tau_1(f) = \inf\{\sum_{i=1}^n \tau(f_i) : |f| = \sum_{i=1}^n |f_i|\}.$$

Then it is easy to see that τ_1 is a function norm with the weak Fatou property, which is equivalent to τ. Let G be the the positive part of the unit ball of L'_{τ_1}, i.e., $G = \{g \geq 0 : \tau'_1(g) \leq 1\}$. It follows from [10, Theorem 112.2] that τ_1 is equivalent to the second associate norm τ''_1, i.e., τ_1 is equivalent to $\sup\{\int |f|g d\mu : g \in G\}$. The conclusion now follows easily, since $\rho(f) = (\tau(|f|^p))^{\frac{1}{p}}$.

We note that the function norm ρ_1 has the Fatou property, so that the assumption that ρ has the weak Fatou property is necessary in the above proposition. If one assumes that the convexity constant M equals one and that ρ has the Fatou property, then $\rho = \rho_1$.

2. MINKOWSKI'S INEQUALITY FOR FUNCTION NORMS

We start with the special case that $\lambda = \| \cdot \|_1$ or $\rho = \| \cdot \|_\infty$.

PROPOSITION 2.1. *Let ρ and λ be function norms with the Fatou property and let $f(x, y)$ be a measurable function. Then we have*

$$\rho(\|f_x\|_1) \le \|\rho(f^y)\|_1$$

and

$$\|\lambda(f_x)\|_\infty \le \lambda(\|f^y\|_\infty).$$

PROOF. Let $0 \le g \in L'_\rho$ with $\rho'(g) \le 1$. Then we have

$$\int \left(\int |f(x, y)| d\nu(y) \right) g(x) d\mu(x) = \int \left(\int |f(x, y)| g(x) d\mu(x) \right) d\nu(y)$$
$$\le \int \rho(f^y) d\nu(y) = \|\rho(f^y)\|_1.$$

By taking the supremum over the collection of all such g we obtain

$$\rho(\|f_x\|_1) \le \|\rho(f^y)\|_1.$$

This proves the first inequality. The second inequality can be proved along similar lines or by a duality argument.

COROLLARY 2.2 (MINKOWSKI'S INTEGRAL INEQUALITY). *Let $0 < r \le s < \infty$ and let $f(x, y)$ be a $\mu \times \nu$-measurable function. Then we have*

$$\left(\int \left(\int |f(x, y)|^r d\nu(y) \right)^{\frac{s}{r}} d\mu(x) \right)^{\frac{1}{s}} \le \left(\int \left(\int |f(x, y)|^s d\mu(x) \right)^{\frac{r}{s}} d\nu(y) \right)^{\frac{1}{r}}.$$

PROOF. Let $p = \frac{s}{r}$ and take $\rho = \| \cdot \|_p$. Then apply the above proposition to $|f(x, y)|^r$ to get

$$\left(\int \left(\int |f(x, y)|^r d\nu(y) \right)^{\frac{s}{r}} d\mu(x) \right)^{\frac{r}{s}} = \rho(\||f_x|^r\|_1)$$
$$\le \|\rho(|f^y|^r)\|_1 = \left(\int |f(x, y)|^s d\mu(x) \right)^{\frac{r}{s}} d\nu(y)).$$

THEOREM 2.3. *Let ρ and λ be function norms with the Fatou property and assume that there exists $1 \le p \le \infty$ such that ρ is p-convex and λ is p-concave. Then there exists a constant C such that for all measurable $f(x,y)$ we have*

$$\rho(\lambda(f_x)) \le M\lambda(\rho(f^y)).$$

PROOF. If $p = \infty$, then ρ is equivalent to $\|\cdot\|_\infty$ and the theorem follows then from Proposition 2.1. Assume therefore that $1 \le p < \infty$. Then λ is an order continuous norm. We shall first prove the theorem under the additional hypothesis that also ρ is order continuous. In that case the product norms $\rho\lambda$ and $\lambda\rho$ are also order continuous and it therefore suffices to prove the inequality for functions in the collection $\mathcal{P} = \{f(x,y) : f(x,y) = \sum_{i=1}^n f_i(x)g_i(y), f_i, g_i \ge 0 \text{ measurable}, g_i \text{ mutually disjoint}\}$. Let $f = \sum_{i=1}^n f_i g_i \in \mathcal{P}$. Let M denote a convexity constant of ρ and a concavity constant of λ. Then we have the following inequalities

$$\rho(\lambda(f_x)) \le \rho(\sum_{i=1}^n |f_i|\lambda(g_i))$$

$$\le \rho((\sum_{i=1}^n |f_i|^p \lambda(g_i)^p)^{\frac{1}{p}})$$

$$\le M(\sum_{i=1}^n \rho(f_i)^p \lambda(g_i)^p)^{\frac{1}{p}}$$

$$\le M^2 \lambda(\sum_{i=1}^n \rho(f_i)g_i) = M^2 \lambda(\rho(f^y)),$$

which proves the inequality in case ρ is order continuous. In the general case we use Proposition 1.2 to pass to the equivalent norm ρ_1. Let $f(x,y)$ be a measurable function. Then by the definition of G we can find $g_n \in G$ so that

$$\rho_1(\lambda(f_x)) = \sup_n(\int \lambda(f_x)^p g_n(x)d\mu(x))^{\frac{1}{p}}.$$

Denote by τ_m the semi-norm $\tau_m(h) = (\int |h(x)|^p g_n(x)d\mu(x))^{\frac{1}{p}}$. Then τ_n is order continuous, p-convex and has the Fatou property. Hence by the first part of the proof there exists a constant C, independent of f and m, such that

$$\tau_m(\lambda(f_x)) \le C\lambda(\tau_m(f^y))$$

for all m. Now $\tau_m(f^y) \leq \rho_1(f^y)$ implies that $\tau_m(\lambda(f_x)) \leq C\lambda(\rho_1(f^y))$ for all m. From this it follows that also $\rho_1(\lambda(f_x)) \leq C\lambda(\rho_1(f^y))$. Since ρ_1 is equivalent with ρ, this completes the proof of the theorem.

REMARK. Note that if $\sigma(L_\lambda) < s(L_\rho)$, then for any $\sigma(L_\lambda) < p < s(L_\rho)$ we have that ρ is p–convex and λ is p–concave.

As a first step in showing that the conditions of the above theorem are also necessary, we prove the following proposition.

PROPOSITION 2.4. *Let ρ and λ be function norms with the Fatou property and assume that there exists a constant C such that for all measurable $f(x, y)$ we have*

$$\rho(\lambda(f_x)) \leq M\lambda(\rho(f^y)).$$

Then $\sigma(L_\lambda) \leq s(L_\rho)$.

PROOF. Let $p = s(L_\rho), q = \sigma(L_\lambda)$ and let n be a positive integer. Then by Krivine's theorem (Theorem 1.1) there exist disjoint $g_1, \ldots, g_n$ in L_ρ and disjoint $h_1, \ldots, h_n$ in L_λ such that for all n–tuples $\{a_i\}$ of real numbers we have

$$\|\{a_i\}\|_p \leq \rho(\sum_{i=1}^{n} a_i g_i) \leq 2\|\{a_i\}\|_p$$

and

$$\|\{a_i\}\|_q \leq \lambda(\sum_{i=1}^{n} a_i h_i) \leq 2\|\{a_i\}\|_q.$$

Let $f(x, y) = \sum_{i=1}^{n} a_i g_i(x) h_i(y)$. Then we have

$$\lambda(\rho(f^y)) = \lambda(\sum_{i=1}^{n} |a_i| \rho(g_i) h_i(y))$$

$$\leq 2\lambda(\sum_{i=1}^{n} |a_i| h_i) \leq 4\|\{a_i\}\|_q$$

and

$$\rho(\lambda(f_x)) = \rho(\sum_{i=1}^{n} |a_i| g_i(x) \lambda(h_i))$$

$$\geq \rho(\sum_{i=1}^{n} |a_i| g_i) \geq \|\{a_i\}\|_p.$$

¿From this it follows that $\|\{a_i\}\|_p \leq 4C\|\{a_i\}\|_q$ for all n–tuples $\{a_i\}$, where C is independent of n. This implies that $q \leq p$, which concludes the proof of the proposition.

We now prove the converse of Theorem 2.3.

THEOREM 2.5. *Let ρ and λ be function norms with the Fatou property and assume that there exists a constant C such that for all measurable $f(x,y)$ we have*

$$\rho(\lambda(f_x)) \leq M\lambda(\rho(f^y)).$$

Then there exists $1 \leq p \leq \infty$ such that ρ is p–convex and λ is p–concave.

PROOF. From the above proposition it follows that $\sigma(L_\lambda) \leq s(L_\rho)$. In case we have that $\sigma(L_\lambda) < s(L_\rho)$, then the theorem will hold for any p such that $\sigma(L_\lambda) < p < s(L_\rho)$. Hence assume that $\sigma(L_\lambda) = s(L_\rho) = p$. By Krivine's theorem (Theorem 1.1) there exist disjoint $h_1, \ldots, h_n$ in L_λ such that for all n–tuples $\{a_i\}$ of real numbers we have

$$\|\{a_i\}\|_p \leq \lambda\left(\sum_{i=1}^n a_i h_i\right) \leq 2\|\{a_i\}\|_p.$$

Let $g_1, \ldots, g_n$ be in L_ρ and put $f(x,y) = \sum_{i=1}^n g_i(x)h_i(y)$. Then we have

$$\lambda(\rho(f^y)) = \lambda\left(\sum_{i=1}^n \rho(g_i)h_i\right) \leq 2\|\{\rho(g_i)\}\|_p$$

and

$$\rho(\lambda(f_x)) \geq \rho\left(\|\{g_i\}\|_p\right).$$

¿From $\rho(\lambda(f_x)) \leq M\lambda(\rho(f^y))$ it follows now that

$$\rho\left(\left(\sum_{i=1}^n |g_i|^p\right)^{\frac{1}{p}}\right) \leq 2M\left(\sum_{i=1}^n \rho(g_i)^p\right)^{\frac{1}{p}},$$

i.e., ρ is p–convex. By a similar argument (or by duality) one can show that λ is p–concave.

The following corollary is due to A.V. Buhvalov ([1]), the special case $\rho = \lambda$ is due to N.J. Nielsen ([7]).

COROLLARY 2.6(GENERALIZED KOLMOGOROV-NAGUMO'S THEOREM).*Let ρ and λ be function norms with the Fatou property and assume that the double norm $\rho(\lambda(f_x))$ is equivalent with the double norm $\lambda(\rho(f^y))$. Then there exists $1 \le p \le \infty$ such that ρ and λ are equivalent to an L_p-norm.*

PROOF. Applying the above theorem twice we get that there exist $1 \le p, q \le \infty$ such that ρ is p–convex and q–concave, and λ is q–convex and p–convex. Hence it follows that $p = q$. If $p < \infty$ the result now follows from [6, Lemma 1.b.13]. In case $p = \infty$ an inspection of the proof of [6, Lemma 1.b.13] shows that ρ and λ are equivalent to an AM-norm.

3. INTEGRAL OPERATORS OF FINITE DOUBLE NORM

Recall that a linear operator T from L_λ into L_ρ is called an integral operator if there exists a $\mu \times \nu$–measurable function $T(x, y)$ on $X \times Y$ such that

$$\int |T(x, y)f(y)|d\nu(y) < \infty$$

a.e. for all $f \in L_\lambda$ and such that

$$Tf(x) = \int T(x, y)f(y)d\nu(y)$$

a.e. for all $f \in L_\lambda$. Such an integral operator is called an integral operator of finite double norm, or Hille-Tamarkin operator, if $\rho(\lambda'(T_x)) < \infty$. Characterizations and compactness properties of such operators are discussed in [9] (see also the references in [9]). An integral operator is called an integral operator of finite inverse double norm if $\lambda'(\rho(T^y)) < \infty$. Integral operators which are of finite double and finite inverse double norm are sometimes called integral operators of complete finite double norm. It is well known that the spaces of integral operators of finite double norm and of finite inverse double norm are Banach function spaces with respect to the product norm $\rho(\lambda'(\cdot))$, respectively $\lambda'(\rho(\cdot))$. Our main results of the previous section imply the following theorem.

THEOREM 3.1. *Let λ and ρ be Banach function norms with the Fatou property. Then the following holds.*

(1) *Every integral operator of finite inverse double norm is an integral operator of finite double norm if and only if there exists $1 \le p \le \infty$ such that ρ is p–convex and λ is p'–convex, where $\frac{1}{p} + \frac{1}{p'} = 1$.*

(2) *Every integral operator of finite double norm is an integral operator of finite inverse double norm if and only if there exists $1 \leq p \leq \infty$ such that ρ is p'-concave and λ is p-concave, where $\frac{1}{p} + \frac{1}{p'} = 1$.*

PROOF. Every integral operator of finite inverse double norm is an integral operator of finite double norm if and only if there exists a constant M such that $\rho(\lambda'(T_x)) \leq M\lambda'(\rho(T^y))$. From Theorem 2.3 and Theorem 2.5. it follows that this inequality holds if and only if there exists $1 \leq p \leq \infty$ such that ρ is p-convex and λ' is p-concave. Now λ' is p-concave implies that $\lambda'' = \lambda$ is p'-convex. This proves (1). The proof of (2) is similar and therefore omitted.

Integral operators of complete finite double norm occur naturally in the study of power summability of eigenvalues of integral operators. In [2] it was e.g. proved that on an order continuous Banach function space the eigenvalues of an integral operator of complete finite double norm are always 4th-power summable. More precise exponents of summability were then given in terms of the lower and upper indices of the Banach function space. The following corollary of Theorem 3.1 shows that in some cases integral operators of finite (inverse) double norm are of complete finite double norm.

COROLLARY 3.2. *Let $X = Y$ and let $\rho = \lambda$ be a Banach function norm with the Fatou property. Then the following holds.*

(1) *Every integral operator of finite inverse double norm from L_ρ into L_ρ is an integral operator of finite double norm if and only if there exists $2 \leq p \leq \infty$ such that ρ is p-convex.*

(2) *Every integral operator of finite double norm from L_ρ into L_ρ is an integral operator of finite inverse double norm if and only if there exists $1 \leq p \leq 2$ such that ρ is p-concave.*

(3) *The collection of integral operators of finite double norm from L_ρ into L_ρ coincides with the collection of integral operators of inverse finite double norm from L_ρ into L_ρ if and only if L_ρ is lattice isomorphic to an L_2-space.*

PROOF. By the above theorem every integral operator of finite inverse double norm from L_ρ into L_ρ is an integral operator of finite double norm if and only if there exists $1 \leq p \leq \infty$ such that ρ is p–convex and ρ is p'–convex. Since at least one of p and p' is greater than or equal to 2, part (1) follows. Part (2) follows similarly. To prove part (3) note first that when L_ρ is lattice isomorphic to an L_2–space then the collections of integral operators of finite double norm, respectively finite inverse double norm both coincide with the collection of Hilbert-Schmidt operators. Now by parts (1) and (2) The collection of integral operators of finite double norm from L_ρ into L_ρ coincides with the collection of integral operators of inverse finite double norm from L_ρ into L_ρ if and only if L_ρ is both 2–convex and 2–concave and the result follows from this as in Corollary 2.6.

REFERENCES

1. A.V. Buhvalov, *Geometrical applications of the Kolmogorov–Nagumo's theorem*, "Qualitative and approximate methods of the investigations of operator equations", Jaroslavl', 1983, pp. 18–29. (Russian)
2. H. König and L. Weis, *On the eigenvalues of orderbounded integral operators*, Integr. Eq. and Oper. Th. **6** (1983).
3. J.L. Krivine, *Sous-espaces de dimension finie des espaces de Banach réticulés*, Ann. of Math. **104** (1976), 1–29.
4. W.A.J. Luxemburg, *On the measurability of a function which occurs in a paper by A.C. Zaanen*, Proc. Netherl. Acad. Sci. (A) **61** (1958), 259–265.
5. ———, *Addendum to "On the measurability of a function which occurs in a paper by A.C. Zaanen"*, Proc. Netherl. Acad. Sci. (A) **66** (1963), 587–590.
6. J. Lindenstrauss and L. Tzafriri, *Classical Banach spaces II*, Ergebnisse der Mathematik und ihrer Grenzgebiete vol. 97, Springer–Verlag, Berlin–Heidelberg–New York, 1979.
7. N.J. Nielsen, *On Banach ideals determined by Banach lattices and their applications*, Diss. Math. **109** (1973), 1-66.
8. Anton. R. Schep, *Krivine's theorem and the indices of a Banach lattice*, Acta Appl. Math. **27** (1992), 111-121.
9. ———, *Compactness properties of Carleman and Hille-Tamarkin operators*, Canad. J. Math. **37** (1985), 921–933.
10. A.C. Zaanen, *Riesz spaces II*, North–Holland Mathematical Library vol. 30, North–Holland Publishing Company, Amsterdam–New York–Oxford, 1983.

DEPARTMENT OF MATHEMATICS,
UNIVERSITY OF SOUTH CAROLINA,
COLUMBIA, SC 29208
E-mail address: schep@math.scarolina.edu

Operator Theory:
Advances and Applications, Vol. 75
© 1995 Birkhäuser Verlag Basel/Switzerland

PROGRAM ANALYSIS SYMPOSIUM

on the occasion of the 80th birthday of

A.C. Zaanen

September 2 and 3, 1993

Thursday 2 September

10.30 - 11.00: Opening address on the Mathematical life of A.C. Zaanen by
W.A.J. Luxemburg (Pasadena).

11.00 - 12.00: *H. König* (Saarbrücken): New aspects after 100
years of measure and integration theory.

14.00 - 15.00: *J. Korevaar* (Amsterdam): Approximating integrals
by arithmetic means and related questions.

15.30 - 16.30: *H. Brezis* (Paris): Ginzburg-Landau vortices and
quantization effects.

Friday 3 September

09.30 - 10.30: *H. König* (Kiel): Vector-valued L_p-convergence
of orthogonal series and Lagrange interpolation.

11.00 - 12.00: *N. Nikolskii* (Bordeaux): 50 years of free interpolation.

14.00 - 15.00: *J. Duistermaat* (Utrecht): Selfsimilarity of "Riemann's
nondifferentiable function".

15.30 - 16.30: *W. Arendt* (Besançon): Evolution and positivity.

Titles previously published in the series

OPERATOR THEORY: ADVANCES AND APPLICATIONS
BIRKHÄUSER VERLAG

59. **T. Ando, I. Gohberg** (Eds.): Operator Theory and Complex Analysis, 1992, (3-7643-2824-X)

60. **P.A. Kuchment:** Floquet Theory for Partial Differential Equations, 1993, (3-7643-2901-7)

61. **A. Gheondea, D. Timotin, F.-H. Vasilescu** (Eds.): Operator Extensions, Interpolation of Functions and Related Topics, 1993, (3-7643-2902-5)

62. **T. Furuta, I. Gohberg, T. Nakazi** (Eds.): Contributions to Operator Theory and its Applications. The Tsuyoshi Ando Anniversary Volume, 1993, (3-7643-2928-9)

63. **I. Gohberg, S. Goldberg, M.A. Kaashoek:** Classes of Linear Operators, Volume 2, 1993, (3-7643-2944-0)

64. **I. Gohberg** (Ed.): New Aspects in Interpolation and Completion Theories, 1993, (3-7643-2948-3)

65. **M.M. Djrbashian:** Harmonic Analysis and Boundary Value Problems in the Complex Domain, 1993, (3-7643-2855-X)

66. **V. Khatskevich, D. Shoiykhet:** Differentiable Operators and Nonlinear Equations, 1993, (3-7643-2929-7)

67. **N.V. Govorov †:** Riemann's Boundary Problem with Infinite Index, 1994, (3-7643-2999-8)

68. **A. Halanay, V. Ionescu:** Time-Varying Discrete Linear Systems Input-Output Operators. Riccati Equations. Disturbance Attenuation, 1994, (3-7643-5012-1)

69. **A. Ashyralyev, P.E. Sobolevskii:** Well-Posedness of Parabolic Difference Equations, 1994, (3-7643-5024-5)

70. **M. Demuth, P. Exner, G. Neidhardt, V. Zagrebnov** (Eds): Mathematical Results in Quantum Mechanics. International Conference in Blossin (Germany), May 17-21, 1993, 1994, (3-7643-5025-3)

71. **E.L. Basor, I. Gohberg** (Eds): Toeplitz Operators and Related Topics. The Harold Widom Anniversary Volume. Workshop on Toeplitz and Wiener-Hopf Operators, Santa Cruz, California, September 20–22, 1992, 1994 (3-7643-5068-7)

72. **I. Gohberg, L.A. Sakhnovich** (Eds): Matrix and Operator Valued Functions. The Vladimir Petrovich Potapov Memorial Volume, (3-7643-5091-1)

73. **A. Feintuch, I. Gohberg** (Eds): Nonselfadjoint Operators and Related Topics. Workshop on Operator Theory and Its Applications, Beersheva, February 24–28, 1994, (3-7643-5097-0)

74. **R. Hagen, S. Roch, B. Silbermann:** Spectral Theory of Approximation Methods for Convolution Equations, 1994, (3-7643-5112-8)

75. **C.B. Huijsmans, M.A. Kaashoek, B. de Pagter**: Operator Theory in Function Spaces and Banach Lattices. The A.C. Zaanen Anniversary Volume, 1994 (ISBN 3-7643-5146-2)

 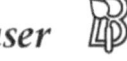